VOLUME 1

APPLIED PHYSICS AND ENGINEERING

An International Series

Relaxation in Shock Waves

Relaxation in Shock Waves

Ye. V. Stupochenko
S. A. Losev
and A. I. Osipov

Translated by
Scripta Technica, Inc.

Translation Editor
Richard Shao-lin Lee
College of Engineering
State University of New York
Stony Brook, New York

SPRINGER-VERLAG • Berlin • Heidelberg • New York • 1967

Originally published under the title, **Relaksatsionnyye Protsessy v Udarnykh Volnakh,** by Nauka Press, Moscow, 1965.

ISBN 978-3-642-48248-9 ISBN 978-3-642-48246-5 (eBook)
DOI 10.1007/978-3-642-48246-5

 Library of Congress Catalog Card Number 67-21459.

Softcover reprint of the original edition 1967

Title No. 3891.

Translation Editor's Preface

This book by Ye. V. Stupochenko, S. A. Losev and A. I. Osipov distinguishes itself as an unusually comprehensive treatment of both the theoretical and experimental aspects of the subject of relaxation processes in shock waves. S. A. Losev has done a considerable amount of work in molecular absorption spectroscopy of gases in relaxation, and, consequently, the chapter on experimental methods of study of nonequilibrium phenomena in shock waves has turned out to be particularly extensive in both scope and depth. A. I. Osipov has contributed extensively to the theoretical understanding of relaxation processes in gases through his work on transition probability in nonadiabatic collisions by quantum-mechanical calculation as can be seen from the many quotations in the chapter on relaxation processes in shock waves. Finally, Ye. V. Stupochenko's heavy contribution to the development of the study of nonequilibrium gas flow must also be mentioned, as it is obviously reflected at various places throughout the book.

It is worth particular mention that the authors have compiled the most extensive bibliography of references (over six hundred in total) in the literature of relaxation gasdynamics to date. The authors' thorough familiarity with the Western (particularly American) work in this area can be seen from the many Western references quoted, about two-thirds of the total number. On the contrary, most of the more than two hundred Russian references are not popularly known in the West and, therefore, should be of value to the English-speaking readers.

A number of misprints found in the original Russian edition of this book have been corrected in the English translation, and, for the convenience of the English-speaking readers, the appropriate information on the origins of quoted Western references or the English versions of quoted Russian references, wherever available, has been added in the bibliography.

Stony Brook, October 1966 RICHARD SHAO-LIN LEE

Foreword

This book surveys the present state of the theory of relaxation processes occurring in shock waves and in gases, and reviews the experimental accomplishments in this area.

When the shock wave was introduced into gasdynamics, it was viewed as the surface of a discontinuity separating two gas regions in thermodynamic equilibrium with each other. This concept of the shock wave has remained valid for the overwhelming majority of problems and is applicable in those cases where the actual thickness of the wave is small by comparison with the flow dimensions. However, it may not hold with high-amplitude shock waves in rarefied gases where the thickness of the shock wave becomes of the same order as the dimensions of the streamlined body or the characteristic dimensions of the fluid flow. In these cases the structure of the shock wave and the processes which take place in it assume fundamental importance.

Until recently, these processes were of only academic interest but are now attracting much attention because of the burgeoning growth of aerospace sciences and resultant demands on high-velocity gasdynamics. The intensive studies of shock waves produced experimental techniques which (as it frequently happens in science) find much wider application than in the original problem of simulating flow past solid bodies at high altitudes. In addition, the results of these specialized studies on ultrafast processes in shock waves have been found pertinent to a great variety of other sciences, from ultrasonics and optics to chemical kinetics.

We now have several recent books on the use of shock waves for studying physicochemical processes (Ya. B. Zel'dovich and Yu. P. Rayzer, *Physics of Shock Waves and of High-Temperature Gasdynamics Phenomena*, 1963 (1966); E. F. Green and J. P. Teonnies, *Chemische Reaktionen in Stosswellen*, Darmstadt, 1959; J. N. Breadley, *Shock Waves in Chemistry and Physics*, London, 1962; A. G. Gaydon and I. R. Hurle, *The Shock Tube in High-Temperature Chemical Physics,* London, 1963). However, none of these has treated relaxation processes in shock waves exhaustively, a lack which this book attempts to remedy. We shall concentrate on the physics of the relaxation phenomena and, in particular, on the relationships governing the individual processes involved in establishing statistical equilibrium at various degrees of molecular freedom, as well as the kinetics of thermal dissociation and ionization. These processes are fundamental to strong shock waves where the temperatures may reach ten to twenty thousand degrees. By contrast, we shall not dwell on the many applications of relaxation phenomena in chemical physics, gasdynamics, etc. We shall discuss radiation of nonequilibrium zones behind a shock front only briefly and shall not deal with propagation of shock waves in plasma, a problem beyond the scope of this book.

The first chapter presents basic theory: the formation and structure of shock waves, the qualitative view of relaxation processes in gases, and the theory underlying experimental techniques for study of these processes. The following three chapters are devoted to relaxation processes in shock waves. Here the authors tried to deal as comprehensively as possible with the main directions, techniques and results of experimental and theoretical studies in this field. The fifth chapter is devoted to the important practical problem of nonequilibrium phenomena behind a shock front in air.

The motion of the gas in the relaxation zone of a shock wave, as any other flow in which the dimensions of the relaxation zone are of the same order as the characteristic flow length, is described by "relaxation" gasdynamics. The full treatment of the problems is beyond the scope of this book, and the sixth (final) chapter is merely a brief survey of the flow properties of a gas undergoing relaxation.

We hope that this volume will be useful to working scientists in physical gasdynamics, high-temperature physics, chemical physics, and students in these disciplines.

The authors are deeply indebted to N. A. Generalov, Yu. P. Rayzer and Ye. V. Samuylov for their valuable suggestions, as well as to all colleagues who have helped in writing this book.

Contents

Chapter 3. Experimental Methods of Study of Nonequilibrium Phenomena in Shock Waves

Chapter 4. Relaxation Processes in Shock Waves

Chapter 5. Nonequilibrium Phenomena in Shock Waves in Air

Chapter 6. Flow of a Gas Undergoing Relaxation

I

Experimental Study of Shock-wave Structure

[1] CREATION AND STRUCTURE OF SHOCK WAVES

[1] Introduction

Phenomena to which this book is devoted are a result of two important properties of moving gases: 1) the "nonlinear" character of the flow, which is expressed in nonlinearity of the basic differential equations of gasdynamics; and 2) statistical inequilibrium, which is peculiar to a moving medium (excluding certain types of motion; for example, uniform translational motion of the system as a whole in a homogeneous field of external forces).

Under certain conditions the nonlinear effects result in the appearance of surfaces at which the flow variables, such as, the velocity, density, pressure, etc., become discontinuous, even if the initial disturbance of the medium was sufficiently smooth. The statistical inequilibrium takes on a particular, sharply expressed character near these discontinuity surfaces and determines the complex structure of shock waves.

In hydronamics of an ideal fluid, shock waves are geometrical surfaces separating two regions of thermodynamically equilibrium states of a medium. It follows from the laws of conservation of mass, momentum and energy, and from the second law of thermodynamics applied to the propagation of such surfaces, that when a medium passes through a shock wave the entropy increases in a jump. Thus, the dissipative processes which transform the medium from one thermodynamic equilibrium state to another are localized in the geometric surfaces of the discontinuity.

This idealized picture of shock waves and the dissipation processes bound up with them is a result of approximations which serve as a basis of Euler's equation for the motion of an ideal fluid. From the molecular-kinetic viewpoint, the state at which a gas is at thermodynamic equilibrium is a state of total statistical equilibrium, which includes equilibrium distribution of the energy over all degrees of molecular, atomic, ionic and electronic freedom, as well as chemical and ionization equilibrium between these particles. (We disregard problems related to radiation.) Dissipative processes in shock waves are a complex process of transition to a new statistical equilibrium state. The time needed for completing this process determines the thickness of the shock wave which is defined as the width of the nonequilibrium transitional layer separating two thermodynamically equilibrium states of the gas.

The structure of this transitional layer is determined by the character and the rate of the relaxation process which takes place in it, the latter process being defined as that of establishing statistical equilibrium. As will be seen below, the structure of sufficiently weak shock waves can be exhaustively described by the Navier-Stokes equations for a viscous and thermally conducting fluid. When the gradients of the flow variables (velocity, density, temperature, etc.), are small, it is sufficient to add the viscosity and thermal conductivity coefficients to equations of hydrodynamics in order to fully take into account the statistical nonequilibrium of the gas, which in this case is expressed in relatively minor corrections to the equilibrium distributions.

However, the picture is entirely different in the case of strong shock waves. When the gas passes through the leading "boundary" of the transitional layer, the state of the former changes so rapidly that strong disturbances to the statistical equilibrium take place. Subsequent flow of the gas is accompanied by complex relaxation processes, *i.e.*, excitation of rotational, vibrational, electronic degrees of freedom, development of chemical transformations (for example, thermal dissociation), and, finally, thermal ionization.

Each of these processes has a characteristic relaxation time. The natural unit of measurement for relaxation times is the time required for a molecule to traverse its mean-free path. By order of magnitude it coincides with the time needed for establishing the (local) Maxwell distribution. The rate of propagation of the shock wave is, by order of magnitude, equal to the thermal velocities of the molecules. Hence, that part of the transition layer in which the Maxwellian distribution is being established has width of the order of the mean-free path, and in macroscopic gasdynamics it should be considered as a geometric surface of discontinuity. Above we have used the term "boundary" of the transition layer to describe precisely this zone. Henceforth, it will be called the front of a (strong) shock wave.

Depending on the conditions, the characteristic times of all the other processes (except for rotational relaxation) can be expressed, in the assumed units, by large numbers; for example, the time of vibrational relaxation of the order of thousands or tens of thousands of mean-free path traverse times (varying within wide limits). The processes for establishing the chemical and ionization equilibrium are slower. However, in our ordinary laboratory scales, complex processes in a gas compressed by a shock wave are very rapid; they occur in time intervals of the order of 0.1 up to 10^2 to 10^3 microseconds. This is responsible for the exacting demands required of experimental methods for studying the shock-wave structure.

The present chapter is introductory in character. Here we consider certain results of the shock-wave theory (Sect. 1), and the elementary theory of relaxation processes in gases (Sect. 2). Also, a short outline of the experimental methods used in the study of shock-wave structure is given (Sect. 3).

[2] Origin of the Discontinuity in an Ideal Fluid. The Hugoniot Curve

The evolution of a smooth disturbance into a discontinuous one was theoretically discovered by Riemann [169], Rankine [480] and Hugoniot [385] in 1860–1890. They also formulated the basic properties of shock waves, which are defined as surfaces of discontinuity of flow variables. Subsequently, in our discussion, the theory of one-dimensional flow and of plane shock waves is of primary importance as the first, and in many cases, sufficient, approximation of the actual processes. It is true that it is then necessary to consider the possible effect of various factors (solid-body surfaces, which make the problem no longer one-dimensional, etc.). In this section we shall restrict ourselves only to the presentation of the most general results of this theory, which serves as a reliable basis in the interpretation of a number of phenomena in gasdynamic flows. (For a more detailed presentation of the shock-wave theory, see [69, 71, 98]).

Following Riemann's presentation, which is ideal because of its simplicity and accuracy, we now consider the formation of the surface of discontinuity in a one-dimensional flow of an ideal fluid with an initially continuous velocity and density distribution.

Let all the quantities under consideration be a function only of the coordinate x and the time t. Then the equation of motion of an ideal fluid and the continuity equation will take on the form:

$$\left.\begin{aligned} \frac{\partial v}{\partial t} + v\frac{\partial v}{\partial x} &= -\frac{1}{\varrho}\frac{\partial p}{\partial x}, \\ \frac{\partial \varrho}{\partial t} + \frac{\partial}{\partial x}\varrho v &= 0. \end{aligned}\right\} \qquad (1\text{–}1)$$

Here v is the gas velocity at all points of the plane intersecting the x-axis at point x, p is the pressure and ϱ is the density.

It is assumed that the flow is isentropic, *i.e.*, the entropy of a unit mass of the gas is regarded as being constant everywhere. Then p and ϱ are related by

$$p = \varphi(\varrho), \tag{1–2}$$

whereupon it is assumed that

$$(\mathrm{d}p/\mathrm{d}\varrho) = \varphi'(\varrho)$$

is positive for all values of ϱ and increases together with ϱ.

Eqs. (1–1) can be rewritten as

$$\left.\begin{aligned} \frac{\partial v}{\partial t} + v\frac{\partial v}{\partial x} &= -\varphi'(\varrho)\frac{\partial \ln \varrho}{\partial x}, \\ \frac{\partial \ln \varrho}{\partial t} + v\frac{\partial \ln \varrho}{\partial x} &= -\frac{\partial v}{\partial x}. \end{aligned}\right\} \tag{1–3}$$

Introducing the notation:

$$\left.\begin{aligned} \int \sqrt{\varphi'(\varrho)}\,\mathrm{d}\ln \varrho &= f(\varrho), \\ 2r = f(\varrho) + v,\ 2s &= f(\varrho) - v, \end{aligned}\right\} \tag{1–4}$$

multiplying the second of Eqs. (1–3) by $\pm\sqrt{\varphi'(\varrho)}$ and adding it to the first, we get

$$\left.\begin{aligned} \frac{\partial r}{\partial t} &= -(v + \sqrt{\varphi'(\varrho)})\frac{\partial r}{\partial x}, \\ \frac{\partial s}{\partial t} &= -(v - \sqrt{\varphi'(\varrho)})\frac{\partial s}{\partial x}, \end{aligned}\right\} \tag{1–5}$$

or, since

$$\mathrm{d}r = \frac{\partial r}{\partial t}\mathrm{d}t + \frac{\partial r}{\partial x}\mathrm{d}x$$

and

$$\mathrm{d}s = \frac{\partial s}{\partial t}\mathrm{d}t + \frac{\partial s}{\partial x}\mathrm{d}x,$$

we get

$$\left.\begin{aligned} \mathrm{d}r &= \frac{\partial r}{\partial x}\{dx - (v + \sqrt{\varphi'(\varrho)})\,\mathrm{d}t\}, \\ \mathrm{d}s &= \frac{\partial s}{\partial x}\{dx - (v - \sqrt{\varphi'(\varrho)})\,\mathrm{d}t\}. \end{aligned}\right\} \tag{1–6}$$

It follows from these equations that the plane which is perpendicular to the x-axis and which moves with the velocity

$$\frac{\mathrm{d}x}{\mathrm{d}t} = v + \sqrt{\varphi'(\varrho)}, \tag{1–7}$$

where v and ϱ are values taken in this plane, is characterized during the entire time when it is in motion by a constant value r; $r=\text{const}$. Similarly, $s=\text{const}$ for the plane which moves with the velocity

$$\frac{dx}{dt} = v - \sqrt{\varphi'(\varrho)}. \tag{1–8}$$

The motion of planes with constant values of r or s can be considered as the motion (or propagation) of these values of r and s. The quantity $\sqrt{\varphi'(\varrho)}$ is equal to the speed of sound. Hence, it follows from Eqs. (1–7) and (1–8) that the values of r and s propagate relative to the medium in opposite directions with the speed of sound (which varies from point to point).

Consequently, each value of r will sooner or later meet each value of s which is ahead of it.

Let us imagine that the disturbance was bounded first along the x-axis by the interval ab ($a<b$). Outside of this interval ϱ and v and, consequently, r and s are constant and equal to

$$\varrho_1, v_1, r_1, s_1 \quad \text{for} \quad x \leqslant a,$$

$$\varrho_2, v_2, r_2, s_2 \quad \text{for} \quad x \geqslant b.$$

We denote the boundaries of the region where r is variable by a' and b' and, correspondingly, the boundaries of the variable values of s by a'' and b''. (It may be assumed that at the initial instant of time these regions are identical with ab, *i.e.*, that then $a' \doteq a'' = a$ and $b' = b'' = b$.)

With time, each of the values of r in the region $a'b'$, propagating in the positive x direction with the velocity (1–7) will meet the value $s=s_2$; starting with this time region $a'b'$ will have a constant value of s which is equal to s_2. Similarly, the region $a''b''$ of variable values of s (each of which propagates with the speed of sound relative to the medium in the negative x direction) starting with a certain time instant will be characterized by a constant value of $r=r_1$.

Initially, the disturbance is divided into two waves which travel (relative to the medium) in opposite directions. The wave moving forward (region $a'b'$) propagates through a medium with $s=s_2$, which does not change on the passing of the wave. On the propagation of the wave moving in the opposite direction (region $a''b''$) it is the quantity $r=r_1$ which remains constant. In the space between them

$$r = r_1, \quad s = s_2.$$

Let us now consider the propagation of a known value of r in the traveling wave $a'b'$. Here $s=s_2$, and, hence, we find from Eq. (1–4) that the velocity v in any point in this wave is related with the density ϱ in the following manner:

$$v = f(\varrho) - 2s_2.$$

The plane which is described by these values of ϱ and v and, consequently, is characterized by the corresponding value of r, moves in the positive x direction with a velocity which is

$$\sqrt{\varphi'(\varrho)} + v = \sqrt{\varphi'(\varrho)} + f(\varrho) - 2s_2 .$$

Similarly, we find that in the wave traveling in the negative x direction, the given values of ϱ and v move in this direction with the velocity

$$\sqrt{\varphi'(\varrho)} + f(\varrho) - 2r_1 .$$

Thus, in traveling waves, high density values have their corresponding high rates of propagation.

If the density distribution along the x-axis is described by a curve in a plane in which the densities are laid off on the ordinate axis and the x-axis serves as the abscissa axis, then the results thus obtained mean that large coordinates move with high velocities. If the traveling wave is a compression wave, *i.e.*, a wave in which the density decreases in the direction of propagation, then the points with the large ordinate [great distance from the origin] should overtake points with smaller ordinates and, consequently, starting with a certain time, the density ϱ in a certain region would not be a unique function of x. Since this is impossible, then it follows that a condition appears which makes the above relationship governing the motion of ordinates invalid. In fact, this relationship is based on the assumption that v and ϱ as functions of x are continuous and that their first derivatives are finite. These assumptions cease to be valid as soon as the tangent to the density distribution curve is found to be normal to the x-axis at any point, which, sooner or later, must happen in a compression wave. A discontinuity surface is produced where a lower density directly precedes a higher density. Thus, the compression waves become increasingly thinner and, finally, are transformed into shock waves.

Starting with the time of the formation of discontinuity, Eqs. (1–1) and (1–2) are no longer sufficient for describing the fluid flow. The flow equations must be supplemented by conditions at the discontinuity surface. It is found that it is sufficient to specify satisfaction of the laws of conservation of mass, momentum and energy fluxes attendant to the passing of the shock wave through an element of the volume of the medium.

Let us consider the flow in a coordinate system moving with the discontinuity surface. The expressions for the mass, momentum and energy flux densities have, correspondingly, the form (one-dimensional flow) [98]:

$$\varrho v,\ p + \varrho v^2,\ \varrho v\left(\frac{v^2}{2} + h\right); \qquad (1\text{–}9)$$

where h is the enthalpy per unit mass.

The conservation laws require that these fluxes be continuous at the discontinuity surface. Hence, conditions at this surface can be written as

$$\left.\begin{aligned} \varrho_1 v_1 &= \varrho_2 v_2 , \\ p_1 + \varrho_1 v_1^2 &= p_2 + \varrho_2 v_2^2 , \\ \frac{v_1^2}{2} + h_1 &= \frac{v_2^2}{2} + h_2 . \end{aligned}\right\} \qquad (1\text{–}10)$$

From Eq. (1–10) we get the following relationships for a shock-wave propagating in a stationary medium:

$$V = \tilde{v}_1 \sqrt{\frac{p_2 - p_1}{\tilde{v}_1 - \tilde{v}_2}}; \qquad (1\text{–}11)$$

$$u = (\tilde{v}_1 - \tilde{v}_2) \sqrt{\frac{p_2 - p_1}{\tilde{v}_1 - \tilde{v}_2}}; \qquad (1\text{–}12)$$

and

$$\varepsilon_1 - \varepsilon_2 = \tfrac{1}{2}(p_2 + p_1)(\tilde{v}_2 - \tilde{v}_1). \qquad (1\text{–}13)$$

Here $\tilde{v}$ is the specific volume, ε is the specific internal energy, u is the velocity of flow behind the shock wave and V is the rate of propagation of the discontinuity. Here, as in the following, the subscript 1 is assigned to the undisturbed medium and the subscript 2 pertains to the disturbed medium.

Other conditions for the discontinuity surface were obtained by Riemann on the basis of the laws of conservation of mass and momentum and on the assumption that the change of state of the gas on the passing of the shock wave is isentropic. This contradicted the law of conservation of energy.

Eq. (1–13) contains only thermodynamic quantities to both sides of the discontinuity surface. The specific internal energy ε is determined by the values of p and ϱ, *i.e.*, $\varepsilon(p, \varrho)$. Hence, for a specified initial state, Eq. (1–13) establishes the relationship between p_2 and $\tilde{v}_2$, which is called the Hugoniot curve or shock adiabatic (which was found independently by Rankin and Hugoniot).

The Hugoniot curve relates the states to various entropy values (in contrast to the isentropic curve*. Its describing curve in the $p_2, \tilde{v}_2$ plane passes [as can easily be seen from Eq. (1–13)] through a point with coordinates $p_2 = p_1$, $\tilde{v}_2 = \tilde{v}_1$. This point divides the Hugoniot curve into two parts, one which corresponds to values of $s_2 > s_1$, where s is the specific entropy, and the other which corresponds to values of $s_2 < s_1$. The conservation laws are symmetrical with respect to the initial and final states of the medium and thus do not determine the direction of propagation of the discontinuity. It is determined by the requirement that the entropy must increase.

* [Transl. Note: termed by the Russians as Poisson's adiabatic.]

A change in the entropy on passing through the discontinuity surface means that conditions at the discontinuity surfaces cannot simply be added to Eqs. (1–1) and (1–2), but also that the latter be changed. With the appearance of a discontinuity the flow is no longer isentropic, and from this instant p must be regarded as a function of two independent variables, for example, ϱ and s. Thus, Eq. (1–2) is replaced by

$$p = \varphi(\varrho, s). \tag{1–14}$$

[3] Dissipative Processes in Shock Waves

In the Riemann scheme of formation and motion of shock waves, the dissipation processes are localized in the geometrical surfaces of discontinuity of the flow variables. This is a result of approximations contained in Eq. (1–1), which is considered to be valid everywhere where the density, pressure and velocity gradients are finite. This equation was written without taking into account the viscosity and thermal conductivity, which is permissible for sufficiently small velocity and temperature gradients. However, as we have seen, in the process of shock-wave formation these gradients can become as large as desired. Hence, the question arises naturally concerning the possible role of these terms in the flow equations which contain the viscosity and thermal conductivity coefficients.

Viscosity and thermal conductivity, playing an increasingly important role as the space nonhomogeneity increases, change the picture of formation of the shock wave as well as its structure. Instead of a geometric discontinuity surface there appears a transition layer of finite thickness. Assuming that the flow in this transition layer is (quasi-) steady, it is possible, as this is entirely obvious from the conservation laws, to obtain Eq. (1–13) for thermodynamic quantities to both sides of the transition layer (let us note the obvious arbitrariness of the concept "thickness" of the transition layer, taking into account the asymptotic nature of the change in the state of the medium at its "boundaries").

Let us estimate the order of magnitude of the thickness Δx of the transition layer, assuming that it is in a steady-state condition, and restricting ourselves, for simplicity, to weak shock waves.

First, let us consider the case when the dissipative processes are limited to thermal conductivity.

The equation describing the variation in the entropy density ϱs in this case has the form [98]:

$$\frac{\partial(\varrho s)}{\partial t} = -\operatorname{div}(\varrho s \boldsymbol{v}) + \frac{1}{T}\operatorname{div}(\lambda \operatorname{grad} T); \tag{1–15}$$

where λ is the coefficient of thermal conductivity. Transforming the last term in the right-hand side, Eq. (1–15) can be written as

$$\frac{\partial(\varrho s)}{\partial t} = -\operatorname{div}(\varrho s \boldsymbol{v}) - \operatorname{div}\boldsymbol{S} + \sigma, \tag{1–16}$$

where use is made of the notation:

$$\boldsymbol{S} = -\lambda\frac{\operatorname{grad} T}{T}; \tag{1–17}$$

$$\sigma = \lambda\left(\frac{\operatorname{grad} T}{T}\right)^2. \tag{1–18}$$

Let us isolate a volume Γ bound by surface Ω, which is stationary, relative to the coordinate system under consideration. Integrating both sides of Eq. (1–16) over this volume and transforming the integrals of the first two terms in the right-hand side over the surface Ω, we get

$$\frac{\partial}{\partial t}\int_\Gamma \varrho s \, d\Gamma = -\int_\Omega \varrho s \boldsymbol{v} \, d\boldsymbol{\Omega} - \int_\Omega \boldsymbol{S} \, d\boldsymbol{\Omega} + \int_\Gamma \sigma \, d\Gamma, \tag{1–19}$$

where $d\Omega$ is a vector whose absolute magnitude is $d\Omega$ and which is directed along the inward normal to the surface element $d\Omega$.

Hence, the total change in the entropy of the medium in the volume thus isolated is broken up into three components. The first two terms in the right-hand side of Eq. (1–19) represent the rate of change of the entropy, which is produced by the flow of the entropy through the bounding surface. The vector $\boldsymbol{S}$ is called the entropy flux density vector. It determines that part of the entropy flux which is not related to the macroscopic flow of the fluid; the total entropy flux includes the convection term $-\varrho s\boldsymbol{v}$. The quantity denotes the local intensity of the entropy source. This quantity, which is of fundamental significance in the thermodynamics of irreversible processes, is called the local entropy generation (or local entropy production) and is defined as the rate of increase in the entropy of a unit volume as a result of irreversible processes which take place in it, in the given case of the heat conduction processes. These concepts (the entropy origin and the entropy flux density vector) did not have to be introduced explicitly into the estimates which are of interest to us here, but they will be needed in considering the propagation of strong shock waves in multiatomic gases (Chapter 6).

Integrating both sides of Eq. (1–16) over the steady-state layer under study, and taking into account the fact the vector $\boldsymbol{S}$ vanishes at the boundaries of the layer, we find the following expression for the total change in the entropy per unit time of the fluid (the magnitude of the change is referred to a unit area perpendicular to the x-axis):

$$\int_{-\infty}^{+\infty} \sigma \, dx = \varrho_1 V (s_2 - s_1), \tag{1–20}$$

where the integral in the left-hand side is practically determined by a certain finite transition layer.

Appropriately selecting the length Δx, this integral can be represented in the form:

$$\int_{-\infty}^{+\infty} \sigma \, dx = \sigma^* \Delta x; \tag{1–21}$$

where

$$\sigma^* = \lambda \left(\frac{T_2 - T_1}{\Delta x} \right)^2 \frac{1}{T_1^2}, \tag{1–22}$$

and, consequently,

$$\lambda \frac{(T_2 - T_1)^2}{T_1^2 \Delta x} = \varrho_1 V (s_2 - s_1). \tag{1–23}$$

Eq. (1–23) should be regarded as one of the possible definitions of the thickness Δx of a shock wave.

Changing to the notation

$$T_2 - T_1 = \Delta T, \quad s_2 - s_1 = \Delta s,$$

etc., and dropping the subscript of quantities pertaining to the undisturbed fluid, we write Eq. (1–13) of the Hugoniot curve in the form:

$$\Delta \varepsilon = - \left(p + \frac{\Delta p}{2} \right) \Delta \tilde{v}. \tag{1–24}$$

Considering ε as a function of $\tilde{v}$ and s, we expand $\Delta \varepsilon$ and Δp in a series in terms of powers of Δs and $\Delta \tilde{v}$, assuming that these quantities are sufficiently small. Since

$$\frac{\partial \varepsilon}{\partial \tilde{v}} = - p, \tag{1–25}$$

then the right-hand side of Eq. (1–24) can be represented as

$$- \left(p + \frac{\Delta p}{2} \right) \Delta \tilde{v} = \frac{\partial \varepsilon}{\partial \tilde{v}} \Delta \tilde{v} + \frac{1}{2} \frac{\partial^2 \varepsilon}{\partial \tilde{v}^2} (\Delta \tilde{v})^2 + \frac{1}{2} \frac{\partial^3 \varepsilon}{\partial \tilde{v}^3} \frac{(\Delta \tilde{v})^3}{2} + O(\Delta \tilde{v} \Delta s), \tag{1–26}$$

where $O(a)$ denotes a quantity of the order of a.

On the other hand,

$$\Delta\varepsilon = \frac{\partial\varepsilon}{\partial s}\Delta s + \frac{\partial\varepsilon}{\partial\tilde{v}}\Delta\tilde{v} + \frac{\partial^2\varepsilon}{\partial\tilde{v}^2}\frac{(\Delta\tilde{v})^2}{2} + \frac{\partial^3\varepsilon}{\partial\tilde{v}^3}\frac{(\Delta\tilde{v})^3}{6} + O((\Delta s)^2) + O(\Delta s\Delta\tilde{v}). \quad (1\text{–}27)$$

Substituting Eqs. (1–26) and (1–27) into (1–24), we see that terms proportional to $\Delta\tilde{v}$ and $(\Delta\tilde{v})^2$ cancel out. This means that Δs is a quantity of the third order of smallness if $\Delta\tilde{v}$ is assumed to be a quantity of the first order of smallness. Hence, dropping quantities with a higher order of smallness, we get

$$T\Delta s = -\frac{1}{12}\left(\frac{\partial^2 p}{\partial\tilde{v}^2}\right)_s (\Delta\tilde{v})^3. \quad (1\text{–}28)$$

The value of ΔT, to within the first order of smallness is equal to

$$\Delta T = \left(\frac{\partial T}{\partial\tilde{v}}\right)\Delta\tilde{v}. \quad (1\text{–}29)$$

It follows from Eqs. (1–11) and (1–28) that in weak shock waves it is possible to assume $V \sim c$, where c is the isentropic speed of sound. Hence, from Eqs. (1–28), (1–23) and (1–29), we get the following expression for the shock-wave thickness:

$$\Delta x = \frac{12\lambda}{T_1\varrho_1 c}\cdot\frac{\left(\frac{\partial T}{\partial\tilde{v}}\right)_{1s}}{\left(\frac{\partial^2 p}{\partial\tilde{v}^2}\right)_{1s}}\cdot\frac{1}{\tilde{v}_1 - \tilde{v}_2}. \quad (1\text{–}30)$$

The thickness of a shock wave in a viscous fluid (with zero thermal conductivity) can be estimated similarly.

The rate of change, in the entropy of a unit volume of a viscous fluid is determined by [54, 54a, 98]

$$\frac{\partial(\varrho s)}{\partial t} = -\operatorname{div}(\varrho s\boldsymbol{v}) + \sigma, \quad (1\text{–}31)$$

where σ the generation of entropy, which is related to irreversible processes of internal friction, has the form:

$$\sigma = \frac{\eta}{2T}\left(\frac{\partial v_i}{\partial x_k} + \frac{\partial v_k}{\partial x_i} - \frac{2}{3}\delta_{ik}\frac{\partial v_l}{\partial x_l}\right)^2 + \frac{\zeta}{T}(\operatorname{div}\boldsymbol{v})^2; \quad (1\text{–}32)$$

η and ζ are the shear and bulk viscosity coefficients, respectively.

For one-dimensional flow, Eq. (1–32) takes on the form:

$$\sigma = \frac{1}{T}\left(\frac{4}{3}\eta + \zeta\right)\left(\frac{\partial v}{\partial x}\right)^2. \quad (1\text{–}33)$$

Eq. (1–20), where σ is determined by Eq. (1–33), can be written as

$$\sigma^* \cdot \Delta x = \varrho_1 V (s_2 - s_1). \tag{1–34}$$

Here

$$\sigma^* = \frac{1}{T_1}\left(\frac{4}{3}\eta + \zeta\right)\left(\frac{v_1 - v_2}{\Delta x}\right)^2, \tag{1–35}$$

and Δx is defined as the thickness of the shock wave in a viscous fluid. It follows from Eqs. (1–11) and (1–12) [since $u = v_1 - v_2$], that

$$(v_1 - v_2)^2 = (\varrho_1 V)^2 (\tilde{v}_1 - \tilde{v}_2)^2. \tag{1–36}$$

Using Eqs. (1–36) and (1–28) we get from Eq. (1–34) [assuming that $V \sim c$]

$$\Delta x = 12\left(\frac{4}{3}\eta + \zeta\right)\frac{\varrho_1 c}{\left(\dfrac{\partial^2 p}{\partial \tilde{v}^2}\right)_{1s}} \cdot \frac{1}{\tilde{v}_1 - \tilde{v}_2}. \tag{1–37}$$

It follows from Eqs. (1–30) and (1–37) that the thickness of shock waves in a thermally conducting and viscous fluid increases beyond bounds with a reduction in the shock-wave strength; also, the gradients of pressure, velocity and other quantities in the transition layer decrease simultaneously. Hence, in describing the flow in a transition layer of a sufficiently weak shock wave, it is possible to use flow equations containing coefficients of thermal conductivity and viscosity. When using this approach it is not necessary to introduce the concept of the discontinuity surface. (It is obvious that in the case when the characteristic lengths in comparison with the transition layer thickness are large, it is possible to disregard the shock-wave structure and approximate it by a geometric discontinuity surface.)

However, the situation becomes different in the case of strong shock waves.

Eqs. (1–30) and (1–37) yield the thickness of shock waves which are characterized by small jumps in the fluid variables. However, they can be used for approximate estimates even for strong shock waves. Let us consider a shock wave in a monatomic gas, where the difference $\tilde{v}_1 - \tilde{v}_2$ is of the same order as the magnitudes of $\tilde{v}_1$ and $\tilde{v}_2$. By order of magnitude, the derivatives in Eq. (1–30) are equal to

$$\frac{\partial T}{\partial \tilde{v}} \sim -\frac{T}{\tilde{v}}, \frac{\partial^2 p}{\partial \tilde{v}^2} \sim \frac{p}{\tilde{v}^2},$$

c is a quantity of the order of magnitude of the mean-thermomolecular velocity, the coefficient of thermal conductivity is $\lambda \sim nkl_0c$, where n is the particle number density, k is the Boltzmann constant and l_0 is the mean-free path.

Using these estimates, from Eq. (1–26) we get

$$\Delta x \sim l_0. \tag{1–38}$$

The same result can be obtained from Eq. (1–37) if we take into account that for a monatomic gas $\eta \sim \varrho l_0 c$ and $\zeta = 0$.

Thus, as the shock waves become stronger, their thickness becomes comparable with the molecular mean-free path.

From the viewpoint of molecular-kinetic concepts, the equations of phenomenological gasdynamics and the concept of a continuous medium as such are valid within the framework of certain approximations. The field of their applicability is determined by the condition that the ratio of the mean-free length to the characteristic macroscopic length L be small:

$$\frac{l_0}{L} \ll 1 . \tag{1–39}$$

Here L is a distance over which the macroscopic quantities vary perceptibly.

This condition is contradicted by Eq. (1–38), according to which the thickness of the transition region is of the order of l_0. This means that the flow equations are inapplicable in this region and its structure should be considered from the molecular-kinetic viewpoint. Within the framework of the phenomenological description, strong shock waves are represented by geometric discontinuity surfaces. To both sides of these surfaces the flow can be described by flow equations containing the viscosity and thermal conductivity, but the principal contribution to the entropy generation is made by dissipation processes, which are localized in the discontinuity "surfaces."

These processes develop in a thin gas layer, the state of which is substantially nonequilibrium. Let us estimate the thickness of this layer. For the sake of simplicity, we shall restrict ourselves to the case of a shock wave in a monatomic gas ("structureless" particles). Then the state of the gas in the layer under study can be described by the distribution function $f(\boldsymbol{c}, \boldsymbol{r}, t)$, where $\boldsymbol{r}$ is the vector of the molecule position,* and $\boldsymbol{c}$ is its velocity. The expression $f(\boldsymbol{c}, \boldsymbol{r}, t)\, d\boldsymbol{r}\, d\boldsymbol{c}$ gives the molecular population in the element $d\boldsymbol{r}$ of the coordinate space and in the element $d\boldsymbol{c}$ of the velocity space in the vicinity of points $\boldsymbol{r}$ and $\boldsymbol{c}$, respectively.

In the case of a strong shock wave, f changes perceptibly across the layer thickness. Under these conditions the thermodynamic definition of entropy is inapplicable to the layer as a whole, as well as to its elements. The concept of entropy generation in the form used in the thermodynamics of irreversible processes is also inapplicable here, since it assumes a thermodynamic description of nonequilibrium systems.

However, both these concepts are made applicable to systems with a nonequilibrium distribution function $f(\boldsymbol{c}, \boldsymbol{r}, t)$ by using Boltzmann's H function.

* In gaskinetic problems, the term "molecule" is also used for monatomic particles.

The local value of the generalized entropy H of a unit volume is determined as follows (see, for example, [72]):

$$H = -k\int \ln f \cdot f d\boldsymbol{c}; \tag{1–40}$$

where k is Boltzmann's constant and the integration is performed over the entire velocity space*. For an equilibrium distribution function H is identical with the thermodynamic entropy.

Differentiating Eq. (1–35) with respect to time, we get

$$\frac{\partial H}{\partial t} = -k\int (1 + \ln f)\frac{\partial f}{\partial t}\, d\boldsymbol{c}. \tag{1–41}$$

Substituting into Eq. (1–41) the expression for $\partial f/\partial t$ from the Boltzmann kinetic equation (in the absence of external forces),

$$\frac{\partial f}{\partial t} + \boldsymbol{c}\frac{\partial f}{\partial \boldsymbol{r}} = \left(\frac{\partial f}{\partial t}\right)_e, \tag{1–42}$$

where $(\partial f/\partial t)_e$ is the collision integral which represents the rate of change of the distribution function as a result of molecular collisions, we get

$$\frac{\partial H}{\partial t} + \operatorname{div}\boldsymbol{S} = \sigma. \tag{1–43}$$

Here we have used the notation:

$$\boldsymbol{S} = -k\int \boldsymbol{c}\ln f \cdot f d\boldsymbol{c}; \tag{1–44}$$

and

$$\sigma = -k\int (1 + \ln f)\left(\frac{\partial f}{\partial t}\right)_e d\boldsymbol{c}. \tag{1–45}$$

The interpretation of Eqs. (1–43), (1–44) and (1–45) is obvious. The vector $\boldsymbol{S}$ is the generalized entropy flux density vector, and σ is the generation of the generalized entropy.

Boltzmann has shown that σ is positive for gases not at equilibrium and zero in the case of statistical equilibrium:

$$\sigma \geqslant 0. \tag{1–46}$$

* The molecules are regarded as being classical particles. For quantum particles, the motion of which is quasi-classical,

$$H = k\int f \ln \frac{em^3}{fh^3}\, dc.$$

Here h is Planck's constant, m is the mass of a molecule and e is the base of natural logarithms.

The definition (1–40) and Eqs. (1–44)—(1–46) together with the conclusion which follows from them on the monotonic increase in the (generalized) entropy of the isolated system make up the content of Boltzmann's fundamental H theorem.

Now let us use these results to estimate the thickness of the transition layer in strong shock waves.

It can be seen from Eq. (1–45) that the intensity of the dissipation processes in the transition layer is determined by the collision integral, *i.e.*, by the effectiveness of collisions as a mechanism which restores the disturbed Maxwellian distributions. This effectiveness can be determined as follows. The collision integral has the form [72]:

$$\left(\frac{\partial f}{\partial t}\right)_e = \int\limits_{(c_1)}\int\limits_0^{2\pi}\int\limits_0^{\infty} |\boldsymbol{c}_1 - \boldsymbol{c}|\cdot(f'f'_1 - ff_1)\, b\, db\, d\varphi\, d\boldsymbol{c}_1 . \qquad (1\text{–}47)$$

Here $\boldsymbol{c}$ and $\boldsymbol{c}_1$ are the molecular velocities before collision, $\boldsymbol{c}'$ and $\boldsymbol{c}'_1$ are the molecular velocities following the collision, $f_1 \equiv f(\boldsymbol{c}_1), f' \equiv f(\boldsymbol{c}'), f'_1 \equiv f(\boldsymbol{c}'_1)$, b is the focal distance, and φ is the angular collision parameter.

When f deviates perceptibly from the Maxwellian distribution, the left-hand side of Eq. (1–47) is, by order of magnitude, equal to f/τ, where τ is the relaxation time (τ is the characteristic time of the process of restoration of the local Maxwellian distribution). We can estimate f as being given by $fc_{av}^3 \sim n$, where c_{av} is of the order of some mean-thermal velocity. Substituting these estimates into Eq. (1–47) and denoting the effective range of intermolecular forces by r_0, we get (dropping numerical multipliers)

$$\frac{n}{\tau c_{av}^3} \sim c_{av} \frac{n^2}{c_{av}^6} r_0^2 c_{av}^3 . \qquad (1\text{–}48)$$

Whence

$$\tau \sim \frac{1}{r_0^2 c_{av} n} . \qquad (1\text{–}49)$$

The denominator in (1–49) is, by order of magnitude, equal to the number of collisions which the molecule undergoes per unit time. Hence by order of magnitude, τ is found to be equal to the mean time for traversing a mean-free path.

This estimate characterizes the extraordinary effectiveness of collisions as a mechanism for the restoration of statistical equilibrium in a monatomic gas and is in agreement with the previously presented estimates of the transition layer thickness in sufficiently strong shock waves.

In fact, assuming that the transition layer is at steady state, and integrating both sides of Eq. (1–43) with respect to x, we get

$$S_2 - S_1 = \sigma^* \Delta x, \tag{1–50}$$

where S_1 and S_2 are values of the entropy flux density at the boundaries of the layer, σ^* is a certain average value of entropy generation and Δx is the thickness of the transition layer.

Estimating both sides of Eq. (1–50) with respect to the order of magnitude using Eqs. (1–44) and (1–45) [assuming that the changes in f, c_{av}, etc., on passing through this layer are quantities of the same order of magnitude as f, c_{av}, etc., respectively], we get

$$kc_{av} \ln f \cdot fc_{av}^3 \sim k \frac{\ln f \cdot f}{\tau} c_{av}^3 \Delta x,$$

i.e.,

$$\Delta x \sim c_{av}\tau \sim l_{0\,av}. \tag{1–51}$$

This result is identical to that given by Eq. (1–38). The width of the transition layer in sufficiently strong shock waves is found to be of the order of the mean-free molecular path. In macroscopic gasdynamics, which is based on the consideration of a gas as a continuous medium,* these lengths are equal to zero. It follows from this that, in the case of sufficiently strong shock waves, the consideration of the macroscopic gasdynamics of the viscosity and heat-conduction processes in equations does not eliminate the necessity of introducing geometric discontinuity surfaces, which approximate the thin transition layer.

[4] Shock Waves in Multiatomic Gases

The estimate given in Eq. (1–38) and the conclusions which follow from it, which is formulated above, pertain primarily to monatomic gases (structureless particles). The formal application of gasdynamic equations to the description of shock-wave structures in multiatomic gases does not change these results, since the bulk viscosity coefficient ζ and the thermal conductivity λ of multiatomic gases are quantities of the same order of magnitude as η and λ of monatomic gases, respectively. However, here we encounter an important feature which is related to the nature of volume viscosity.

From the viewpoint of molecular-kinetic concepts, volume viscosity is related to relaxation processes which occur in the gas, attendant to changes in its volume. Retarded energy transfer between various molecular degrees of freedom, displacement of the chemical equilibrium, etc., can serve as

* [Transl. Note: continuum, as commonly called.]

these processes. In connection with this, in the theory of the coefficient of volume viscosity with the time τ_0 of traversing a mean-free path, another characteristic time appears, *i.e.*, the relaxation time τ, and in the theory of gasdynamic flows appears the characteristic length $l \sim c\tau$. If $\tau \gg \tau_0$, then l is a macroscopic quantity (as is τ).

It is assumed in gasdynamic equations that the change in the states of a fluid in gasdynamic flows can be described by the equation of state of a thermodynamically equilibrium medium. Obviously, this assertion is valid only for sufficiently smooth changes of state. The quantitative criterion for the applicability of the equations of "equilibrium" gasdynamics is the condition:

$$L \gg l \sim c\tau; \qquad (1\text{–}52)$$

where L is the previously described characteristic macroscopic length. It is known from the theory of volume viscosity [98, 99, 119] that the condition given by Eq. (1–52) is also needed so that it will be possible to consider the bulk viscosity coefficient ζ as a "constant" of the fluid. Condition (1–52) is, evidently, more rigorous than (1–39). Thus in view of these considerations, let us evaluate the situation which arises in the Riemann picture of formation and motion of shock waves.

A traveling compression wave (this concept is rigorously defined for an ideal fluid, while here it is valid only as an approximate description) is characterized by an increase in gradients with time, which results in the fact that the spatial nonhomogeneity becomes perceptible over a length of the order of l. From this "instant" (the fact that it cannot be defined precisely is obvious), the criterion of Eq. (1–52) is not satisfied. Flows arise and the description of them requires substantial changes in the equations of macroscopic gasdynamics. These equations lose their validity at three points: 1) the equation of state of a fluid in thermodynamic equilibrium is no longer valid; 2) the contribution of viscous stresses made to the tensor by the term containing the volume viscosity coefficient loses its meaning; and 3) a change takes place in that part of the thermal flux which is due to the transfer of energy of internal ("relaxational") degrees of freedom. These three points have a common basis, *i.e.*, substantial nonequilibrium of the states which are produced in gasdynamic flows under certain conditions.

Thus, the evolution of a smooth disturbance into a strong shock wave in a multiatomic gas passes through a stage at which the gasdynamics equations should take into account the lack of thermodynamic equilibrium of the states of the gas and the relaxation processes which take place following this. The same situation arises in the steady-state problem in the description of the structure of a strong shock wave in a multiatomic gas.

As was pointed out, the discontinuity surface, which represents a shock wave in a monatomic gas, approximates the transition layer, the thickness of which is of the order of a mean-free path. In this layer the molecular velocity distribution function changes rapidly from the Maxwellian distribution, which corresponds to the gas ahead of the shock wave, to another Maxwellian distribution for the gas behind the shock wave. Let us imagine that a moderate amount of diatomic molecules has been added to a monatomic gas which practically does not change (by virtue of its moderateness) the fluid properties. If $\tau \gg \tau_0$, where τ_0 is the time required for traversing a mean-free path and τ is the time for establishing an equilibrium distribution of energy between the internal and translational degrees of molecular freedom, then, on passing through the shock wave the internal energy of the molecules remains practically unchanged. Its equilibrium distribution, which corresponds to the gas temperature behind the shock wave, will be established after a time of the order of τ. Hence, a region with a nonequilibrium energy distribution exists behind the discontinuity surface. Its width is of the order of $v_2\tau$, where v_2 is the rate of propagation of the shock wave relative to the disturbed fluid.

Turning now to a shock wave in a multiatomic gas, we will note that the inequality $\tau \gg \tau_0$ means that the overwhelming majority of collisions are elastic. Hence, the discontinuity surface in a multiatomic gas approximates a thin transition layer, where the gas behaves practically in the same manner as a monatomic gas. Henceforth, this layer will be called the shock-wave front*. Immediately after passing the shock front, the velocity distribution of the translational molecular motion is found to be Maxwellian and corresponds to such values of the temperature, density and flow velocity, which are related to the flow parameters ahead of the shock wave by the Hugoniot relationship for a monatomic gas. However, in practice, the energy distribution with respect to other degrees of freedom directly behind the shock front does not have sufficient time to change.

The relative slowness of the process of establishing the equilibrium energy distribution between the internal and translational degrees of freedom ($\tau \gg \tau_0$) results in the fact that the shock front is followed by a region of incomplete statistical equilibrium, that is, by a "relaxation zone," the width l of which is large in comparison with l_0; $l \gg l_0$. Hence, the structure of this region can be studied by methods of macroscopic gasdynamics. However, the system of gasdynamic equations must be supplemented by equations which take into account the relaxation processes in the fluid.**

* [Transl. Note: shock front.]

** The effect of the retarded energy transfer between the internal and translational degrees of freedom on the shock-wave structure was considered in the monograph by Zel'dovich [69].

At distances greater than l the relaxation processes practically end, and if the flow takes place under steady-state conditions, then the state of the gas immediately behind the relaxation zone satisfies the Hugoniot curve, in which the thermodynamic variables are assumed to be equal to their values in the state of complete thermodynamic equilibrium. Obviously, the specification of (quasi-)steady-state conditions is necessary in order to be able to use conservation laws which relate the states of the gas ahead of the shock wave and behind the relaxation zone.

The problem of the relaxation times for various degrees of freedom will be considered in more detail in the following section. Here it will only be noted that the time required for establishing equilibrium between the translational and rotational degrees of freedom is usually (by the order of magnitude) close to τ_0, *i.e.*, to the time needed for traversing a mean-free path (being slightly greater than the latter). Here the applicability [equations] of macroscopic gasdynamics is problematic. Hence, it is advantageous to approximate by the discontinuity surface that layer where both the Maxwellian distribution and the equilibrium between the translational and rotational degrees of freedom are established. The states of the gas directly to both sides of this surface are, evidently, connected by the Hugoniot relationship, where it has been considered that the remaining degrees of freedom are practically excluded when passing through it. This has an effect (not the same) on the relaxation zone and beyond its boundaries.

Thus, in the theory of shock waves in multiatomic gases, it becomes necessary to generalize the flow equations in a manner such that they would take into account the relaxation processes in the fluid. As we have seen, this is a result of two basic problems: 1) the evolution of a smooth disturbance into a shock wave; and 2) the structure of the steady-state shock wave. However, the second problem is of primary interest to us. We note concurrently that at present, there is no analytic solution of the problem of shock-wave formation which would be a generalization of the results obtained by Riemann or Hugoniot for the relaxation media and which would include the description of all the successive stages: the stage of the Navier-Stokes flow, appearance of a region which is described by higher-approximation equations (Barnett-type equations [220], etc.)—this stage is of short duration and of little consequence, and, finally, the stage of explicitly expressed relaxation processes which directly precede the appearance of the "geometrical" discontinuity surface. This surface approximates a thin layer the structure of which cannot be studied by methods of macroscopic gasdynamics.

[2] RELAXATION PROCESSES IN GASES (ELEMENTARY THEORY)

When the equilibrium is disturbed, relaxation processes arise in the system which attempt to return it to the state of the total statistical equilibrium. In di- and multiatomic gases these are processes which result in the establishment of equilibrium with respect to individual degrees of molecular freedom: translational; rotational; vibrational; electronic; and also in the establishment of a chemical and ionization equilibrium.

Let us consider each of these elementary processes separately, hence, we shall simplify the problem as much as possible. We assume that the gas with a nonequilibrium distribution with respect to any one given degree of freedom (which undergoes relaxation) comprises a moderate addition to a monatomic gas which is at equilibrium. Also, it will be assumed that the system as a whole is at rest. To estimate the order of magnitude of the relaxation time, we shall assume that the process of relaxation of the translational, rotational and vibrational degrees of freedom is described by

$$\frac{\mathrm{d}E(t)}{\mathrm{d}t} = -\frac{E(t) - E_0}{\tau}, \qquad (2\text{–}1)$$

where $E(t)$ is the energy of the appropriate degrees of freedom per unit volume, E_0 is the equilibrium value of this energy, and τ is the relaxation time. Eq. (2–1) determines the relaxation time τ, which can be written in the form:

$$\tau = \frac{E_0}{\left(\dfrac{\mathrm{d}E}{\mathrm{d}t}\right)_{E=0}}. \qquad (2\text{–}2)$$

Eq. (2–1) is the simplest relaxation equation. For states close to that of equilibrium, Eq. (2–1) can be obtained for, example, by expanding $\partial E/\partial t$ in a series with respect to the powers of the deviation $E - E_0$, which is assumed to be small, and by restricting ourselves here to the first nonvanishing term. As will be shown in Chapter 4, in states far from equilibrium, Eq. (2–1) is not always valid. However, even in this case it is possible to use Eq. (2–2) to estimate the relaxation time.

In general, a more rigorous determination of the relaxation time requires the use of another approach. First, one must solve the dynamic problem and find the probabilities of various elementary actions. Second, the probabilities thus obtained should be used to set up gas-kinetic equations, the solution of which will yield the relaxation time. The problem thus stated is considered in Chapter 4. However, here we restrict ourselves only to elementary considerations, making no attempt to go beyond the scope of classical physics.

A comparison of the results obtained in this manner, with the more exact quantum-mechanical calculations and the determination of the limits of their applicability, will be given in Chapter 4.

[1] Establishment of the Maxwellian Distribution

The fastest process in a shock wave which is produced by collisions is the energy transfer between the translational degrees of freedom. The relaxation time of this process can be estimated by using Eq. (2–2). Let the monatomic gas A comprise a moderate admixture in gas B which is at the temperature T. As a result of collisions between molecules of A and B (collisions between the molecules of A can be neglected), the gas A is transformed from a certain initial state into a state with temperature T. Here $E_0=\frac{3}{2}kTn_A$, where n_A is the number density of molecules of gas A.

To evaluate $\partial E/\partial t$ we will use a model of solid elastic spheres, *i.e.*, we shall assume that the gas molecules can be represented in the form of elastic spheres with diameter d_A for gas A and with diameter d_B for gas B. In addition, we shall assume that all the collisions take place along the lines of centers (head-on collisions), *i.e.*, we shall consider conditions which are most favorable for transfer of translational energy. In this case

$$\left.\frac{\mathrm{d}E}{\mathrm{d}t}\right|_{E=0} = n_A \int \Delta\varepsilon Z^{(0)}(v)\,\mathrm{d}v\,, \tag{2–3}$$

where $Z^{(0)}(v)\,\mathrm{d}v$ is the number of collisions which a molecule of gas A (with a mass m_A) undergoes with molecules of gas B (with a mass m_B) which have a velocity of absolute values in the interval between v and $v+\mathrm{d}v$; $\Delta\varepsilon$ is the energy transferred to a stationary molecule m_A (according to Eq. (2–2), $\mathrm{d}E/\mathrm{d}t$ is calculated at $E=0$) on a single collision with molecule m_B. In head-on collisions of elastic spheres

$$\frac{\Delta\varepsilon}{\varepsilon} = \frac{4m_A m_B}{(m_A+m_B)^2}, \quad \varepsilon = \frac{m_B v^2}{2}. \tag{2–4}$$

Since it is assumed that gas B has a Maxwellian velocity distribution, then $Z^0(v)\,\mathrm{d}v$ has the form:

$$Z^{(0)}(v)\,\mathrm{d}v = n_B d^2 \left(\frac{2\pi kT}{m_B}\right)^{\frac{1}{2}} \left(\frac{m_B}{kT}\right)^2 e^{-\frac{m_B v^2}{2kT}} v^3\,\mathrm{d}v; \tag{2–5}$$

where n_B is the number density of molecules of gas B, and $d=(d_A+d_B)/2$. Substituting Eqs. (2–4) and (2–5) into (2–3), we get

$$\left.\frac{\mathrm{d}E}{\mathrm{d}t}\right|_{E=0} = \frac{4m_A m_B}{(m_A+m_A)^2} Z2kT\,, \tag{2–6}$$

where Z is the total number of collisions which molecules of gas A undergo with molecules of gas B per unit time. This number is equal to

$$Z = 2d^2 n_B \left(\frac{2\pi kT}{m_B}\right)^{\frac{1}{2}}. \tag{2–7}$$

If now we substitute the expression for E_0 and dE/dt into Eq. (2–2), then the relaxation time can be written as

$$\tau_{\text{trans}} = \frac{3}{4} \frac{(m_A + m_B)^2}{4m_A m_B} \cdot \frac{1}{Z}. \tag{2–8}$$

Eq. (2–8) becomes very simple in the case of $m_A \approx m_B$. In this case

$$\tau_{\text{trans}} \sim \frac{1}{Z} \equiv \tau_0\,, \tag{2–9}$$

where τ_0 is the mean time of traverse of a mean-free path and τ_{trans} is the time for establishing the Maxwellian distribution (for the translational degrees of freedom).

The equilibrium with respect to the translational degrees of freedom in gas A is thus established during a time of the order of the average time between two collisions. In this form this result is independent of the molecular model and is applicable also for a single-component system. For shock waves this result means that the process of establishing the Maxwellian distribution in the first approximation takes place over a distance of the order of a mean-free molecular path.*

Since, as we shall show in the following, all the remaining processes related to collisions and resulting in the establishment of total statistical equilibrium take place during a time greater than τ_0, the introduction of the concept of a shock-wave front as a region becomes clear. In this region, in the first approximation, a Maxwellian distribution is established. The isolation of the shock-wave front is also justified by the fact that the concept of temperature can be introduced for processes behind the front. From the viewpoint of gasdynamic considerations (see Sect. 1), the shock front is equivalent to the concept of the discontinuity surface.

Eq. (2–8) makes it possible to analyze one particular case which is of interest. Let us consider the process of establishing the equilibrium with respect to the translational degrees of freedom in the heavier component A

* The relaxation time τ_{trans} thus found, generally speaking, corresponds to the time needed for establishing the Maxwellian distribution in a region with energies of the order of kT. It is not *a priori* evident that τ_{trans} coincides with the relaxation time of particles which are located in the tail of the Maxwellian distribution. However, these quantities are close to one another, and in a number of cases, are identical.

($m_A \gg m_B$) which is a small addition to the light gas B. The relaxation time of such a process has the form

$$\tau_{\text{trans}} \sim \frac{m_A}{m_B} \tau_0 \gg \tau_0 ; \tag{2–10}$$

and, consequently, the relaxation of the translational energy of particles of the heavier component placed in the incubator made up of the light gas particles takes place appreciably slower than in a single-component system. A similar phenomenon is observed also on relaxation of the lighter component of a heavier gas. This result is obvious. Transfer of translational energy between the light and heavy components is made difficult by the large difference in masses which is directly responsible for increasing the relaxation time. The latter effect can be observed in I_2–He, Xe–He and Xe–H_2 mixtures.

[2] Excitation of Rotational Degrees of Freedom

The energy transfer translational and internal degrees of molecular freedom (which usually include rotational, vibrational and electronic degrees of freedom) takes place comparatively slowly. The only exception to this is rotational relaxation of heavy molecules.

The time of rotational relaxation can be estimated by using Eq. (2–2) and the results above. Let us consider a system consisting of a monatomic gas containing a moderate admixture of molecules which are approximated by the model of rigid rotators. It is assumed that the velocities of the centers of gravity of all the particles have a Maxwellian distribution corresponding to the temperature T of the translational degrees of freedom. The initial distribution of the rotational energy of the rotators is assumed to be arbitrary. Under conditions of total equilibrium at temperature T, the average energy devolved on the rotational degrees of freedom of the rotators is

$$E_{\text{or}} = kTn ,$$

where n is the molecular number density. It is assumed that T is higher than the characteristic rotational temperature, which comprises several °K* for heavy gases. The rate of energy transfer from the translational to the rotational degrees of freedom $(dE_{\text{rot}}/dt)_{E_{\text{rot}}}=0$ will be determined from the problem of a rotator-atom collision. The rotator-rotator collisions can be neglected because of their low concentration. For simplicity, we shall assume that

* The term "characteristic rotational temperature" is used to denote the rotational constant B_e in °K.

all the collisions take place under conditions which are most favorable for the transformation of translational into rotational energy. In these collisions the moving atom follows a straight line which passes through the rotator-atom perpendicular to the rotator axis (Fig. 1). According to Eq. (2–2), the rotational energy increment is calculated on the assumption that the rotator does not rotate before the collision ($E_{rot}=0$) and that, in addition, the center of gravity of the rotator is at rest. The collision for the selected model and configuration takes place so rapidly that, at the instant of collision, the right-hand atom does not acquire a velocity. Hence the collision between the atom and rotator can be regarded as a head-on collision of the arriving particle (with a mass m_1) and the rotator-atom (with a mass $m_2/2$, where m_2 is the mass of the rotator).

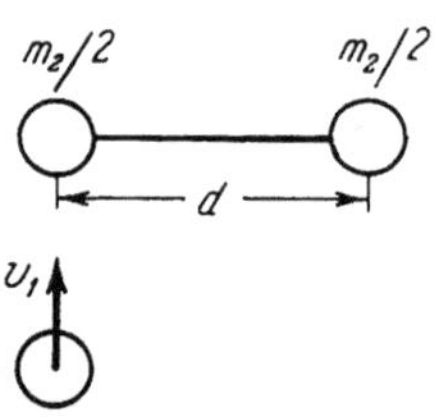

Fig. 1. Collision of an atom with a rotator.

In such a collision the moving atom imparts to the rotator an energy:

$$\frac{\Delta\varepsilon}{\varepsilon} = \frac{4m_1\left(\frac{m_2}{2}\right)}{\left(m_1 + \frac{m_2}{2}\right)^2};$$

which is distributed between its translational and rotational degrees of freedom. In the case of equal masses ($m_1 = m_2 \equiv m$), this energy is divided equally between the rotational and translational degrees of freedom of the rotator.

In fact, when $m_1 = m_2$, the velocity of the rotator immediately after the collision is $v = \frac{4}{3}v_1$, where v_1 is the velocity of the moving atom before the collision. The distribution of energy between the rotational and translational motions of the rotator can be found by determining the translational velocity v and the angular velocity Ω of the rotator. At the time of collision, the instantaneous axis of rotation of the rotator passes through the right atom; hence,

$$d\Omega = v = \frac{4}{3}v_1$$

(d is the distance between the atoms in the rotator); and the rotational energy is

$$E_{rot} = \frac{I\Omega^2}{2} = \frac{2mv_1^2}{9}$$

(I is the moment of inertia). The energy of the translational degrees of freedom is $E_{trans} = \Delta\varepsilon - E_{rot} = (2mv_1^2)/9$. The result thus obtained can be formalized in the following form: on collision between a rotator with a particle of comparable mass, the rotational degrees of freedom of the rotator acquire an energy which is, by order of magnitude, comparable to the energy obtained by the translational degrees of freedom of the rotator. This result is true also for a rotator-rotator collision. Since the average energy of the rotational and translational degrees of molecular freedom, as well as the energy transferred to the rotational and translational degrees of freedom in each collision, is of the same order of magnitude, then the rotational and translational relaxation times are found also to be of the same order of magnitude. In general, τ_{rot}, the time of rotational relaxation, is slightly greater than the relaxation time for the translational energy, since not every collision is accompanied by rotational excitation. In particular, in head-on collisions between a rotator and an atom, which is shown schematically in Fig. 2, the rotational energy of the rotator does not change. In connection with this we can write:

$$\tau_{rot} > \tau_{trans};$$

where τ_{rot} also is slightly greater than τ_{trans}.

Thus, rotational relaxation takes place during a time of the order of the average time of traverse of a mean-free path, which is somewhat greater than the relaxation time for the translational energy. Experimental data which were obtained by ultrasonic methods for O_2 and N_2 at room temperature yield values of $\tau_{rot} \sim 5\tau_0$. The rotational relaxation time for light molecules is appreciably greater than τ_0. Thus, for H_2 at room temperature, $\tau_{rot} \sim 300\tau_0$, and for D_2, it is $\sim 200\tau_0$, etc. The reduction in the rotational relaxation rate in light gases is due to the high energy of the rotational quantum of light molecules. While the characteristic rotational temperature for N_2 is 2.9 °K, for H_2, it is 87.7 °K. In practice, this means that light molecules have a higher angular velocity. On the other hand, the scheme (which was developed above) for calculating the rotational relaxation times is valid only for sufficiently high velocities of relative motion, for which the duration of the collision is so short that the molecule does not succeed in turning during the collision. For heavy molecules, this condition, as a rule, is sufficiently satisfied and, in practice, does not impose limitations on the velocities of the moving particles; for light molecules this condition is not so well satisfied. The molecular interaction which produces a rapid rotation of one or both molecules is more complicated, since in this case multiple collisions are possible during the interaction time. The result of such a complex interaction will, first of all, be energy transfer between the centers of mass of the colliding molecules, and also, a small fraction of the translational energy will be converted to rotation. In quantum-mechanical terms this means that the most probable result of such an interaction will be elastic scattering of molecules which is not accompanied by changes in their internal state. In addition, a small part of the total number of collisions will result in the excitation of rotational degrees of molecular freedom. The increase in the rotational relaxation time is in this case a result of the low efficiency of the process of energy transfer between the translational and rotational degrees of freedom following the collisions of light molecules.

[3] Excitation of Vibrational Degrees of Freedom

The equilibrium with respect to vibrational degrees of molecular freedom at ordinary temperatures is established relatively slower than the rotational relaxation equilibrium. The comparatively high frequency of vibrational motion of diatomic molecules (for O_2 $h\omega/k = 2240\,^{\circ}K$, for N_2 it is $3354\,^{\circ}K$, for H_2 it is $5910\,^{\circ}K$, etc.) is responsible for the fact that, under ordinary conditions, the time of molecular collision is appreciably greater than the period of vibrations. The average energy which is transferred to the vibrational degrees of freedom in these "adiabatic" collisions is low in comparison with the translational energy, and decreases with an increase in the frequency.

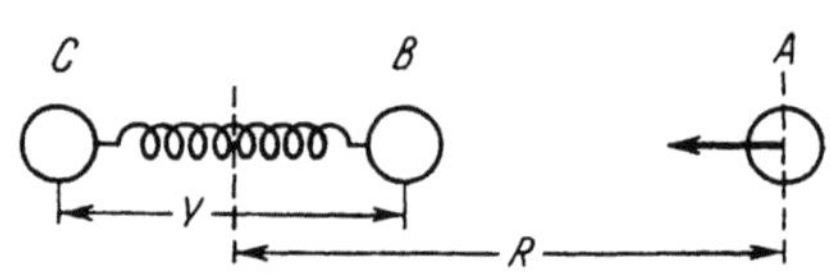

Fig. 2. Collision between an atom and an oscillator.

We will determine the vibrational relaxation time from Eq. (2–1) written for vibrational energy. We shall consider the elementary cases of vibrational relaxation of diatomic molecules which comprise a small admixture in a monatomic gas. We shall assume that the velocities of translational motion of the centers of mass of all the particles have a Maxwellian distribution with the temperature T. The initial distribution of the vibrational molecular energy is arbitrary. The vibrational relaxation process under ordinary conditions is accompanied by excitation of only several first vibrational levels, hence, the molecules can be considered to be harmonic oscillators. The average value of the vibrational molecular energy at temperature T is

$$E_{0\,\text{vib}} = \hbar\omega \frac{e^{-\frac{\hbar\omega}{kT}}}{1 - e^{-\frac{\hbar\omega}{kT}}} n, \tag{2–11}$$

where n is the molecular number density.

To determine the vibrational relaxation time in accordance with Eq. (2–2), it is also necessary to find the vibrational energy increment per unit time $(\mathrm{d}E_{\text{vib}}/\mathrm{d}t)_{E_{\text{vib}}=0}$. Let us consider the collision of an oscillator with a monatomic particle. We shall assume that the collision takes place under conditions most favorable to the excitation of vibrations. The configuration of such a collision is shown in Fig. 2, and corresponds to a head-on atom-oscillator collision in which the atom moves along a straight line connecting the nuclei of the diatomic molecule. To evaluate $(\mathrm{d}E_{\text{vib}}/\mathrm{d}t)_{E_{\text{vib}}=0}$, we must find the energy $\Delta\varepsilon$, which is transformed from the translational degrees of freedom of the colliding particles into the vibratory energy of the oscillator in a single collision, and to average the latter over all the velocities of relative motion of the moving particles.

The equation of motion of the oscillator under the action of a driving force $F(t)$ in the system of the mass centers is written as follows:

$$\ddot{y} + \omega^2 y = \frac{1}{m} F(t); \tag{2–12}$$

where y is the displacement of the vibrational coordinate from the equilibrium value; ω is the angular frequency of the oscillator; $F(t)$ is the force exerted on the oscillator by the atom; and m is the reduced mass of the oscillator. According to Eq. (2–2), we assume that initially the oscillator is at rest:

$$y = (-\infty) = 0, \quad \dot{y}(-\infty) = 0. \tag{2–13}$$

The transferred energy $\Delta\varepsilon$ is determined by the expression:

$$\Delta\varepsilon = \frac{m}{2}(\dot{y}^2 + \omega^2 y^2)\Big|_{t=\infty} \tag{2–14}$$

Eq. (2–12) can be integrated in the general case. We rewrite it in the form:

$$\frac{d}{dt}(\dot{y} + i\omega y) - i\omega(\dot{y} + i\omega y) = \frac{1}{m} F(t),$$

or

$$\frac{d\xi}{dt} - i\omega\xi = \frac{1}{m} F(t); \tag{2–15}$$

where $\xi = \dot{y} + i\omega y$.

The solution of Eq. (2–15) together with the initial conditions (2–13) has the form:

$$\xi = e^{i\omega t} \int_{-\infty}^{t} \frac{1}{m} F(t) e^{-i\omega t}\, dt. \tag{2–16}$$

Substituting Eq. (2–16) into (2–14) we get

$$\Delta\varepsilon = \frac{m}{2}|\xi|^2 = \frac{1}{2m}\left|\int_{-\infty}^{\infty} F(t) e^{-i\omega t}\, dt\right|^2. \tag{2–17}$$

The physical meaning of the solution thus found is obvious. The magnitude of the transferred energy is determined by the square of the modulus of the Fourier component of force $F(t)$ with a frequency equal to the natural frequency of the system. To calculate $\Delta\varepsilon$, it is necessary to specify the form of the intermolecular interaction potential, *i.e.*, the form of $F(t)$. We shall assume that the intermolecular interaction potential is described by the sum of exponential expressions:

$$V = Ce^{-\alpha r_{ij}};$$

which correspond to the interaction between the moving atom A with each of the atoms of the molecule BC (r_{ij} is the distance between the center of masses of the moving atom A and one of the atoms of molecule BC). This is the general form of the potential of short-duration forces; here $\alpha^{-1} \sim 0.2-0.5$ Å. Since $\alpha Y_0 \gg 1$ (Y_0 is the equilibrium value of the intramolecular coordinate Y), we do not have to consider the interaction between atoms A and C. The intermolecular interaction potential in this case can be written in the form:

$$V(R, Y) = Ce^{-\alpha r_{AB}};$$

where R is the distance between the centers of mass of the colliding particles, and r_{AB} is the distance between the centers of mass of atoms A and B. It can be seen from Fig. 2 that $r_{AB} = R - \lambda Y$, where $\lambda = m_C/(m_B + m_C)$. In our further discussion it is convenient to change from R and Y to the coordinates r and y, where $r = R - R_0$ and $y = Y - Y_0$. The value of R_0 corresponds to the coordinate of the point at which the atom A reverses when $Y = Y_0$. In terms of the new variables the potential has the form:

$$V(r, y) = W_0 e^{-\alpha(r - \lambda y)}. \tag{2–18}$$

Under ordinary conditions the vibrational amplitude at the lower vibrational levels is small in comparison with the range of intermolucelar forces. Hence, $\alpha y \ll 1$, and

$$V(r, y) = W_0 e^{-\alpha r}(1 + \lambda\alpha y). \tag{2–19}$$

The condition $\alpha y \ll 1$ means that vibrations of the molecule *BC* do not appreciably affect the trajectory of relative motion of particles *A* and *BC* and in the first approximation they can be neglected. In this case the trajectory of relative motion of the centers of mass of the colliding particles can be found from the classical equation of motion:

$$\mu \frac{d^2 r}{dt^2} = -\frac{dV(r,0)}{dr}; \tag{2–20}$$

or

$$\frac{dr}{dt} = \left\{\frac{2}{\mu}[W_0 - V(r)]\right\}^{\frac{1}{2}}; \tag{2–21}$$

where μ is the reduced mass of particle *A* and oscillator *BC*, while $W_0 = \mu v^2/2$, v being the relative velocity of particles *A* and *BC* before collision. The solution of Eq. (2–21) can be obtained as follows: we introduce a new variable $z = \exp(\alpha r/2)$. In terms of z, Eq. (2–21) has the form:

$$\left(\frac{dz}{dt}\right)^2 = \frac{W_0 \alpha^2}{2\mu}(z^2 - 1). \tag{2–22}$$

Differentiating Eq. (2–22) with respect to z, we get

$$\frac{d^2 z}{dt^2} = \frac{W_0 \alpha^2}{2\mu} z. \tag{2–23}$$

The solution of Eq. (2–23) is known. With the initial conditions $z = 1\,(r = 0)$ when $t = 0$ it has the form:

$$z = \mathrm{ch}\left(\frac{W_0 \alpha^2}{2\mu}\right)^{\frac{1}{2}} t.$$

Taking into account the fact that $(W_0 \alpha^2/2\mu) = \alpha^2 v^2/4$, we get, finally,

$$e^{-\alpha r} = \mathrm{sch}^2\left(\frac{\alpha v t}{2}\right). \tag{2–24}$$

Using Eqs. (2–19) and (2–24), it is easy to find the expression for the force exerted on the oscillator by the atom:

$$F(t) = -\frac{\partial V}{\partial y} = -\alpha W_0 \lambda\, \mathrm{sch}^2\left(\frac{\alpha v t}{2}\right). \tag{2–25}$$

The magnitude of the transferred energy which is determined by Eq. (2–17) on substitution of Eq. (2–25), is written as

$$\Delta\varepsilon = \frac{(\alpha W_0 \lambda)^2}{2m} \left| \int_{-\infty}^{\infty} \frac{e^{-i\omega t}}{\mathrm{ch}^2\left(\frac{\alpha v t}{2}\right)} dt \right|^2. \tag{2–26}$$

The integral contained in the right-hand side of Eq. (2–26) is calculated by the known equation:

$$\int_0^\infty \frac{\cos px}{\text{ch}^2 x}\,\text{d}x = \frac{1}{2}\,\frac{\pi p}{\text{sh}\dfrac{\pi p}{2}}.$$

Thus, we obtain

$$\Delta\varepsilon = \frac{(\alpha W_0 \lambda)^2}{2m}\left(\frac{\dfrac{4\pi\omega}{\alpha^2 v^2}}{\text{sh}\dfrac{\pi\omega}{\alpha v}}\right)^2. \tag{2–27}$$

Under ordinary conditions the collision time τ_{col} for the majority of molecules is appreciably greater than the period of vibrations of the molecule atoms, which allows us to write $\omega\tau_{\text{col}} \gg 1$. By order of magnitude $\tau_{\text{col}} \sim (1/\alpha v)$, hence, $\omega/\alpha v \gg 1$ (condition specifying that the collisions are adiabatic). (2–28) Considering Eq. (2–28), the expression for the transferred energy can be written as

$$\Delta\varepsilon = \frac{8\pi^2\omega^2\mu^2\lambda^2}{\alpha^2 m}\, e^{-\frac{2\pi\omega}{\alpha v}}. \tag{2–29}$$

In this form $\Delta\varepsilon$ is identical to its value obtained in Chapter 4 by quantum-mechanical methods. It is now easy to find

$$\left.\frac{dE_{\text{vib}}}{dt}\right|_{E_{\text{vib}}=0}.$$

Substituting Eq. (2–29) into (2–3) and taking into account the fact that the relative velocities have a Maxwellian distribution, we get

$$\left.\frac{dE_{\text{vib}}}{dt}\right|_{E_{\text{vib}}=0} = n\hbar\omega P_{01} Z, \tag{2–30}$$

where

$$P_{01} = \frac{1}{Z}\int_0^\infty \frac{\Delta\varepsilon}{\hbar\omega} Z^{(0)}(v)\,\text{d}v, \tag{2–31}$$

here $Z^{(0)}(v)\,\text{d}v$ is given by Eq. (2–5), in which m_B is replaced by the reduced mass of the colliding particles μ, and Z and n are the same as in Eq. (2–6). The quantity $\hbar\omega P_{01}$ has the meaning of the average energy which is transferred to the vibrational degree of freedom in one collision. In making quantum-mechanical calculations (see Sect. 17) and making the same approximation, the exact same value is obtained for $\hbar\omega P_{01}$, in which case P_{01} has the meaning of the probability of transition of the molecule from the zero vibrational level to the first, referred to one collision. Usually in the vibrational relaxation theory, the quantity P_{10} is employed, which is the probability of transition of the molecule from the first vibrational level to the zero level on a single collision. The probabilities P_{10} and P_{01} are related by the detailed balancing condition:

$$P_{01} = P_{10} e^{-\frac{\hbar\omega}{kT}}. \tag{2–32}$$

The magnitude of P_{10} is calculated in Sect. 17. Here we shall restrict ourselves to qualitative estimates only. According to Eqs. (2–31), (2–5) and (2–7), we can write the following expression for P_{10}:

$$P_{10} \sim \frac{1}{2}\left(\frac{\mu}{kT}\right)^2 \int_0^\infty \frac{\Delta\varepsilon}{h\omega} e^{-\frac{\mu v^2}{2kT}} v^3 \, dv;$$

or [see Eq. (2–29)]

$$P_{10} \sim \int_0^\infty e^{-\frac{2\pi\omega}{\alpha v} - \frac{\mu v^2}{2kT}} v^3 \, dv. \tag{2–33}$$

The quantity P_{10} can be evaluated by using the fact that the integrand of Eq. (2–33) has a sharply defined maximum. The position of the maximum is determined from the condition that $(2\pi\omega/\alpha v) + (\mu v^2/2kT)$ be at minimum. From this condition we can find the velocity (v^*) to which corresponds the maximum of the integrand:

$$v^* = \left(\frac{2\pi kT\omega}{\mu\alpha}\right)^{\frac{1}{3}}. \tag{2–34}$$

The main contribution to the integral of Eq. (2–33) is made by the range of velocity in the vicinity of v^*. Hence,

$$P_{10} \sim e^{-\frac{2\pi\omega}{\alpha v^*} - \frac{\mu v^{*2}}{2kT}} \equiv e^{-3\chi}, \tag{2–35}$$

where

$$\chi = \left(\frac{\pi^2 \mu\omega^2}{2\alpha^2 kT}\right)^{\frac{1}{3}}.$$

In the temperature range of practical interest (from room temperature and up) $\chi = 5$–10 for various gases, hence $P_{10} \ll 1$. Thus, for example, when $T = 288\,°K$,

$$P_{10}(O_2 - O_2) \sim 10^{-8},\ P_{10}(N_2 - N_2) < 10^{-12},$$
$$P_{10}(CO_2 - CO_2) \sim 10^{-5} \quad \text{etc.},$$

P_{10} increases sharply with an increase in the temperature, thus, at $T=2000\,°\mathrm{K}$,

$$P_{10}(O_2 - O_2) \sim 10^{-4},$$

at $T=3000\,°\mathrm{K}$,

$$P_{10}(O_2 - O_2) \sim 10^{-3} \quad \text{etc}.$$

Substituting Eqs. (2–30) and (2–11) into Eq. (2–2) and using Eq. (2–32), we obtain for τ_{vib} the following vibrational relaxation time:

$$\tau_{\mathrm{vib}} = \frac{1}{ZP_{10}\left(1 - e^{-\frac{\hbar\omega}{kT}}\right)}. \tag{2–36}$$

Eq. (2–36) is identical to the expression for the vibrational relaxation time which is obtained in quantum mechanics [see Sect. 17, Eq. (17–52)], and is known as the Landau-Teller formula. Since $P_{10} \ll 1$, then τ_{vib}, the vibrational relaxation time, is appreciably greater than the relaxation times for the translational and rotational degrees of freedom, the order of which is Z^{-1}. As the temperature increases, τ_{vib} is reduced perceptibly. The temperature dependence of τ_{vib} has the form:

$$\ln \tau_{\mathrm{vib}} Z\left(1 - e^{-\frac{\hbar\omega}{kT}}\right) \sim T^{-\frac{1}{3}}.$$

This relationship is satisfied for the majority of cases in a wide temperature range (for more details, see Sect. 17).

Thus, the vibrational relaxation process takes place much slower than relaxation processes related to the translational and rotational degrees of freedom. The latter is due to the low efficiency of heat transfer between the translational and vibrational degrees of freedom on molecular collisions.

[4] Molecular Dissociation and Ionization

The slowest processes of establishing equilibrium are those of dissociation, ionization and electronic excitation. The physics of the latter process are very close to those of ionization. Ionization can be regarded as the limiting case of electronic excitation, when the electron acquires enough energy to go from the bound to the free state. Hence, electronic excitation will not be considered separately. In principle, dissociation can also be considered as a limiting case of vibrational excitation. However, substantial differences exist here which are due to the different relation between the energy of the vibrational photon and the dissociation energy. The electronic excitation energy is of the order of the ionization energy while the energy of dissociation exceeds appreciably the energy of the vibrational photon. Hence, in the case of electron excitation, each of the elementary processes which leads to the excitation of bound electrons can also result in ionization if sufficient energy is available. A different situation arises on excitation of vibrational levels. Thus, for example, on molecular collisions in the ground vibrational states, practically not a single molecule is dissociated, although the vibrational excitation process takes place with one or another degree of effectiveness. Here, however, dissociation must be considered separately. The process of dissociation in a system where the disintegrating molecules are a small admixture is described by a kinetic equation of the form:

$$\frac{dn}{dt} = -k_d n N; \tag{2–37}$$

where n is the number density of the dissociating particles, N is the number density of molecules of the neutral gas, and k_d is the dissociation rate constant. Eq. (2–37) describes the initial stage of dissociation, since it does not take into consideration the recombination processes. The time required for establishing equilibrium dissociation τ_d is, by order of magnitude:

$$\tau_d \sim (k_d N)^{-1}.$$

An actual calculation of the rate constant k_d requires that the dissociation mechanism be known. However, for qualitative estimates one may use certain general considerations. Under ordinary conditions, for example, in a gas heated by a shock wave, thermal dissociation takes place due to molecular collisions. However, as a rule, the number of molecules which react per unit time is very small in comparison with the total number of collisions. Dissociation is produced by collisions with particles whose energy (internal plus kinetic) is greater than or equal to the energy of dissociation. On equilibrium at the temperature T, the fraction of these high-energy collisions is proportional to $\exp(-D/kT)$, where D is the dissociation energy. The magnitude of the dissociation rate constant can be estimated assuming that the dissociation process does not disturb the system equilibrium. In this case:

$$k_d = PZ^0 e^{-\frac{D}{kT}}; \tag{2–38}$$

where P is a proportionality coefficient (usually designated by the term steric factor) and Z^0 is the number of collisions which the dissociating molecule undergoes per unit time when $N=1$. Moreover, P is expressed differently for each specific reaction mechanism. Thus, for example, if the only quantity of substance in the reaction is the energy of relative translational motion along the line of centers, $P=\gamma$, where γ is the probability of dissociation on collision of particles possessing sufficient energy. If the entire relative kinetic energy of moving particles is of substance in a reaction, then $P=\gamma[(D/kT)+1]$, etc (see Sect. 18). Knowing the value of k_d it is easy to estimate the time for establishing equilibrium dissociation:

$$\tau_d \sim \frac{e^{\frac{D}{kT}}}{PZ}, \quad (Z = Z^0 N). \tag{2-39}$$

Eq. (2–39) gives an estimate of the lower limit of the time needed for establishing equilibrium dissociation since it was assumed in calculating k_d that the dissociation process does not disturb the equilibrium energy distribution with respect to the different degrees of freedom. Comparing Eqs. (2–39) and (2–30), we find that

$$\tau_d \gg \tau_{\text{vib}}, \tag{2-40}$$

if

$$Pe^{-\frac{D}{kT}} \ll P_{10}. \tag{2-41}$$

Under ordinary conditions, Eq. (2–41) and, consequently, also Eq. (2–40), are well satisfied. Thus, for example, for dissociation of O_2 in an O_2-Ar mixture at 5000 °K, $\tau_d/\tau_{\text{vib}}=60$. At low temperatures this ratio is even larger.

A similar expression can also be written for the time of establishing the equilibrium ionization τ_i. The process of ionization of a monatomic gas behind a shock front takes place primarily by electron-atom collisions (see Sect. 19). The first ionization stage, which produces the necessary number of priming electrons, has not been studied up to the present. Atom-atom collisions, collisions with the admixture atoms, optical excitation, photoionization, etc., can be found to be of importance at this stage. We shall estimate the time τ_i by making the following general assumptions. We assume that the thermal ionization of gas A is governed by the relationship

$$A + e \rightarrow A^+ + e + e,$$

where e is an electron.

The kinetic equation which describes the initial ionization stage (when the recombination can be disregarded) has a form similar to Eq. (2–37):

$$\frac{dn_e}{dt} = k_i n_e N;$$

where n_e and N, respectively, are the electron and atom number density, and k_i is the ionization rate constant. The time for establishing equilibrium ionization:

$$\tau_i \sim (k_i N)^{-1}.$$

The ionization rate constant is defined as the number of collisions of particles A (with electrons) per unit time which lead to ionization. Let us assume that the electron gas has a Maxwellian velocity distribution with a temperature T_e which, in general, can differ from T, the gas temperature (see Sect. 19). In this case the ionization rate constant can be written as

$$k_i = PZ^0 e^{-\frac{I}{kT_e}}, \tag{2–42}$$

where I is the ionization constant and the steric factor P depends on the ionization cross section. Thus, for the time of establishing equilibrium ionization τ_i, we can write the following:

$$\tau_i \sim \frac{e^{\frac{I}{kT_e}}}{PZ} \quad (Z = Z^0 N). \tag{2–43}$$

Since the dissociation energy is of the same order as the ionization energy, then these processes can take place concurrently in a certain temperature region. At high temperatures (for air of the order of 20,000 °K), the slowest process is that of establishing ionization equilibrium; hence, in this temperature region $\tau_i > \tau_d$.

Similarly, it is also possible to estimate the relaxation time for establishing the equilibrium distribution of electronically excited atoms. It has been calculated that when atoms are excited by electronic collisions, this time is perceptibly smaller than the ionization time. The relaxation time for electronic excitation can be found to be comparable to the ionization time only at the beginning of the ionization process, when the electron number density is perceptibly smaller than that of equilibrium. (For more details, see [71].)

[5] Sequence of Relaxation Processes in Shock Waves

The considerations made above enable the sequence of relaxation processes in shock waves to be established. First, during a time of the order of τ_{trans}, the translational degrees of molecular freedom are excited and the Maxwellian equilibrium is established. The section of shock wave which corresponds to this process is regarded as a shock front, and it is possible to introduce the concept of temperature for processes behind the shock front (for translational degrees of freedom). Second, rotational equilibrium is established during the time τ_{rot} and the equilibrium with respect to the vibrational degrees of freedom is established during the time τ_{vib}. The slowest processes are the establishing of equilibrium dissociation (the relaxation time τ_d) and ionization (the relaxation time τ_i). The following inequality is satisfied under actual conditions for pure gases:

$$\tau_{\text{tr}} < \tau_{\text{rot}} \ll \tau_{\text{vib}} \ll \tau_d < \tau_i \,. \tag{2–44}$$

This condition simplifies substantially the experimental study of processes and facilitates the theoretical consideration of the problem. Moreover, when Eq. (2–44) is satisfied, most of the aforementioned elementary processes can be considered separately, *i.e.*, assuming in the first approximation that the preceding process has ended and that the following one has not as yet begun. Thus, for example, when considering vibrational relaxation it is possible to assume in the first approximation that at each time instant (or, correspondingly, at each point in space) there exists a Maxwellian velocity distribution and a Boltzmann distribution of rotational levels. This justifies the use of the Maxwellian function for calculating the increment in the vibrational energy per unit time (2–31). On the other hand, the vibrational relaxation terminates before the molecular dissociation starts. Hence, dissociation can be disregarded entirely in the study of vibrational relaxation. When studying dissociation, one can assume that, at any time instant (in any point in space), there exists a quasi-stationary [steady-state] distribution of molecules with respect to vibrational levels (in general, not an equilibrium distribution), which is a result of the terminated process of vibrational relaxation. The case of rotational relaxation is slightly more complicated. This process takes place at a substantially higher rate than vibrational relaxation, hence, it can be studied without [at all] considering the excitation of vibrations. On the other hand, the rotational relaxation can be rigorously separated from the process of establishing the Maxwellian distribution only in the case of light molecules. For all the other molecules, the equilibrium with respect to the vibrational and rotational degrees of freedom is established almost simultaneously, and in a number of cases it is difficult to separate one from another. The situation is the same for the dissociation and ionization processes. Let us note that the distribution of the relaxation times of various processes which is given by Eq. (2–44) is characteristic only of pure gases at moderate temperatures. In gas mixtures as well as in pure gases at sufficiently high temperatures, the various relaxation processes start to overlap. Thus, for example, the time of vibrational relaxation of N_2 in air is of the same order of magnitude as the time for establishing the equilibrium dissociation of O_2. On the other hand, at very high temperatures (of the order of tens of thousands degrees for air), the dissociation process leads the establishment of quasi-steady-state distribution of molecules with respect to vibrational level. In addition, in gas mixtures one must take into account the processes of transfer between vibrational and rotational photons, as well as the transfer of charges between various components, which complicates the picture of establishing the complete statistical equilibrium still further.

[3] EXPERIMENTAL STUDY OF THE SHOCK-WAVE STRUCTURE

From the viewpoint of the experimenter, shock waves have two peculiar features which make their study difficult. First, the small thickness of the relaxation zones, which is due to the high rate at which the processes take place and, second, the high velocity of motion. According to estimates in the preceding section, the thickness of the relaxation zone behind the shock front is contained within the limits of several mean-free paths to several hundred thousands of these paths, which comprises, at atmospheric pressure, a distance from fractions of a centimeter to several centimeters. The shock wave moves at supersonic speed, hence the gas volume under study, contained in the nonequilibrium region, flows past the stationary instrument during a time from fractions of a microsecond to several hundreds of microseconds. The great speeds also greatly heat up the gas behind the shock front. The gas temperature in the nonequilibrium region comprises tens and hundreds of thousands degrees. Thus, the experimenter is faced with the difficult task of studying the complex structure of a rapidly moving volume of gas with small dimensions. Before we consider in detail the experimental methods used in studying phenomena in shock waves, we must consider the following basic points: 1) how to obtain shock waves under laboratory conditions; 2) what to measure in a shock wave in order to determine its structure; and 3) how to make these measurements. A short answer to these questions is given in the present section, which serves as an introduction for the following two chapters, where a more detailed analysis of the experimental method for studying nonequilibrium phenomena in shock waves is given.

[1] Obtaining the Shock Wave. Simplified Theory of the Shock Tube

Several methods are available for producing shock waves, including explosion (shock wave produced by a point source) and supersonic flow past bodies. However, nonequilibrium phenomena in gases are best studied by a shock wave which propagates in a tube, since in this case the shock front is sufficiently plane and the flow behind the front is one-dimensional in the first approximation. This permits maximum simplification of the gasdynamic picture and frees the experimenter to concentrate on the physicochemical aspects of the problem at hand. There are a large number of such installations which differ according to their purpose, arrangement, auxiliary equipment, overall dimensions, etc. However, all of them are constructed according to a single principle and can be designated by one name – shock tubes. The basic operational principle of all the shock tubes is quite simple: the shock wave is formed following the motion of some given piston in a tube filled by the test gas. The nature and character of this piston varies depending on the type of shock tube (compressed gas, products of explosion, plasma cloud, etc). The gas volume under test is heated by the shock wave and moves in the tube at a high speed. Consequently, the development of physicochemical phenomena here is closely related to the gasdynamic conditions under which they take place. On one hand, the presence of the gasdynamic flow makes it possible to obtain a shock wave and thus rapidly and uniformly heat the gas to a high temperature. On the other hand, the development of these physiochemical processes in a number of cases, substantially changes the gasdynamic characteristics of the flow. Hence, successful setting up of experiments for the study of kinetics of nonequilibrium phenomena in shock waves and correct treatment of the results thus obtained requires a knowledge of the processes responsible for the formation and propagation of shock waves in shock tubes.

The experimental study of the formation and propagation of shock waves in tubes began as early as the nineteenth century.* However, the basic results of these studies were obtained only during the last ten years when

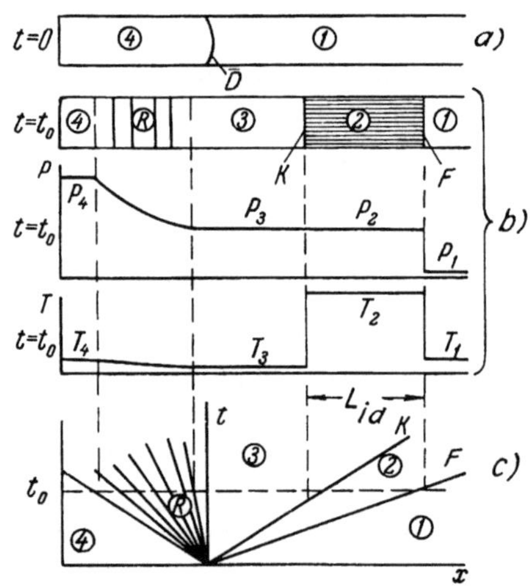

Fig. 3. Simplified diagram of processes in a shock tube. a) Before the experiment starts ($t = 0$); b) at the time $t = t_0$ (distribution of the pressure and temperature in the tube); c) the x-t diagram of the flow in the shock tube. D is the diaphragm; F is the shock front; K is the contact surface; R is the bunch of rarefaction waves, L_{id} is the length of the hot plug.

* The history of the development of shock tubes with reference to applicable sources is presented briefly in [168].

shock tube studies were developed to an extraordinary degree. In the simplest shock tube, the shock waves are formed in the tube with a constant cross section after bursting a diaphragm which separates the low-pressure chamber (filled by the test gas at a pressure of several mm Hg) and the high-pressure chamber (which is filled with compressed gas under a pressure of several atmospheres).

Now we will consider the fundamental processes which take place in an elementary shock tube. An increase in the pressure of the driver gas in the high-pressure chamber breaks the diaphragm which separates the two chambers. The diaphragm can also be broken by some impact mechanism. After the diaphragm bursts, the gas from the high-pressure chamber rushes into the low-pressure chamber and compresses the test gas. A shock wave propagates through the test gas. Simultaneously, a rarefaction wave travels in the high-pressure chamber in the opposite direction. A schematic representation of phenomena which take place in a shock tube after the diaphragm is removed is shown in Fig. 3 in the form of an *x–t* diagram which corresponds to the time development of processes at various points of the tube. In the figure, Region 1 corresponds to the state of the test gas before the shock wave arrives, Region 2 corresponds to a gas heated by the shock wave, Region 3 corresponds to the driver gas between Region 2 and the rarefaction wave, and Region 4 corresponds to the initial state of the gas in the high pressure chamber before the arrival of the rarefaction wave. The surface K which separates the test and driver gases (between Regions 2 and 3) is called the contact surface.* The gas pressure and the flow velocity to both sides of the contact surface are the same ($p_2=p_3$, $u_2=u_3$). Region 2 is sometimes called the plug, and the distance from F the shock front, to the contact surface K is called the plug length L. The gas compressed by the shock wave is heated to a high temperature T_2 while the driver gas is cooled in the rarefaction wave to a temperature T_3. As the shock wave and the rarefaction wave propagate further, they are reflected from the end walls of the shock tube, they begin to interact with one another, etc. A simple and graphic consideration of these processes is given by Gaydon and Hurle [328].

The main problem in shock-tube theory is to establish relationships between quantities which characterize the state of the gas in the tube before the experiment, and the velocity of the shock wave which is produced there. This relationship is obtained easiest by making a number of simplifying assumptions; such a simplified theory of the shock tube is developed in [364, 332, 486]. The approximations used reduce to the following: 1) the gases used in experiments are assumed to be ideal with a constant value of

* [Transl. Note: Also called interface.]

the ratio of specific heats γ**; 2) the diaphragm bursts instantaneously and the shock wave is formed immediately following the opening of the diaphragm; 3) the shock wave in the low pressure chamber moves at a constant speed; 4) the flow in the shock tube is described as if it were one-dimensional; 5) the effect of viscosity, thermal conductivity and other dissipative processes on the properties of the flow, as well as intermixing of the gases at the contact surface is disregarded.

Based on these assumptions, the initial temperature and pressure of the gas can be related to the shock-wave velocity. To do this we must find the relationship between the characteristics of the gas in Regions 1 and 4 after the diaphragm bursts (these designations are standard in many works on shock tubes and thus will be used here; see Fig. 3). The gas in the rarefaction wave expands isentropically; hence, the pressure changes in it can be determined by the expression of the isentrope thus obtaining

$$\frac{p_4}{p_3} = \left(\frac{c_4}{c_3}\right)^{\frac{2\gamma_4}{\gamma_4 - 1}}, \tag{3–1}$$

where p is the pressure and c is the speed of sound. When passing through the rarefaction wave, the quantity $[2c/(\gamma - 1)] + u$, which is called the Riemann invariant, remains constant; then, since $\mu_4 = 0$,

$$\frac{2}{\gamma_4 - 1} c_4 = \frac{2}{\gamma_4 - 1} c_3 + \mu_3 . \tag{3–2}$$

As was pointed out, at the contact surface $p_3 = p_2$ and $u_3 = u_2$. Eliminating c_3, we find that

$$\frac{p_4}{p_2} = \left[\frac{c_4}{c_4 - \dfrac{\gamma_4 - 1}{2} u_2}\right]^{\frac{2\gamma_4}{\gamma_4 - 1}} . \tag{3–3}$$

The pressure ratio in the chambers before the experiment can be written as

$$\frac{p_4}{p_1} = \frac{p_4}{p_2}\frac{p_2}{p_1}, \tag{3–4}$$

where p_2/p_1 is the pressure ratio across the shock front. According to the assumptions made, this ratio is equal to

$$\frac{p_2}{p_1} = \frac{2\gamma_1}{\gamma_1 + 1} \mathsf{M}^2 - \frac{\gamma_1 - 1}{\gamma_1 + 1}, \tag{3–5}$$

where M is the Mach number of the shock wave, which is equal to the ratio of the rate of propagation of the shock front to the speed of sound in the gas

** The quantity γ is equal to the ratio of the specific heat at constant pressure to the specific heat at constant volume.

ahead of the shock front. Using the expression for the flow velocity u_2 in terms of M

$$u_2 = \frac{2}{\gamma_1 + 1} \frac{\mathsf{M}^2 - 1}{\mathsf{M}} c_1, \tag{3-6}$$

and Eqs. (3–3) and (3–5), we get

$$\frac{p_4}{p_1} = \left(\frac{2\gamma_1}{\gamma_1 + 1} \mathsf{M}^2 - \frac{\gamma_1 - 1}{\gamma_1 + 1} \right) \left[\frac{1}{1 - \frac{\gamma_4 - 1}{\gamma_1 + 1} \frac{c_1}{c_4} \left(\mathsf{M} - \frac{1}{\mathsf{M}} \right)} \right]^{\frac{2\gamma_4}{\gamma_4 - 1}}. \tag{3-7}$$

This is the equation which relates the Mach number and the specified values of p_4/p_1 and c_4/c_1; the ratio c_4/c_1 for a selected gas composition is determined by the temperatures T_1 and T_4. Eq. (3–7) is the basic equation of the simplified shock-tube theory. To clarify the basic factors which determine the magnitude of M, we now consider the limiting case of a very large pressure ratio; here $1/\mathsf{M}$ is small in comparison with M and it follows from Eq. (3–7) that

$$\mathsf{M}_\infty = \frac{\gamma_1 + 1}{\gamma_4 - 1} \frac{c_4}{c_1} \equiv \frac{\gamma_1 + 1}{\gamma_4 - 1} \sqrt{\frac{\mu_1 \gamma_4 T_4}{\mu_4 \gamma_1 T_1}}, \tag{3-8}$$

where μ is the molecular weight.

According to Eq. (3–8), the Mach number of a shock wave increases with an increase in the temperature ratio and reduction in the ratio of the molecular weights in the high- and low-pressure chambers. Certain numerical values of M_∞ for a number of gases are given in Table 1. In the general case of any values of p_4/p_1, the values of the Mach number can be determined by constructing graphs of M as a function of p_4/p_1 on the basis of Eq. (3–7) and assuming that c_4/c_1 is an independent parameter. These graphs are shown in

Table 1. Values of c_4/c_1 and M_∞ for various gases when $p_4/p_1 \to \infty$

Driving Gas	Test Gas	c_4/c_1	M_∞
Air	Air	1.0	6.1
Helium	Air	2.93	10.7
Helium	Argon	3.16	12.7
Helium	Oxygen	3.08	11.2
Helium	Carbon Dioxide	3.75	13.0
Helium	Xenon	5.73	22.9
Hydrogen	Air	3.80	21.4
Hydrogen	Argon	4.10	26.8
Hydrogen	Oxygen	4.0	23.6
Hydrogen	Carbon Dioxide	4.86	27.5
Hydrogen	Xenon	7.41	48.6

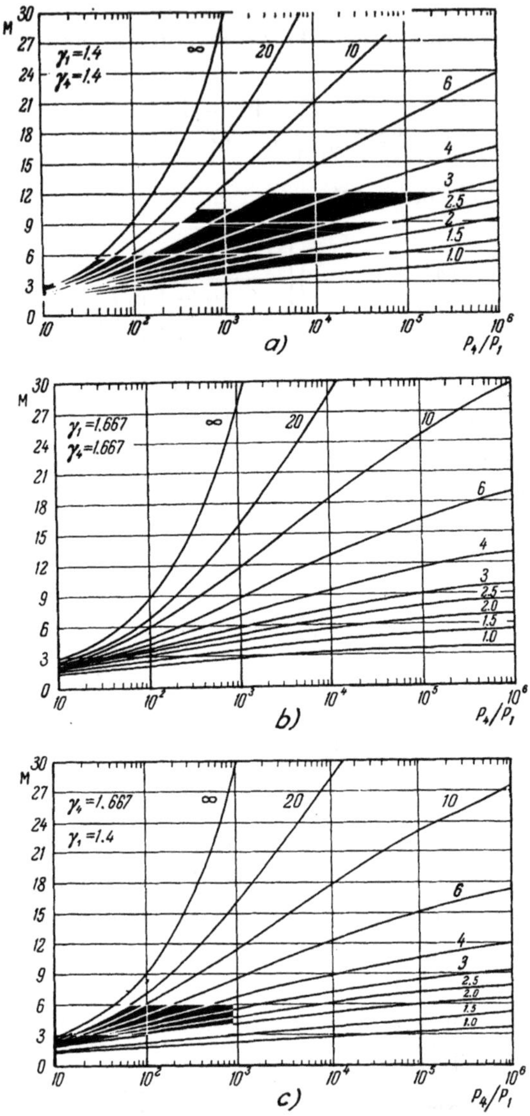

Fig. 4. Dependence of the Mach number of a shock wave in a shock tube on the initial pressure ratio p_4/p_1 at the diaphragm for different values of c_4/c_1 (given by numbers on the graphs) and different means of filling the shock tube. a) $\gamma_1 = \gamma_4 = 1.4$; b) $\gamma_1 = \gamma_4 = 1.667$; c) $\gamma_1 = 1.4$, $\gamma_4 = 1.667$.

Fig. 4. The values of c_4/c_1 used in experiments with shock tubes are given in Table 1 ($T_1 = T_4$).

The relationships thus obtained characterize the simplified operating conditions of a shock tube, however, the actual processes which take place are more complicated. Phenomena taking place in actual shock tubes will be considered in the next chapter. But, in most cases, it may be assumed that the simplified shock-tube theory presented above yields roughly correct results, which is important in selecting conditions for the experiment and estimating the shock-wave velocities on the basis of the initial state.

[2] The Quantities Being Measured

Studies are being made on a volume of heated gas in a shock tube which is contained between the shock front and the contact surface, *i.e.*, the plug. Special attention is given to the nonequilibrium flow zone behind the shock front.

Now we consider what should be measured and how to do this in order to obtain the information characterizing the development of nonequilibrium phenomena in shock waves.

The most direct method for obtaining information on the development of these processes is to measure the distribution of particle concentration in a given state. Thus, when studying vibrational relaxation or the process of excitation of electronic levels, it is desirable to study the concentration of vibrationally excited molecules* or the population of the given electronic states of atoms and molecules. When studying chemical processes (disintegration and formation of molecules), it is desirable to measure the concentrations of the various gas components behind the shock front. When studying ionization phenomena one must know the distribution of the electron (or ion) concentration.

The excitation, disintegration, and ionization of molecules (or atoms) can be accompanied not only by a change in the concentration, but in other gas characteristics (temperature, pressure, density, flow velocity). Hence, measuring these quantities could also yield data on the kinetics of nonequilibrium phenomena. Here the quantity being measured must be "sensitive" to changes in the physicochemical state of the fluid. The dependence of the density, pressure and temperature on the state of the gas behind the shock front can be evaluated easily using the conservation laws for one-dimensional steady-state flow and the equation of state. Assuming that at a certain stage an equilibrium is established with respect to the rotational degrees of freedom, then with respect to vibrational degrees of freedom, and, finally, that equilibrium dissociation is established, etc., we obtain various values of ϱ, p and T. The differences $\Delta\varrho$, Δp and ΔT between the values of these quan-

* Sometimes, when molecular disintegration does not occur, it is possible to determine the concentration of vibrationally excited molecules by measuring the changes in the concentration of molecules in the ground state (see, for example [277]).

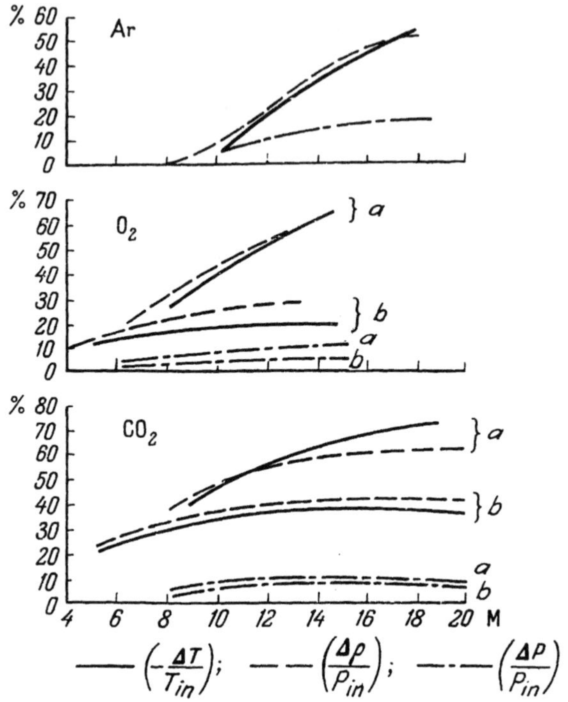

Fig. 5. Relative changes in the temperature, density and pressure behind a shock front in the process of establishing equilibrium as a function of the Mach number of a shock wave (in argon upon atomic ionization, in oxygen and carbon dioxide a) on molecular dissociation; b) on excitation of molecular vibrations). T_{in}, ϱ_{in} and p_{in} are the initial temperature, density and pressure before the ionization (Ar) or excitation of vibrations (O_2, CO_2) start.

tities at the beginning and end of each stage which arise on these assumptions are then referred to the absolute magnitudes of these quantities behind the shock front. As an example of the most typical mon- di- and triatomic gases which are studied in shock tubes, it is most convenient to consider argon, oxygen and carbon dioxide (Fig. 5). It can be seen that the temperature and density behind the shock front undergo the most significant changes in the process of excitation of vibrations, dissociation and ionization of molecules; here the pressure does not change significantly. Consequently, the study of the pressure distribution behind the shock front is barely sufficient to give reliable information on the course of relaxation processes in shock waves. In fact, not a single work in this field has made use of pressure measurements for obtaining kinetic characteristics; hence the problem of pressure measurement methods is not considered in this book.**

** Some information on the methods of pressure measurement in shock waves can be found in [65, 189].

On the other hand, the study of the density and temperature distributions behind the shock front (see Chapter 3) has yielded a large number of interesting results. If the test gas comprises only a small admixture in an inert diluting fluid, then the temperature and density of the diluting fluid do not change appreciably. In this case, in order to get information on processes in the test gas one must measure the concentration of the given gas components (most frequently, by spectroscopy). A particular complication is the study of establishing the Maxwellian distribution of molecular velocities across the shock front and of the rotational relaxation. For this the front thickness is usually measured (density variation in the shock front).

[3] Measuring Methods

Certain measurement methods must be established for measuring the temperature, gas density, concentration of the given components and other quantities. Under laboratory conditions, as a rule, the wave moves at a constant speed past a stationary instrument. Thus, we can consider two methods for recording the changes which take place in the shock wave: 1) obtaining an "instantaneous photograph" of the distribution of flow variables in the shock wave (across the wave front or in the region behind the front); and 2) continuous recording of the state of the gas with respect to time as the shock wave moves past the instrument. In the first case one of the main criteria for the suitability of a given measuring method is the spatial resolving power of the arrangement, and in the second case, it is the resolving power in time. The second approach is used more often.

It was noted above that the development of relaxation processes requires times from certain intervals between molecular collisions to hundreds of thousands (and more) of such intervals. The resolving power of the devices used should be selected in accordance with this; the presently available nano- and microsecond equipment is perfectly suitable for this purpose. The maximum duration of the recorded time intervals is determined by the dimensions of the plug and the flow velocity behind the shock front, and usually comprises from tens to hundreds of microseconds.

The second requirement – that the effect of the methods used on the object under study be small – is satisfied best by optical methods of study, which do not distort the gas flow behind the shock front. The greatest successes in the study of nonequilibrium phenomena have been achieved by using interferometric and spectroscopic methods (in a wide range of wavelengths from X-rays to radio frequencies), *i.e.*, by measuring quantities which are related to the refraction, reflection and absorption of electromagnetic waves in the gas.

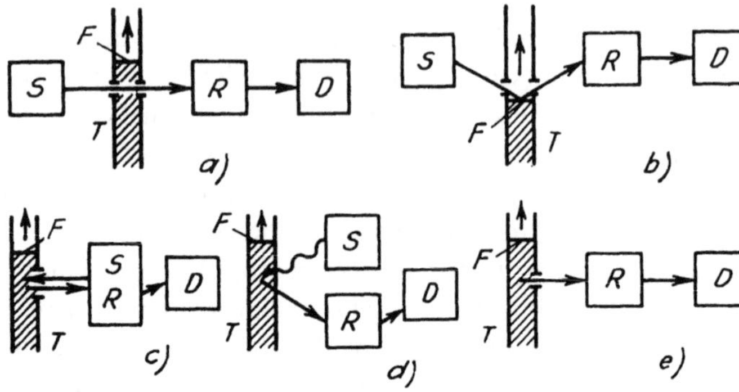

Fig. 6. Principal block diagrams of various measuring methods in shock tubes. a) Measuring the absorption, refraction or scattering coefficients; b) measuring the coefficient of reflection from the shock front; c) measuring the coefficient of reflection from the gas behind the front; d) measuring the electromotive force and the current flowing through a probe; e) measuring the radiation intensity of the gas behind the shock front. S = source; T=shock tube; F=shock front; R=receiver; D=recording device.

Most of the existing measuring arrangements are based on the principle of sounding [probing] the volume under study. These schemes include a source, a receiver and a recording device (Fig. 6). To measure the absorption (in some cases the refraction or scattering) coefficient the probing beam is directed normal to the shock-tube axis (direct sounding method, Fig. 6a). This method is used for measuring the distribution of component (including electron) concentrations and the gas density behind the shock front using spectral absorption analysis, absorption of X-rays and microwaves, scattering of the electron beam, as well as using the interferometer or the Tepler schlieren system (see Sects. 10, 11, 13). When measuring the reflection coefficient the beam is directed either at an acute angle to the shock front (measuring the density distribution across the shock front, Fig. 6b), or along a normal to the shock-tube axis (measuring the electron concentration by the reflection of microwaves, Fig. 6c). The distribution of the electron concentration in a shock wave is measured by displacement of the external magnetic field and recording the EMF which then appears in a special measuring coil, and also by using Langmuir probes (Fig. 6d) [see Sect. 13]. Finally, several arrangements are based on measuring the radiation intensity of the gas proper behind the shock front (Fig. 6e). These are the methods by which one can obtain the distribution of the concentration of excited particles and electrons, as well as the gas temperature distribution (emission spectroscopy, see Sect. 12). There are other methods of studying phenomena in shock waves, however, we wish to note the method of studying the final results of processes in the entire volume of gas behind the shock front (the "freezing" method, measuring the speed of the reflected shock wave, measuring the

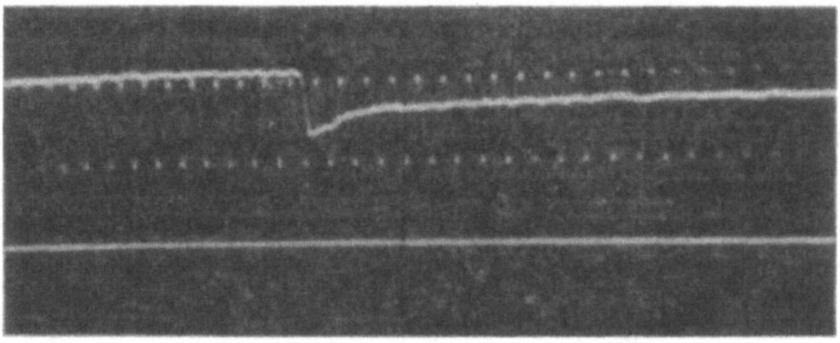

Fig. 7. Oscillogram characterizing the change in intensity of ultraviolet radiation which has passed normally through a shock wave as a function of time, for a shock wave propagating through a mixture of 21% O_2 and 79% Ar at a speed of 3 km/sec. Below is the zero line. The calibrating sinusoidal curve has a frequency of 1 Mc/s.

distance by which the step change [discontinuous jump] is removed from the surface of a blunt body, etc., see Sect. 14). All of these systems are discussed in detail in Chapter 3.

An example of the direct probing arrangement is a system for measuring the distribution of the molecular concentration behind the shock front by spectral absorption analysis (see Fig. 61). Here, use is made of a light source, spectral instrument, radiation receiver (most frequently a photomultiplier) and oscillograph, which is used for recording the received signals. An example of such an oscillogram is shown in Fig. 7. When the shock front appears in the observation section, the signal drops sharply, *i.e.*, the hot gas starts to absorb light. Further, the absorption is smoothly decreased, which corresponds to the dissociation of oxygen molecules. On the basis of such

Table 2. Vibrational Relaxation Time for $O_2(\tau_{vib})$ and the Characteristic Dissociation Time for $O_2(\tau_d)$ in a Mixture of 21% O_2 and 79% Ar (p_2 is the pressure behind the shoch front)

V, km/sec	p_2, atm	τ_{vib} microsec	τ_d microsec	τ_d/τ_{vib}
2.28	0.65	2.6	21.2	8
2.33	1.36	1.8	26.8	15
2.34	3.4	0.8	10.6	13
2.36	3.5	0.7	17.8	25
2.38	1.4	1.3	19.5	15
2.44	0.75	1.1	8.4	8
2.57	0.83	1.8	24.8	14
2.64	0.88	1.1	13.8	13
2.70	1.84	0.7	13.8	20
2.78	0.97	0.5	7.5	15
2.86	1.02	0.4	6.3	16
3.09	1.18	0.4	5.8	14

oscillograms, it is easy to determine the characteristic time τ_d, which is required for molecular dissociation (by the order of magnitude it is equal to the time needed for reducing the difference between the initial and the equilibrium concentration by a factor of *e*). The time τ_{vib}, which is needed for exciting molecular vibrations, is measured similarly. Table 2 gives examples of determining τ_d and τ_{vib} for oxygen molecules behind the shock front in a mixture of 21% O_2 and 79% Ar. It can be seen from the table that when the wave speeds is relatively low τ_d exceeds τ_{vib} appreciably.

2

The Shock Tube

[4] METHODS OF OBTAINING STRONG SHOCK WAVES

To study the excitation of dissociation and ionization of molecules in a shock tube, one must establish a temperature at which these phenomena could take place. What conditions are necessary for this to happen?

As we know, rotational excitation of molecules takes place at very low temperatures. At room temperature and above, the rotational degrees of freedom of all the molecules are practically [already] excited and make a contribution to the specific heat which is equal to $R/2$ for each degree of freedom (R is the universal gas constant). Excitation of molecular vibrations requires higher temperatures. The contribution made by the vibrational degree of freedom of a diatomic molecule to the molar specific heat is determined by

$$C_{\text{vib}} = R\left(\frac{\theta'}{T}\right)^2 \frac{e^{-\frac{\theta'}{T}}}{\left(1 - e^{-\frac{\theta'}{T}}\right)^2}, \tag{4–1}$$

and is illustrated by the graph of Fig. 8a, where θ' is the characteristic vibrational temperature ($\theta' = h\nu/k$, ν is the molecular vibration frequency, h and k are the Planck and Boltzmann constants). It can be seen from Fig. 8a that appreciable excitation of vibrations takes place only at temperatures comparable to θ'. The characteristic vibrational temperature for molecules, such as, N_2, CO, O_2, is several thousand degrees, while iodine molecules are practically excited at temperatures only slightly higher than room temperature. High temperatures are needed for molecular dissociation, as is illustrated by the graphs in Figs. 8b and 8c. Finally, even higher temperatures are needed for ionizing the atoms and molecules of a gas (Fig. 8d).*

* We wish to point out that the degree of dissociation is equal to the ratio of the number of free atoms, which have appeared as a result of dissociation to the total number of free and bound atoms per unit volume. For singly-ionized atoms, the degree of ionization α is defined as the ratio of the number of the ionized atoms to the total number of atoms and ions per unit volume.

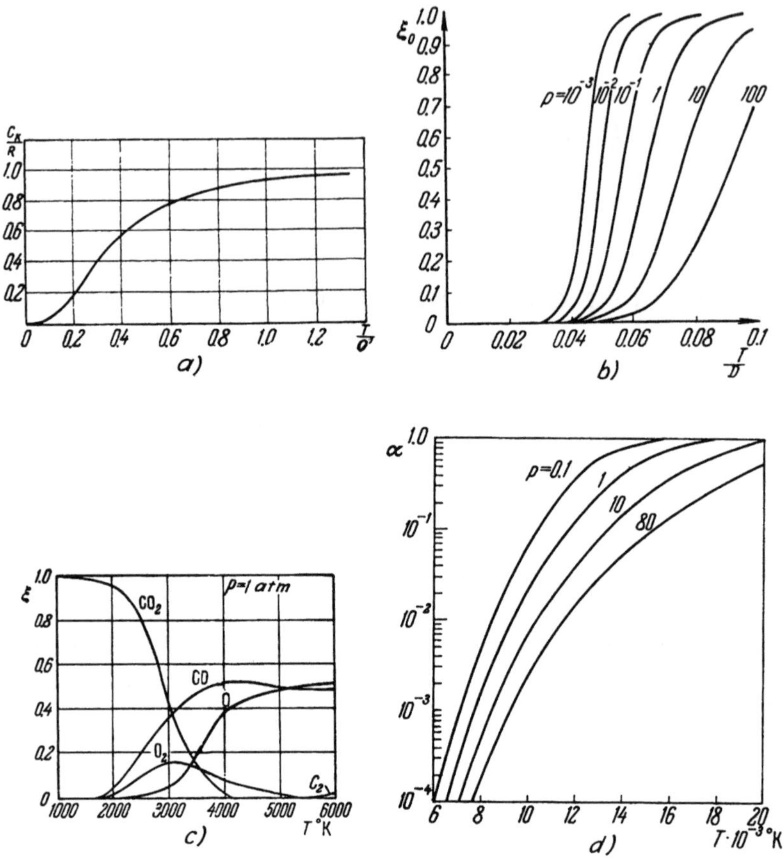

Fig. 8. Equilibrium properties of gases at high temperatures. a) The contribution of molecular vibrations to the specific heat as a function of the ratio of the translational temperature to the characteristic vibrational temperature [Eq. (4.1)]; b) the degree of dissociation of oxygen as a function of the ratio of the translational temperature to the dissociation energy D, expressed in °K. The numbers on the graphs show the gas pressure in atm. Curves for other diatomic gases are close to those for oxygen; c) composition of carbon dioxide at atmospheric pressure as a function of the temperature; d) degree of ionization of argon as a function of the temperature. The gas pressure is given in atm.

The temperature, pressure (and other flow variables) behind the shock front as a function of the shock-wave velocity can be found by using relationships which follow from the conservation laws, the equation of state and a number of supplementary equations. These calculations are made on various assumptions with respect to the completeness of the physicochemical process behind the shock wave (on the assumption of equilibrium, either only in relation to the vibrational degrees of freedom, or in relation to the rotational and vibrational degrees of molecular freedom with and without taking into account the electronic excitation, or assuming a complete static equilibrium, etc.). According to this, the molar fraction of the components either does not change (there is no dissociation) and no corresponding energy expenditures take place, or, in accordance with the law of mass action (or the Saha equation in case of ionization), the changes in the molar fractions of components with corresponding attendant energy expenditures are taken into account. It is assumed then that the sum of partial pressures of the components is equal to the total pressure (Dalton's law). In mixtures of gases of differing chemical nature, consideration is given to the fact that the ratio of the total atomic population of one gas to that of the other is constant. In the case of substantial ionization consideration must be given to the reduction in the ionization potential in plasma. The methods of this kind of calculations are given in detail in a number of works (see, for example [198, 172, 71, 476]).

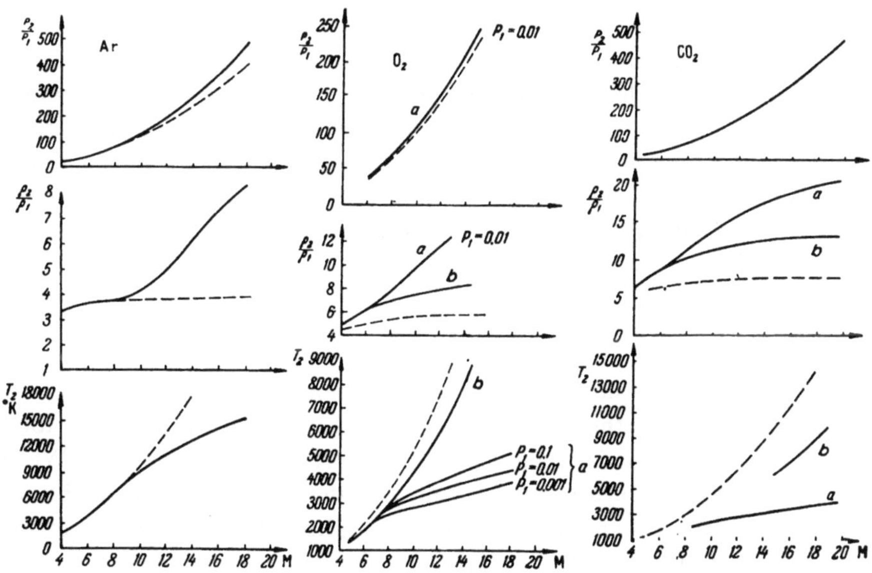

Fig. 9. Equilibrium values of the pressure and density with respect to values ahead of the shock front and the temperature behind the shock front as a function of the Mach number of its propagation in argon, oxygen and carbon dioxide. Designations: the dashed line denotes a gas with a constant ratio of specific heat equal to $\gamma = 1.66$ in Ar, $\gamma = 1.4$ in O_2, $\gamma = 1.33$ in CO_2; the solid line denotes an actual gas (in argon ionization is taken into account, in oxygen and carbon dioxide — a) taking dissociation into account; b) only taking into account the vibrational excitation). The initial pressure in argon is $p_1 = 10$ mm Hg, in oxygen it is $p_1 = 0.1$, 0.01 and 0.001 atm. The data for equilibrium dissociation of carbon dioxide correspond to atmospheric pressure behind the shock front.

As an example, we calculate characteristics of the gas behind a shock front in a monatomic gas (argon), diatomic gas (oxygen) and triatomic gas (carbon dioxide) [Fig. 9]. It follows from the graphs that the required temperatures can be obtained in a shock wave propagating with high velocities only (with $M > 5$).

As was pointed out in the preceding section, the shock tube is most convenient for obtaining shock waves under laboratory conditions. Therefore, it follows that, in order to study thermal excitation, dissociation and ionization in a shock tube, it is necessary to obtain sufficiently strong waves. The present techniques of shock-tube operation makes these conditions possible by using various methods.

The simplest approach is to increase the pressure ratio at the diaphragm and select the gases used in a standard tube with a constant cross section as was described in Sect. 3. However, even when $(p_4/p_1) \to \infty$, the shock-wave velocity cannot exceed the maximum values given in Table 1. On the other hand, the proper selection of gases used for filling the shock-tube chambers is more effective. According to Eq. (3–8), in order to increase the velocity of the shock wave, it is desirable to use a light gas in the high-pressure chamber and a heavy gas in the low-pressure chamber. One of the possibilities is adding a light component (let us say, hydrogen) to the driver gas and a heavy component to the test gas. The latter should preferably be an inert gas, for example, xenon [104]. As we know, the molecular weight of the mixture is

$$\mu = \sum \mu_i \xi_\mu, \tag{4–2}$$

where μ_i is the molecular weight of the ith component $\xi_i = n_i/n = p_i/p$ is the molar fraction of this component, n_i is the number density of particles of the ith component, p_i is the partial pressure of this component, n is the total number density of particles and p is the total pressure of the mixture. When xenon is added to oxygen or nitrogen, and if the ratio of the number of xenon atoms to the number of oxygen (or nitrogen) atoms ahead of the shock front in the low-pressure chamber is β, then the speed of sound in this nonreacting ideal mixture is

$$c_1 = \sqrt{\frac{7 + 10\beta}{5 + 6\beta} \cdot \frac{1 + 2\beta}{\mu_{O_2} + 2\beta\mu_{Xe}}}, \tag{4–3}$$

i.e., it decreases with an increase in β, which aids in obtaining higher Mach numbers for the shock wave for the same pressure ratio p_4/p_1 at the diaphragm. The calculated equilibrium values of the flow variables behind the shock front in an oxygen- (or nitrogen-) xenon mixture are given in [104].

Selecting gases and increasing the pressure ratio are not methods which are generally used for obtaining strong shock waves. The pressure increase is limited by the strength of the chamber walls and a reduction of p_1 involves disturbing the normal operating regime of the tube due to the increasing influence of the boundary layer. The Mach number can be increased by adding the lightest gas, *i.e.*, hydrogen, in the high-pressure chamber, however, the choice of a gas in the low-pressure chamber is dictated by the purpose of the study made. Hence, we shall consider other, more effective methods for increasing the speed of the shock wave.

It is simplest to heat the driver gas. An increase in its temperature increases c_4 and the rate of outflow from the high-pressure chamber. External heating of the entire high-pressure chamber is not too effective and, hence, most extensive use is made of a combustible mixture. This method is used for obtaining very strong shock waves (up to $V=16.8$ km/sec in air [459]). Three methods for heating the gas by combustion in the high-pressure chamber are as follows:

a) *Combustion at a constant volume, i.e., under conditions when the diaphragm is broken after combustion is terminated.* This facilitates the calculation of final (after termination of combustion) characteristics of the gas in the high-pressure chamber, if we disregard the energy losses due to dissocia-

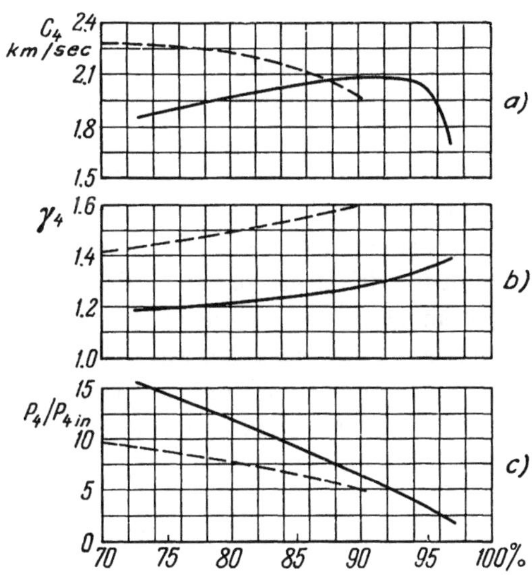

Fig. 10. Theoretically calculated values of the speed of sound c_4 (*a*), the ratio of specific heats γ_4 (*b*) and the final to initial pressure ratio p_4/p_{4in} (*c*) as a result of burning various mixtures in the high-pressure chamber. Designations: solid line—oxygen-hydrogen mixtures, dashed line—mixtures with the hydrogen and oxygen in stoichiometric proportions diluted by helium. The numbers on the abscissa axis pertain to hydrogen for the former mixture and to helium for the latter mixtures.

tion of the starting and final products of combustion and heat transfer through the walls. The results of these calculations, performed by Wittliff, *et al.* [612], are presented in Fig. 10. The most satisfactory results are obtained for an oxygen-hydrogen-helium mixture with the oxygen and hydrogen in stoichiometric proportions and with 70–80% helium. Such a mixture is preferable to a simple oxygen-hydrogen mixture even in considerations for operating safety. According to calculations, for the same peak value of the ratio p_4/p_1 (after combustion), which is equal to 3000, the Mach numbers in air which are obtained in the shock tube should be [612]:

Mixture O_2+H_2+He				Mixture O_2+H_2			
% He	90	79	70	97	92	73	% H_2
M	10.2	11.8	12.4	10.7	12.8	12.7	M

The above experiments show that the actual maximum pressure after combustion at a constant volume can be 30% lower than that theoretically calculated.

b) *Combustion at constant pressure.* In this case it is necessary to use a device which breaks the diaphragm immediately after the mixture is ignited. This causes the mixture that is already burning to expand. This method makes it possible to obtain stronger shock waves due to nonlinear addition of compression waves and shock waves moving in the same direction. The shock-wave velocity thus obtained can be estimated by letting c_4/c_1 go to zero in accordance with Eq. (3–7) [367]. The convenience of using this method is reduced due to the rapid deceleration of the shock front resulting from the rapid drop in the driver gas pressure. This can result in substantially disturbing the homogeneity of the gas in the plug.

c) *Heating the driver gas by a detonation wave.* This method is used in a long chamber in which the combustion front can become a detonation wave. However, a detonation wave introduces appreciable disturbance into the flow and can disturb the homogeneity of the test gas volume.

Various methods are used for igniting the mixture in the high-pressure chamber, for example, ordinary spark plugs, spark dischargers, dischargers with meltable wires, etc. To obtain the most uniform combustion and eliminate detonation, several dischargers (sometimes as many as eighteen, see [458]), located in different parts of the chamber, are sometimes used; good intermixing of the mixture before starting the test is obtained by supplying gas simultaneously through several inlets.

Methods of electrically heating the driver gas are also used. The use of

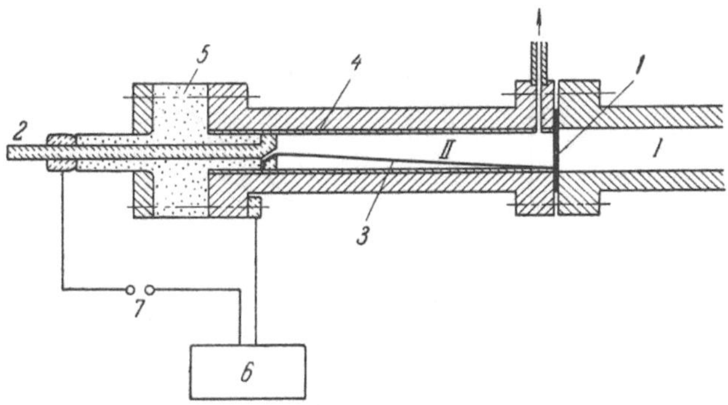

Fig. 11. Construction of a high-pressure chamber where the gas is heated by an electrical discharge. Designations: I) Low-pressure chamber; II) high-pressure chamber; 1) diaphragm; 2) electrode; 3) melting wire for discharge initiation; 4) insulating insert 5) insulator; 6) capacitor bank; 7) discharger.

resistance wires inside the chamber involves great technical difficulties. Pulsating electrical discharges in the high-pressure chamber are a more convenient method. A schematic of such a tube used for studying nonequilibrium phenomena in shock wave is shown in Fig. 11 [281]. Figure 12 gives Mach numbers of a shock wave in argon which is obtained by electrically heating helium contained in the high-pressure chamber; the thermal ionization of helium was taken into account in the calculations [354]. The Mach numbers were obtained as a function of the factor $\eta E/v$, where E is the stored electrical energy supplied to the spark gap, v is the volume of the heated gas, and η is the fraction of the energy which is transferred to the driver gas. The value of η can vary within wide limits (from several hundredths [237] to 0.8–0.9 [281]).

The development of the electrical heating method has resulted in the development of new kinds of shock tubes. In these devices the high-pressure chamber has been dispensed with and the discharge takes place directly in the test gas. The energy generated by the discharge greatly increases the pressure and temperature in the gas volume in the vicinity of the discharge zone; this gas, by expanding, acts as a piston and produces the shock wave. In certain designs the shock-wave velocity is increased by the interaction between the magnetic field of the discharge and the plasma thus produced. The basic processes which take place in these devices have been considered in [591, 55, 409] and in a number of other works. Many experiments (specifically [323, 257, 83, 5, 398, 28, 176, 222, 218, 219, 95 and 442]) have shown that this method can be used to obtain shock-wave velocities of several tens of km/sec. However, in this case the volume of heated gas behind the shock front being studied, is highly insignificant (see, for example [117a]). An-

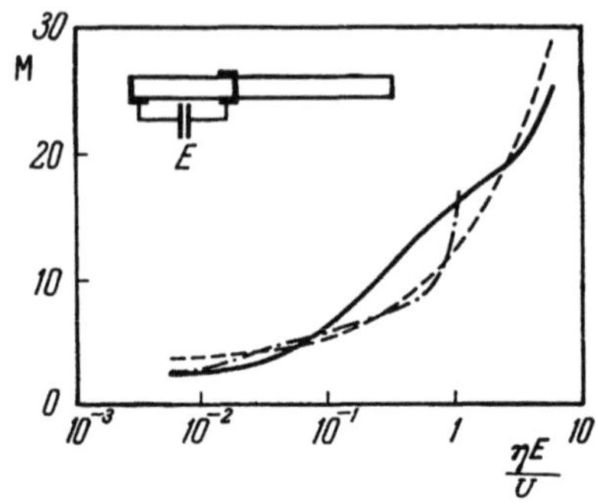

Fig. 12. Theoretically calculated value of the Mach number of a shock wave in a shock tube where the gas is electrically heated in the high-pressure chamber, as a function of $(\eta E/v)$, where E is the stored electrical energy in joules, v is the volume of the heated gas in cm^3, and η is that fraction of the energy which is transferred to the driver gas, *i.e.*, helium. The dashed line pertains to a pressure of 0.7 atm, the solid line is for a pressure of 0.07 atm, and the dash-dot line corresponds to a pressure of 0.007 atm. The low-pressure chamber contained argon under a pressure of 0.007 atm.

other feature of all these experiments is the fact that the shock-wave velocity is not uniform along the tube where the rapid velocity increase is replaced by a substantial deceleration due to the drop off in pressure in the driver piston after the discharge has ceased (see Fig. 18). Ladyzhenskiy and Chernikova [95] have shown that a smoother variation in the wave velocity can be obtained by changing the shape of the electrodes and increasing the discharge

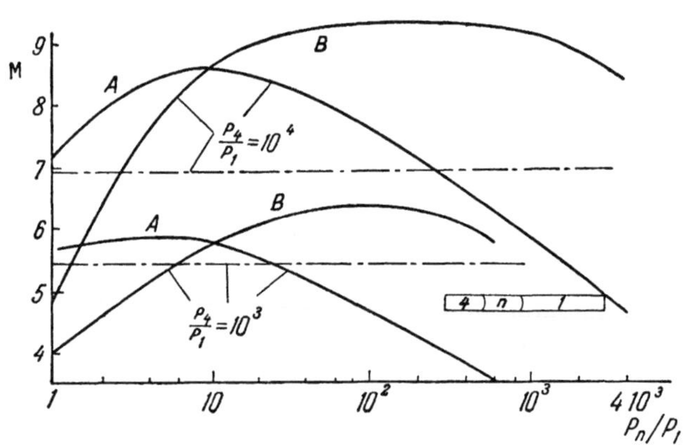

Fig. 13. Calculated values of the Mach number of a shock wave in a two-diaphragm shock tube as a function of the pressure ratio in the intermediate chamber and the low-pressure chamber p_n/p_1 for an overall pressure ratio (p_4/p_1) of 10^3 and 10^4, with the high-pressure chamber filled with helium and the low-pressure chamber filled with air. In the intermediate chamber *a*) is air, *b*) is helium. The curve for the Mach number as a function p_n/p_1 in the one-diaphragm arrangement without an intermediate chamber is denoted by -·-·-·-·-·-

time, however, this reduces the absolute velocity. Despite the above difficulties, successful studies were performed in these installations of nonequilibrium phenomena in very strong shock waves (30–40 and more km/sec), for example, in helium [442]. A detailed consideration of problems concerned with the operation of discharge devices is beyond the scope of this book. The reader is referred to the above original works: to the work by Wright [617] and the survey by Kolb and Griem [410].

Not only can the gas in the high-pressure chamber be heated by combustion of combustible mixtures or by electrical discharge, but also by a shock wave. This, naturally, increases not only the temperature, but also the pressure. The driver gas is heated by a shock wave in a shock tube which has several chambers separated by diaphragms and filled with different gases at various pressures (for multidiaphragm versions of shock-tube operation see [251, 513, 523, 363 and 370]). The two-diaphragm arrangement is the simplest and most frequently used (Fig. 13). This arrangement can be operated in two ways. First, the diaphragm separating the intermediate and the low-pressure chambers is made sufficiently strong and breaks only some time after the shock wave reflects from it. Due to this delay, a highly heated gas at a high pressure is formed near this diaphragm by the passing of the reflected shock wave. Second, this diaphragm is broken by the incident shock wave and the flow is expanded under nonsteady-state conditions into the low-pressure chamber. The first of these methods, all other conditions being the same, makes it possible to obtain stronger waves. However, the difference is highly insignificant, being of the order of only several percent.

In comparison with the simple shock tube, the convenience in using two-diaphragm systems depends on the kind of gases used and on the pressure ratio in the chambers. Using hydrogen at a pressure of 5 atm in the intermediate chamber for the same pressure ratio in the high- and low-pressure chambers, it is possible to obtain shock-wave velocities which are 15–20% greater than those obtained using only one diaphragm [205].

Very strong shock waves can be produced by using solid explosives [126, 293, 535]. By combining explosives with different rates of propagation of detonation waves and of different shapes, it is possible to form a plane detonation wave which, passing out from the volume occupied by the explosives, sets a plate into motion thus generating a plane shock wave with a maximum velocity corresponding to $\mathsf{M}=30$ at a high initial pressure [293, 535]. An interesting method for obtaining strong shock waves under laboratory conditions by using cumulative streams was employed by Novikov [137]. In his experiments cumulative charges of a special shape had a cylindrical hollow lined with iron; a glass shock tube divided into two chambers by an aluminum diaphragm was connected to the lining (Fig. 14). The chamber

Fig. 14. Schematic diagram of a shock tube using the cumulative effect of detonating a solid explosive. 1) Test gas chamber; 2) driver gas chamber; 3) charge.

connected to the charge was filled with helium to a pressure of 0.5 atm; the test gas (argon, air) was placed in the low-pressure chamber. The explosion of the cumulative charge forced a stream from the lining material which moved at a high velocity [138]. This stream, acting as a piston, heats and compresses the gas in Chamber 2, then the diaphragm is ruptured and the shock wave is formed in the test gas. In each experiment the glass tube was broken and had to be replaced. In this device shock waves were obtained propagating in argon with a velocity of 10.7 km/sec (for a $p_1 = 40$ mm Hg), and in air with a velocity of up to 9 km/sec (for a $p_1 = 25$ mm Hg).

An increase in the Mach number of a shock wave for the same pressure ratio p_4/p_1 can be obtained by changing the cross-sectional area of the tube, in particular, by reducing the tube cross section at the diaphragm section, *i.e.*, in the region where the gas flows out from the high-pressure chamber. As is known, there are two types of gas outflow from a high-pressure to a low-pressure region: one being the nonsteady-state (for example, outflow immediately after sudden removal of the diaphragm); and the other, steady-state (for example, steady flow in nozzles, orifices, etc). As was shown [69, 486] in the case of steady-state outflow from the diaphragm section, it is possible (in sub- or transonic flows) to obtain higher driver gas velocities.

Table 3. Maximum Values of M_∞ of a Shock Wave (when $(p_4/p_1) \to \infty$) for Different Ratios of Shock-Tube Cross-Sectional Areas in the Diaphragm Section

Driver Gas	Test Gas	S_4/S_1			
		1	1.51	2.25	∞
Helium	Nitrogen	10.9	11.3	11.6	12.5
Helium	Argon	12.8	13.3	13.7	14.7
Hydrogen	Nitrogen	22.6	23.2	23.6	24.7
Hydrogen	Argon	27.5	28.3	28.8	30.1

To establish steady-state outflow from the high-pressure chamber, the diaphragm cross section is narrowed down. Simple calculations can be used to determine the effectiveness of this method of increasing the Mach number of the shock wave; the maximal values of M obtained when $(p_4/p_1) \to \infty$ and for different ratios of the cross-sectional areas of the high-pressure chamber S_4 and the low-pressure chamber S_1 are given in Table 3. The experimental

Fig. 15. Photograph of the radiation of gas behind a shock front propagating in air with $M = 19$ in a 10×10 mm^2 duct directly following a 45° emergence from a round duct 50 mm in diameter.

results, obtained by Alpher and White [255, 226, 597], are in good agreement with these calculations.

A perceptible gain in the shock-wave velocity can be obtained by reducing the tube diameter in the low-pressure chamber, through which it is an already-formed wave that travels. Consideration of the problem of changing the shock-wave velocity on passing through a converging nozzle has shown that the following relation is obtained for relatively moderate cross-sectional area changes [290, 292]:

$$S(z-1)^{\frac{1}{K(z)}} = \text{const}; \tag{4–4}$$

where S is the cross-sectional area, $z=(p_2/p_1)$, and $K(z)$ is a slowly varying function of z. For $\gamma=1.4$ $K(z)\sim 0.4$ for all z; then for sufficiently strong shock waves ($z \gg 1$) and $\gamma=1.4$, we get

$$SM^5 = \text{const}. \tag{4–5}$$

One of the conditions for satisfying this relationship is a sufficiently smooth transition from one cross section to another. Bird's experiments [253] have shown that such a transition makes it possible to obtain stronger shock waves in a narrow cross section than when the tube duct undergoes a sudden expansion. It was found that a smoothed taper with cone angle of about 15° is most convenient. Despite the fact that Eq. (4–5) is approximate, satisfactory agreement with experimental results is obtained even when the cross section is reduced by more than a factor of 30 (the Mach number of the shock wave is increased by a factor of two) [254]. However, reducing the shock-tube duct diameter results in disturbing the homogeneity of the gas flow behind the shock front (see Sect. 6). A highly significant deceleration of the shock wave as a result of development of the boundary layer can be observed in very narrow ducts. This can be illustrated by a photograph of the radiation of the gas behind a shock front (Fig. 15). The photograph shows the deceleration of the shock front and inhomogeneity of the structure of the luminescent gas plug.

There are other ways of increasing the temperature of the test gas, however, we wish to note the method of reflecting the wave from the end wall of the shock tube. Kholev [217] has obtained an appreciable temperature increase by reflecting the shock waves from a wall with a 140° angle of convergence rather than from a flat wall (Fig. 16). It is difficult to study the performance in this case because of the appreciable nonhomogeneity of the gas.

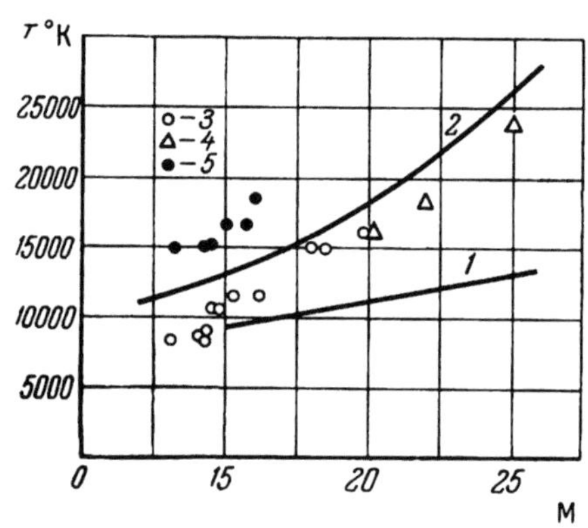

Fig. 16. Gas temperature on reflection of a shock wave from a straight wall of a shock tube and one with a 140° angle of convergence as a function of the Mach number of the incident shock wave in xenon. 1) Theoretically calculated value behind the front of the incident wave; 2) same as above, for a wave reflected from a flat wall; the points denote measured values; 3) straight wall; $p_1 = 3$ mm of Hg; 4) same as above, $p_1 = 2$ mm Hg; 5) a 140° angle, $p_1 = 3$ mmHg.

By using the methods listed for increasing the shock-wave velocities, it is possible to obtain shock temperatures and pressures behind the shock front which will produce excitation, disintegration and ionization of atoms and molecules of practically any gas. We know from experience that the most effective and convenient means for increasing the shock-wave velocity is the use of hydrogen and light combustible mixtures in the high-pressure chamber as well as the use of the two-diaphragm arrangement of shock-tube operation; also other methods for obtaining strong shock waves can sometimes be useful. Thus, the use of shock tubes creates the necessary conditions for a successful study of homogeneous processes in gases in a wide temperature range.

[5] GASDYNAMIC FLOWS IN SHOCK TUBES

Equation (3.7) obtained in the preceding chapter and relating the initial state of the gas in the shock tube and the velocity of the shock wave produced in it, follows from the simplified shock-tube theory. In this theory it is assumed (as was pointed out in Sect. 3) that the ratio of specific heats γ of the gases which are used remains constant, the bursting of the diaphragm and the formation of the shock wave take place instantaneously, the gas flow in the shock tube is steady-state and one-dimensional, and that it is possible to disregard dissipative phenomena and the intermixing of gases at the interface between the test and driver gases. Now we will consider the actual processes which take place in shock tubes.

Experiments in shock tubes show that in many cases Eq. (3–7) yields correct estimates of the dependence of the shock-wave velocity on the initial state of the gases. As an example, we can cite the results shown in Fig. 17.

The disagreement between the calculated and measured values can be

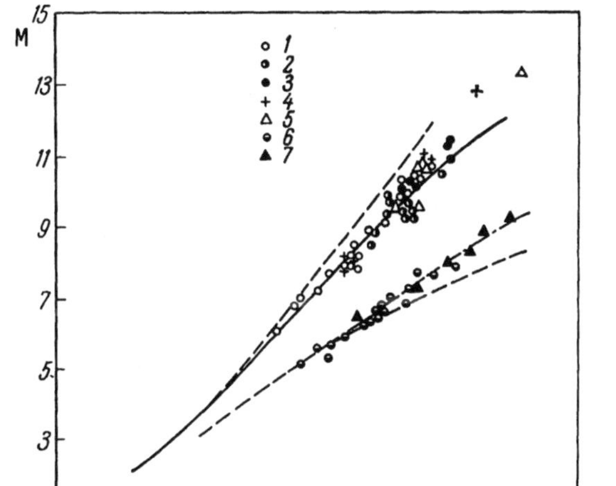

Fig. 17. Measured values of Mach numbers for a shock wave in the air as a function of the initial pressure ratio p_4/p_1 at the diaphragm for a hydrogen-(1–5) and helium-(6–7) filled high-pressure chamber. Dashed line corresponds to calculations according to the simplified shock-tube theory [Eq. (3.7)]; solid line pertains to calculations where the actual properties of air and hydrogen have been considered; the dash-dot line refers to calculations in which the nonlinear superposition of compression waves which form on gradual opening of the diaphragm have been taken into account. For experiments with hydrogen the thickness of diaphragms (made of copper) was: 1) 0.2 mm; 2) 0.3 mm; 3) 0.4 mm; 4) 0.47 mm; 5) 0.5 mm. Jones' results [397] are denoted by the number 6, and those due to White [597] by 7.

reduced by omitting [some] assumptions which were used as a basis for the simplified shock-tube theory. The first refinement of the theory can be accomplished by neglecting the assumption of a constant γ and taking into account the actual physicochemical transformations which take place in the gas. To consider the actual properties of the dissociating gas in the low-pressure chamber, the ratio p_2/p_1 given by Eq. (3–5) is replaced by a quantity which takes into account the actual temperature dependence of the specific heat and the change in the gas composition. On the other hand, the ratio p_4/p_3 should be represented by magnitudes which take into account the actual change in the specific heat following the cooling of the gas which flows out from the high-pressure chamber. It is easy to show that this cooling is quite substantial. Thus, for example, when the hydrogen is accelerated in the rarefaction wave to a velocity of $u_3 = 3.9$ km/sec, the temperature drops from 300 to 20 °K (when the actual properties of hydrogen are taken into account); here the magnitude of p_4/p_3 is reduced by more than a factor of 3. Mach numbers calculated as a function of the ratio p_4/p_1 for the hydrogen-air combination taking into account the real properties of air and hydrogen (solid curve in Fig. 17) give a better agreement with experimental data.

The data presented in Fig. 17 show that the actual maximum velocity of a shock wave for sufficiently high Mach numbers exceeds somewhat the theoretically calculated values (for the actual case of hydrogen-air). The same is observed when helium is the driver gas; in this case the experimental results exceed even the ideal theoretical values. These deviations are due to the fact that actually the diaphragm does not open instantaneously, and the shock wave does not form near the diaphragm opening, but at some distance from it. Detailed studies of the process of shock-wave formation in a tube after bursting the diaphragm were performed by Henshall [364], Glass and Patterson [332] and White [597]. It was found that several hundreds of microseconds are needed to break the diaphragm. When the diaphragm begins to open a jet breaks out from the high-pressure chamber, which forms a chain of compression waves in the test gas; the superposition of these waves creates the shock wave. The shock-wave strength continues to increase by the addition of compression waves which form during the subsequent stage of diaphragm opening and which then overtake the shock wave. The shock front is distorted at the beginning of the formation section, and then as a result of interaction with the tube walls, the front is straightened out. The nonlinear superposition of the nonsteady-state compression waves results in the formation of a shock wave moving at a speed higher than one formed by the instantaneous removal of the diaphragm. This is substantiated by experiments in which helium was used for driving air (see Fig. 17).

The character of the velocity distribution along the tube depends on the nature of the driving piston as well as on the dissipation processes which are related to friction between the gas and the tube walls, heat transfer, etc. Numerous measurements of the distribution of shock-wave velocities along the tube have shown that three sections exist in a shock wave: 1) the *acceleration* section, or the shock-wave *formation* section, over which the wave velocity is increased; 2) section of *uniform* motion, over which the velocity remains practically constant; and 3) wave *deceleration* section, where its velocity is reduced. This is seen clearly from the results plotted in Fig. 18,

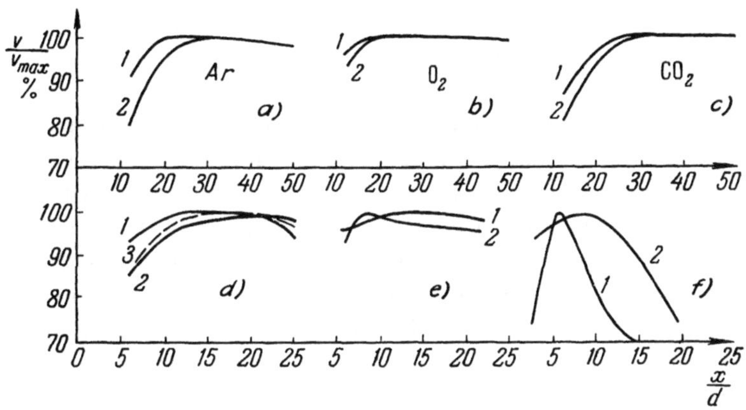

Fig. 18. Relative change in the shock wave as a function of the distance from the diaphragm, referred to the shock-tube diameter, in various shock tubes (V_{max} is the maximum shock-wave velocity, d is the shock-tube diameter). a), b) c) in a tube with $d=50$ mm, $V_{max} \sim 3$–3.5 km/sec with hydrogen used as the driver gas [1) $p_1=7.6$ mm Hg; 2) $p_1=3$ mm Hg]; d) in a two-diaphragm shock tube [205], $d=92$ mm, $V_{max} \sim 3$ km/sec [1), 2), are results of two different experiments with oxygen in the low-pressure chamber, $p_1=30$ mm Hg, 3) nitrogen, $p_1=12.7$ mm Hg]; e) combustible mixture is the driver gas [422], $d=610$ mm, air is the test gas, $p_1=0.02$ mm Hg [1) is an $O_2+H_2+N_2$ mixture as the driver gas, $V_{max}=6$ km/sec, 2) is a H_2+O_2 mixture with $V_{max}=8$ km/sec]; f) pulsating discharge tube [95], $d=40$ mm, air is the test gas, $p_1=0.2$ mm Hg [1) is for a capacitance of $C=300$ microfarads, inductance of the discharge circuit $L=1.66$ microhenries, discharge time $T=140$ microseconds, $V_{max}=23$ km/sec; 2) $C=300$ microfarads, $L=4.86$ of $C=300$ microfarads, inductance of the discharge circuit $L=1.66$ microhenries, discharge time $T=140$ microseconds, $V_{max}=23$ km/sec; 2) $C=300$ microfarads, $L=4.86$ microhenries, $T=240$ microseconds, $V_{max}=14.5$ km/sec].

which characterize the velocity distribution for a shock wave in different fluids and for different kinds of shock tube operation. The reduction in the shock-wave velocity is more perceptible when using lighter gas as the driver, *e.g.*, the shock wave in air is decelerated more perceptibly when it is driven by hydrogen than in the case when helium is the driver (Fig. 19). The most

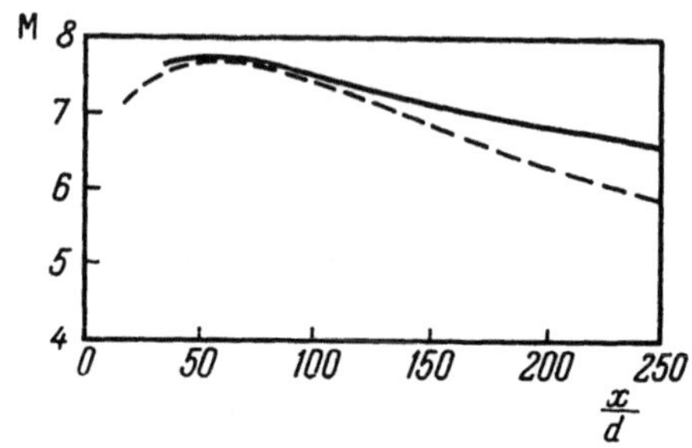

Fig. 19. Effect of the driver gas on the change in the Mach number of a shock wave in air along the shock tube. Solid line pertains to the use of helium as the driver gas; the dashed line corresponds to the use of hydrogen for driving the test gas.

appreciable change in the wave velocity is observed in pulse-type installations when using a discharge with a very short duration (Fig. 18f).

Studies have shown that the dimensions of the shock-wave formation and acceleration (to the maximum velocity) section depend on the initial pressure ratio p_4/p_1 and are practically independent of the kind of gas used (Fig. 20). The scattering which is then obtained is due (only) to differences in the character of the diaphragm bursting. The formation and acceleration of a shock wave is more complicated when a combustible mixture is used as the driver gas or when using several diaphragms.

The dimensions of the shock-wave formation section, the shock-wave velocity and the shape of the interface are related to the intensive intermixing

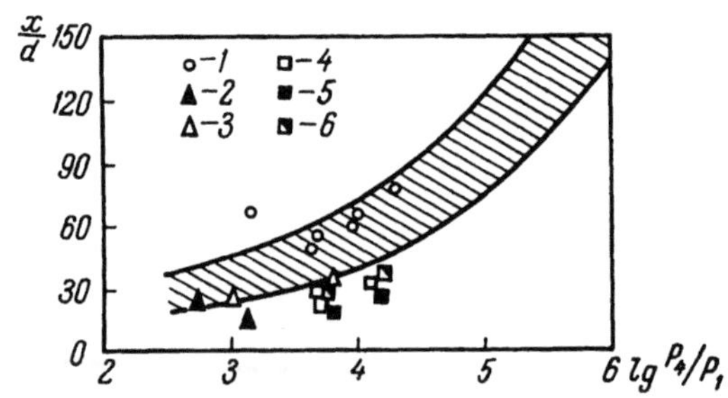

Fig. 20. Relative dimensions of the shock-wave formation section as a function of the initial pressure ratio p_4/p_1 for certain methods of filling the high-n and low-pressure chambers. The shaded area represents results obtained by White [597], in a tube of 82.5×82.5 mm^2 with helium as the driver gas and nitrogen, oxygen or air as the test gas, and also for the case when the driving gas is hydrogen and argon is the test gas; 1) Jones' results [397] for a tube with $d = 95$ mm, with hydrogen as the driver gas and air as the test gas; 2), 3) results due to Fayzullov et al [205], $d = 92$ mm, hydrogen is the driver gas and argon (2) and nitrogen (3) are the test gases; 4)–6) results due to Losev et al, $d = 50$ mm, hydrogen is the driver gas while oxygen (4), argon (5) and carbon dioxide (6) are the test gases.

of the driver and test gases at the end of the plug. The most substantial effect on the intermixing is exerted by the development of the boundary layer at the tube walls. The gas which was previously in Region 2 and has then flowed into the boundary layer is retarded, it deforms the interface and is intermixed with the driver gas. The cooling of the test gas and the heating of the driver gas in the intermixing zone results in density changes, formation of additional compression (or rarefaction) waves and in the final result changes the shock-wave velocity. This can manifest itself most perceptibly when combustion takes place at the interface (for example, when oxygen or air is driven by hydrogen). An instantaneous photograph of this region (Fig. 21) shows that the interface is actually intricate in shape, and is extended in the direction of the shock front. Intermixing of the gas results in the formation of a contact region the length of which, according to Hooker [379], depends little on the wave velocity, initial pressure and the time needed for opening the diaphragm. The intermixing of the gas in the contact region reduces the homogeneous gas volume behind the shock front.

It was discovered in the early researches on shock tubes [364, 332], that the velocity of the leading part of the interface exceeds substantially the magnitude predicted by the simplified shock-tube theory, which is equal to the theoretical value of the flow velocity u_2 (for a measured shock-wave velocity). Subsequently, this phenomenon was observed by all the investigators who have made simultaneous measurements of V the shock velocity and V_i the interface velocity. Thus measurements performed by these authors of V and V_i at a distance of 50 diameters from the point of diaphragm rupture for a shock wave propagating in oxygen in a 50-mm diameter tube have shown that in the majority of cases the velocity of the leading part [nose] of the interface is practically equal to the shock-wave velocity. Similar results were obtained for argon and carbon dioxide (Fig. 22). Henshall, Dem'yanov and others [364, 57, 546, 496] have shown that the acceleration of the interface is due to the development of a boundary layer at the tube walls.

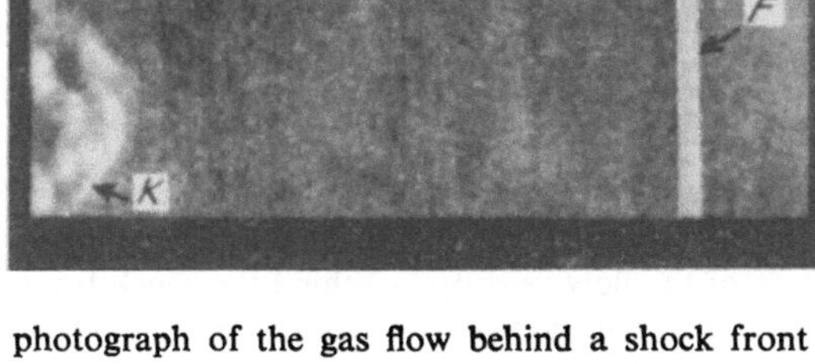

Fig. 21. Shlieren photograph of the gas flow behind a shock front propagating in a shock tube. F = shock front, K = interface (nitrogen, $p_1 = 0.9$ mm Hg, M = 10.6) [73].

The boundary layer which forms at the shock-tube walls has an important effect on the character of the shock-wave motion. Direct observation of the boundary layer in the flow behind the shock front in a tube is a very complex problem. Results obtained by optical (shadow, interferometer) methods (for example, film transducers, adding special solid admixtures to the flow, etc., see, for example [33, 360, 435, 436, 359, 289]), were obtained primarily in weak shock waves. However, the results which were obtained in these

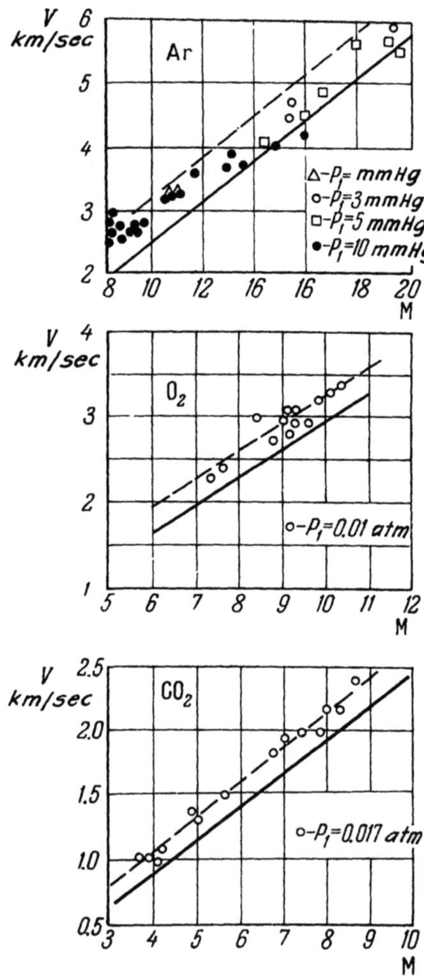

Fig. 22. Measured values of the velocity of the leading part of the interface following the propagation of shock waves in argon, oxygen and carbon dioxide. Solid line denotes the theoretical values of the flow velocity u_2 behind the shock front on the assumption of total static equilibrium; the dashed line denotes the shock-front velocity. Data for carbon dioxide were taken from [9]. 1) mm Hg.

experiments make it possible to determine the reliability of available theoretical results.

The first theoretical works devoted to the study of the boundary layer in the shock tube and its effect on the shock-wave propagation were by Donaldson and Sullivan [315], Henshall [364] and Hollyer [378]. Donaldson and Sullivan have estimated the deceleration of the shock wave assuming that the flow is incompressible, which is not valid in the case of strong shock waves. They have proposed that the length of the plug be estimated by equating the "losses" of the gas mass from the main body of the flow due to outflow into the boundary layer in he region of the interface to the "refilling" of the gas mass due to [the] flow [of gas] through the shock front; this approach is sometimes used when considering the problem of the length of a hot plug [231, 317, 379]. Later, Henshall estimated the thickness of the boundary layer in the tube by assuming it to be laminar and incompressible. He also obtained a qualitative agreement with experimental data, referring to the relationship between the acceleration of the interface and the formation of the boundary layer. The next important development was made by Mirels [448, 449], and by Trimpi and Cohen [575]. The theories reported by these authors postulate continuous generation of nonsteady-state compression and rarefaction waves as a result of the formation of a boundary layer at the shock-tube walls. These disturbances propagate in all directions, interact with the shock front and the interface, and change the character of the flow by disturbing the homogeneity. Trimpi and Cohen assume that these waves are produced by frictional stresses and heat transfer at the tube walls. Similar to Hollyer, they average these effects over the entire flow cross section, reducing the problem to one-dimensional flow. Mirels assumes that the boundary layer which thus appears acts as an elastic wall in displacing the flow from the region adjoining the wall and produces at the external boundary of the layer a nonzero component of the flow velocity. Both these theories predict deceleration of the shock wave which depends on the wave velocity, distance from the diaphragm, type (laminar or turbulent) and shape of the boundary layer, initial conditions and the tube diameter. In the case of comparatively thin boundary layers with an isentropic main body of the flow use should be made of the Mirels theory, while the theory of Trimpi and Cohen is applicable for thicker boundary layers. Both these theories have been developed for a gas with a constant ratio of specific heats and thus cannot give correct quantitative results when considering strong shock waves. Since successful study of nonequilibrium phenomena is possible only when the boundary layer is relatively thin, it is more expedient to use the Mirels theory for estimating the effect of the boundary layer on the velocity distribution along the tube. One of the results of such calculations (for a

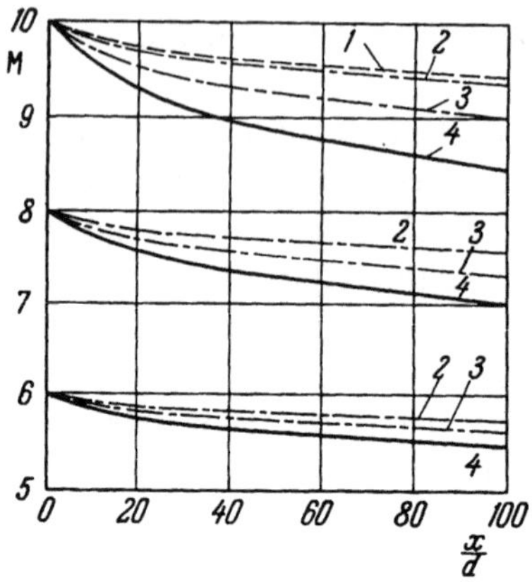

Fig. 23. Theoretically calculated distributions of the Mach number of a shock wave along the shock tube in the presence of a laminary boundary layer at the tube walls for different gases ($d = 50$ mm, atmospheric pressure behind the shock front). 1) Helium or hydrogen in the high-pressure chamber, carbon dioxide is the test gas; 2) hydrogen-oxygen (on the assumption that the gases have their real properties); 3) same as above assuming the gases to be ideal; 4) hydrogen-argon.

laminar boundary layer) is presented in Fig. 23. It follows from these calculations that the deceleration of a shock wave for the same starting Mach number is most perceptible when a wave is propagated in argon, and least when the wave travels in carbon dioxide; this has been substantiated experimentally (see Fig. 18). It can be seen from the case when oxygen is driven by hydrogen that considering the actual properties of the gases used results in decreasing the theoretically calculated deceleration. Calculations made according to the Mirels theory show that replacing the hydrogen by helium somewhat reduces the deceleration, which is also in agreement with experimental results.

Solutions of gasdynamic equations for the entire region behind the shock front assuming either a laminar or turbulent boundary layer were obtained by Dem'yanov [56, 57], and Spence and Woods [546], who have obtained the acceleration of the interface and the deceleration of the shock front and also found the distribution of the flow variables behind the shock front in the presence of a boundary layer.

In a number of shock-tube experiments it is desirable to know the dimensions of the gas volume between the shock front and the leading part of the interface, *i.e.*, the length of the plug. According to simple considerations of conservation of the mass of the gas which flows through the shock front assuming the one-dimensional flow of the simplified shock-tube theory, we find that the length of the plug in the ideal case of constant shock wave and interface velocities (without intermixing and with no effect exerted by the boundary layer) is

$$L_{id} = l\frac{\varrho_1}{\varrho_2}, \tag{5-1}$$

where l is distance from the point of observation to the diaphragm. In an ideal gas, the ratio ϱ_2/ϱ_1 increases due to excitation, dissociation and ionization of the gas behind the shock front; according to Eq. (5–1), this results in reducing the length of the plug. The acceleration of the leading part of the interface with a simultaneous deceleration of the shock wave results in the fact that the length of a plug of gas heated in a shock wave in an actual shock tube will be smaller than that given by Eq. (5–1). It was shown experimentally that the distance between the shock front and the leading part of the interface can actually be by more than a factor of two shorter than that theoretically calculated. This is due to the fact that a part of the test gas flows into the boundary layer, is decelerated, remains behind the leading part of the interface surface and is then intermixed with the driver gas.

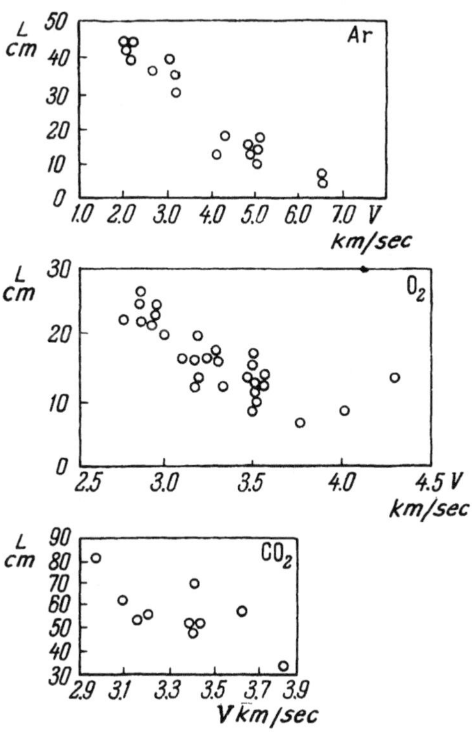

Fig. 24. Measured values of the plug, *i.e.*, of the distance from the shock front to the leading part of the interface, in argon, oxygen and carbon dioxide at a distance of about 50d from the diaphragm and for an initial pressure of ~ 0.01 atm, as a function of the shock-wave velocity.

According to the experimental results, the rates of propagation of the shock front and the leading part of the interface can be quite close; then the length of the plug, unlike Eq. (5–1), will cease to increase upon reaching a maximum value of L_{max}. Roshko [498], Hooker [379] and Mirels [450, 405a], analyzing the results of measuring the plug length in different gases and under different initial conditions, concluded that the velocity of the leading part of the interface approaches the shock-front velocity as early as a distance from the diaphragm of $l \sim 5(\varrho_2/\varrho_1)\, L_{max}$. The theoretically calculated L_{max} for argon and air, as a function of the Mach number of the shock wave (in the presence of a laminar boundary layer) is given in Fig. 25. For the same Mach number the magnitude of L_{max} is proportional to d^2p_1 (for a laminar layer) and $d^{\frac{1}{4}}p_1$ (for a turbulent layer)*. Compared with ideal values of Eq. (5–1), the reduction in the plug length is most pronounced in the case when a turbulent boundary layer is developed at the wall [231]; in this case L_{max}/L_{id} can be as low as 0.1–0.15 [450a].

The most appreciable deviations from the simplified shock-tube theory are observed for a low initial pressure in small-diameter tubes [379, 317]. Measurements of plug length under these conditions ($p_1 \leqslant 0.5$ mm Hg, $d \leqslant 30$–40 mm) have shown that the plug length is greatly dependent on the initial pressure, which is in complete disagreement with the simplified theory. It was discovered [379, 450, 518], that the L_{max} for a given Mach number is

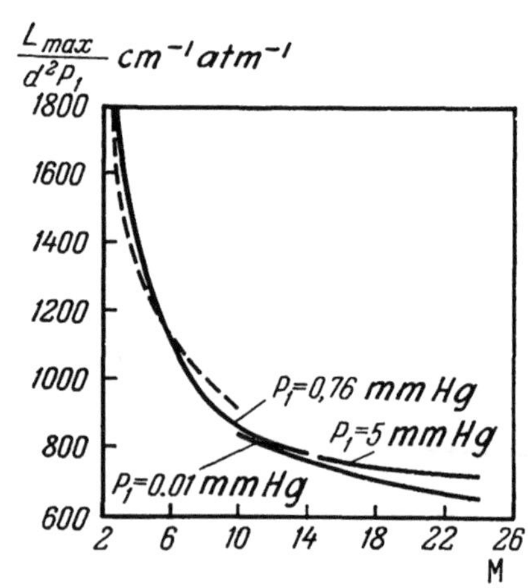

Fig. 25. Theoretically calculated ratios of the maximum plug length L_{max} and the product of the square of the tube diameter and the initial pressure as a function of the Mach number of a shock wave for argon (dashed line) and air (solid line) [450]. 1) cm^{-1} atm^{-1}; 2) mm Hg.

* d is the shock-tube diameter.

proportional to the initial pressure to the square of the shock-tube diameter (see Fig. 25). Thus, all other conditions remaining the same, a reduction in the initial pressure decreases the length of the plug substantially, so that for very small p_1 it comprises a negligible part of the length predicted by the simplified theory (Fig. 26). When the pressure is decreased in a small diameter tube, the obtaining of the same shock-wave velocity requires an appreciably higher pressure ratio p_4/p_1 [317]. Thus, the simplified theory becomes invalid under these conditions since the boundary layer occupies a perceptible part of the tube cross section.

However, the study of nonequilibrium phenomena under low initial pressures (less than 1 mm Hg) is of great interest, since the mean-free molecular path is large in this case, and, consequently, the dimensions of the nonequilibrium regions of the flow are also large. This is most important in the study of the structure of strong shock waves, when the gas temperature behind the front is high and the equilibrium is established relatively rapidly. Only a substantial pressure reduction (increase in the dimensions of nonequilibrium zones) makes it possible in this case to successfully set up an experiment for the study of relaxation processes. The need of performing experiments with low initial pressure is also determined by problems of high-

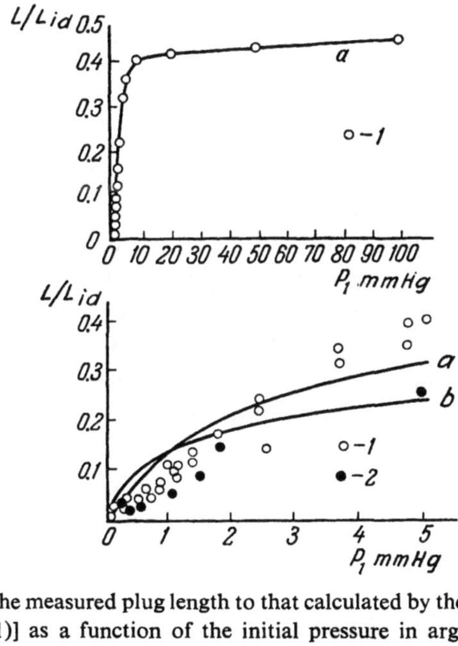

Fig. 26. Ratio of the measured plug length to that calculated by the simplified shock-tube theory [by Eq. (5.1)] as a function of the initial pressure in argon. 1) Results due to Hooker, $d = 39.4$ mm [379]; 2) results due to Duff, $d = 33.4$ mm [317]; the solid line denotes theoretically calculated L/L_{id}, obtained by Hooker for conditions of his experiments (a) and of those by Duff (b). A) mm Hg.

altitude flight. Thus, the air pressure in the earth atmosphere at an altitude of 50 km is 0.6 mm Hg, and at an altitude of 80 km it is about 8.10^{-3} mm Hg. Under these conditions the nonequilibrium region of the flow can be found to be quite substantial (see Chapter 5).

Specially made large-diameter shock tubes are used for experiments at low initial pressures. The shock tube described by Lin and Fyfe [422] can serve as an example of such an installation. The internal diameter of the low-pressure chamber of this tube is 610 mm, the length is about 15 m. An oxygen-hydrogen mixture, with a nitrogen admixture, at an initial pressure of up to 30 atm was used in the high-pressure chamber (diameter 125 mm, length 2 m). The initial pressure of the test gas in the low-pressure chamber was 0.2–0.02 mm Hg, the shock-wave velocity was from 4 to 7 and more km/sec. The length of the plug at a distance of 15 m from the diaphragm, under these conditions, was about 30 cm. As shown by the experiments in this case all the main conclusions of the simplified shock-tube theory are satisfied even at a pressure of $p_1 = 0.02$ mm Hg. The use of such tubes in the study of relaxation processes at high temperatures makes it possible to obtain many interesting results.

[6] INHOMOGENEITY OF THE FLOW BEHIND THE SHOCK FRONT

Usually in the study of nonequilibrium phenomena in shock waves propagating in shock tubes, it is assumed that: 1) the shock wave propagates in the shock tube in a steady-state manner; 2) the flow in the shock tube is one-dimensional; 3) the effect of viscosity, thermal conductivity, diffusion and energy removal by radiation can be neglected. If any inhomogeneities are observed in the flow behind the shock front in the shock tube, then the cause for it must be sought in nonequilibrium phenomena. Conformance to the above conditions then makes it possible to use gasdynamic equations for the steady-state and one-dimensional flow taking into account the physico-chemical transformations.

Obviously, the above conditions actually prevail only with a given degree of approximation, which is responsible for possible errors in determining the rates of the nonequilibrium process and may also be the cause for making its determination entirely impossible. This requires that the satisfaction of the above conditions under actual conditions be determined. Such a consideration was started in the preceding section, where the main causes for disturbing the simplified picture of shock-wave propagation in a shock tube were pointed out. Here this consideration is continued from the viewpoint of determining the possibility of studying nonequilibrium phenomena in shock tubes.

[1] Inhomogeneity of the Flow Along the Plug

The study of nonequilibrium phenomena in shock tubes is based on the study of the distribution of the gas characteristics in the shock wave along the tube axis in the direction of the wave propagation, *i.e.*, along the plug. Hence, first of all, we shall consider the effect of the flow unsteadiness and of dissipation processes on the distribution of flow variables along the plug.

The most appreciable disturbance of the homogeneity of the flow behind the shock front is observed when the wave is formed at a section with substantial acceleration, since the character of the diaphragm bursting and the outflow of gas from the high-pressure chamber (or the action of the discharge in the pulse system) have a pronounced effect on the flow under study in this part of the tube, when the plug is affected by strong longitudinal and transverse compression waves. The fact that large gradients of the flow variables exist in the plug at that section of the tube is substantiated by results obtained by Busygin and Tumakayev [33] who have measured the density distributions in the flow behind the front of a forming shock wave in argon at different distances from the diaphragm (Fig. 27). Despite the absence of any changes in the physical state of the fluid, the density of the gas in the plug increases substantially with an increase in the distance from the wave front. This change is most pronounced for tube sections near the diaphragm and becomes less perceptible when moving away from the diaphragm; at a distance of 60–70 diameters, this density increase becomes insignificant (within the limits of the experimental error, *i.e.*, less than 4%). As the Mach number increases, the inhomogeneity of the plug in the formation zone increases in a manner such that at a distance of 24 diameters from the diaphragm the pressure rise along the flow between the front and the leading part of the interface is 10–15% for M$\sim$3, while for M$\sim$10 it is as high as

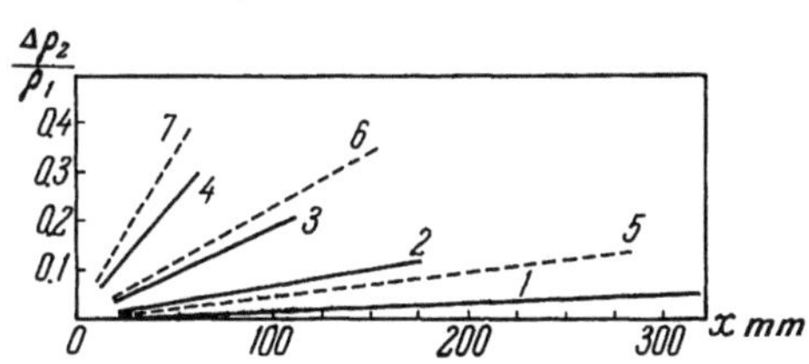

Fig. 27. Relative density change behind a shock front in argon obtained by Busygin and Tumakayev [33] as a function of the distance from the shock front different Mach numbers and distances l from the diaphragm. (1–4) correspond to M $=3$, (1) $l=47d$; (2) $l=24d$; (3) $l=14d$; and (4) $l=8d$; (5–7) correspond to M $=4$, (5) $l=47d$; (6) $l=24d$; (7) $l=8$d.

50–100%. These facts point to the difficulties in studying nonequilibrium phenomena in very short shock tubes, where the shock wave does not succeed in finally forming before reaching the test section.

We now turn to the distribution of flow variables in the plug at greater distances from the diaphragm, by first presenting results of experimental investigations. Rose and Nelson [497] have found that the gas-pressure change in a plug behind the shock wave in air ($M \leqslant 10$) is less than ten percent. On this basis they have assumed it to be approximately constant and calculated the distribution of other flow variables. The study of the pressure distribution performed by Bazhenova, *et al.* [15], also points to the smallness of the pressure variation in the flow behind a shock wave in a shock tube. Duff [317] has used an electron beam to study the density distribution in argon and has found that, for a low initial pressure ($p_1 = 0.5$ mm Hg) in a narrow tube the measured density can increase toward the end of the plug by 50%. In experiments by Busygin and Tumakayev [33] with $p_1 = 3$–10 mm Hg, at a sufficient distance from the diaphragm (60–70 diameters), as was pointed out above, no substantial density changes were discovered.* A study of the temperature distribution in the flow behind the front of a passing shock wave which was performed by Fayzullov, Sobolev and Kudryatsev [205, 206] using a pyrometer has shown that the gas temperature in the plug is practically constant and drops off only toward the end of the plug (in nitrogen with $p_1 = 10$ mm Hg and $V = 3.2$ km/sec this reduction was 12%). A study, made by these authors, of the distribution of molecular oxygen concentration along the plug by the absorption spectroscopy method (see Sect. 11) has shown that, for an initial pressure of several mm Hg the O_2 concentration is practically constant up to the time when the leading part of the interface arrives (Fig. 28). Finally, as was pointed out above, it was established in the majority of works that the leading part of the interface moves

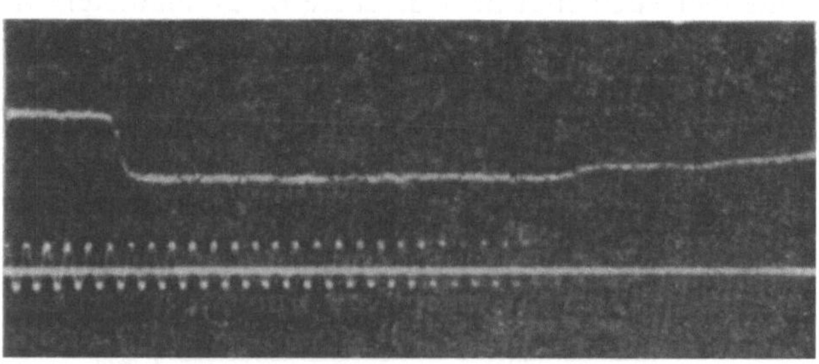

Fig. 28. Oscillogram describing the absorption of light in oxygen between the shock front and the interface ($V = 2.77$ km/sec).

* These authors claim that Duff's results have been distorted by the outflow of the gas from the test section onto the electron gun.

at a velocity which greatly exceeds the calculated flow velocity behind the shock front. This shows that the flow velocity increases as one moves farther away from the front. Unfortunately, the enumerated facts do not give a complete idea of the character of the flow-variable distribution in the plug; a comprehensive and systematic study of this problem has not as yet been made. The quantitative aspect of the above results is also not too reliable. Thus, for example, it was the "total" density in the main body of the flow and in the boundary layer which was recorded by Duff when he measured the gas density by sounding the gas flow across the tube axis by the electron beam. As is pointed out below, the mass of gas contained in the boundary layer when the initial pressure at the end of the tube is low can be substantial. Hence Duff's results cannot be referred to the main body of the flow only. Apparently, the most reliable postulate (qualitatively as well as quantitatively) at the present time is that the velocity of the main body of the flow (in the laboratory coordinate system) increases toward the end of the plug and that it approaches the shock-wave velocity. According to what was said in the preceding section, this takes place in different gases at a sufficient distance from the diaphragm (at the point where nonequilibrium phenomena are usually studied).

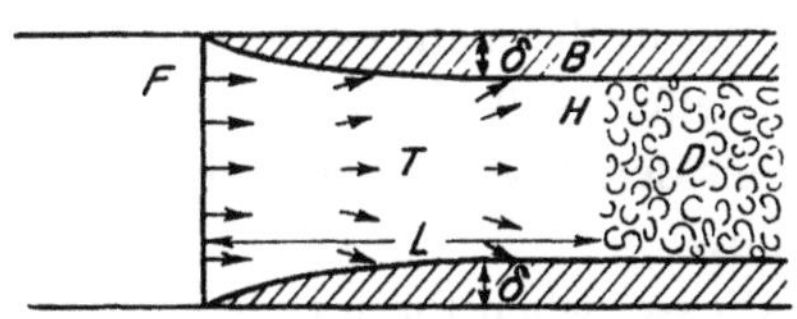

Fig. 29. Schematic representation of the flow behind a shock front in the presence of a boundary layer at the shock-tube walls. *F* is the shock front; *H* is the leading part of the interface; *T* is the test gas in the main body of the flow; *B* is the boundary layer; δ is an arbitrarily assumed boundary-layer thickness; *D* is the driver gas and *L* is the plug length.

In this case, Duff [317] has suggested that the motion of the gas in the main body of the flow from the front to the interface be assumed to be isentropic, one-dimensional and steady-state (in the coordinate system moving with the front), since the inflow of mass through the wave front is equal to the outflow of mass into the boundary layer behind the leading part of the interface (Fig. 29). This model was developed further by Roshko [498], Hooker [379] and Mirels [450], who have assumed that a laminar boundary layer is present at the tube walls. An estimate of the distribution of flow variables in the main body of the flow in this case can be obtained from the relationship for the mass flow

$$\frac{\varrho v}{\varrho_2 v_2} = 1 - \sqrt{\frac{x}{L_{\max}}}, \qquad (6\text{–}1)$$

where ϱ_2 and v_2 pertain to variables directly behind the shock front. Thus, in the coordinate system moving with the shock front, the flow in the main body of the flow behind the shock front corresponds to the subsonic flow of gas in an expanding channel, *i.e.*, the flow velocity drops to zero (with respect to the front), while the density, pressure and temperature increases as the distance from the front increases. A second analogy of the gas flow in the plug in this case is the flow behind a detached shock wave ahead of a blunt body, whose role is played by the leading part of the interface.* Simple estimates (assuming that the test gas is ideal) show that when $\mathsf{M}=8$ the gas density in the main body of the flow toward the end of the plug increases by 11% in Ar, by 8% in O_2 and by 7% in CO_2, the pressure increases by 18% in Ar, 11% in O_2 and 8% in CO_2; while the temperature increases by 7% in Ar, by 3% in O_2 and by 2% in CO_2. The flow velocity undergoes the most appreciable changes, *i.e.*, in the laboratory coordinate system it increased by 35% in Ar, by 22% in O_2 and by 17% in CO_2. In real gases and in the case of higher velocities all these changes are less perceptible. The distribution of the above variables along the plug is easily obtained by using Eq. (6–1) and equations which characterize one-dimensional isentropic deceleration of a gas (Fig. 30). It can be seen from this figure that the more perceptible gradients of the variables in the main body of the flow which are produced by the boundary-layer effect will be observed in the vicinity of the shock front; this must be considered when analyzing nonequilibrium phenomena in this region of the flow.

In the more general case, when the interface velocity is not equal to the shock front velocity, use should be made of results obtained by Dem'yanov [57] and Spence and Woods [546]. Dem'yanov has solved a system of flow equations for the entire plug (assuming that the boundary layer is laminar) and found the distribution of flow variables along the plug. If we assume that the velocity of the shock front remains constant (as in Duff's model) then it

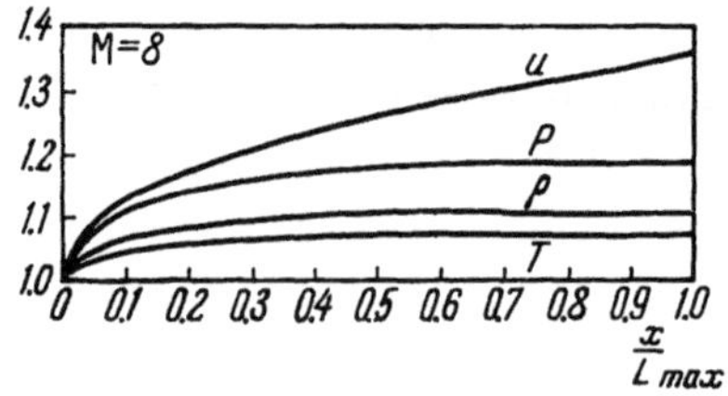

Fig. 30. Relative variation in the flow velocity, pressure, density and temperature in a plug of maximum length (argon, $\mathsf{M}=8$), which are due to the development of a laminar boundary layer at the shock-tube walls (calculated by Mirels [450]).

* This analogy is incomplete due to the intermixing of gases at the interface.

also follows from [57] that the velocity, pressure, density and temperature of the gas for a specified time instant increase as the distance from the wave front becomes larger. However, if it is assumed that the shock-front velocity is appreciably reduced, then using Dem'yanov's calculations we will find that the pressure density and temperature along the length of the plug decrease and the flow velocity increases. The results of quantitative estimates of the changes of variables in the plug obtained by Dem'yanov's formulas for the case of a constant velocity of 3.76 km/sec and deceleration of the shock wave given by $V=3.76\cdot(1-0.001\,l')$ km/sec (where l' is the distance from the point of shock-wave formation in cm) are given in Fig. 31. Thus the

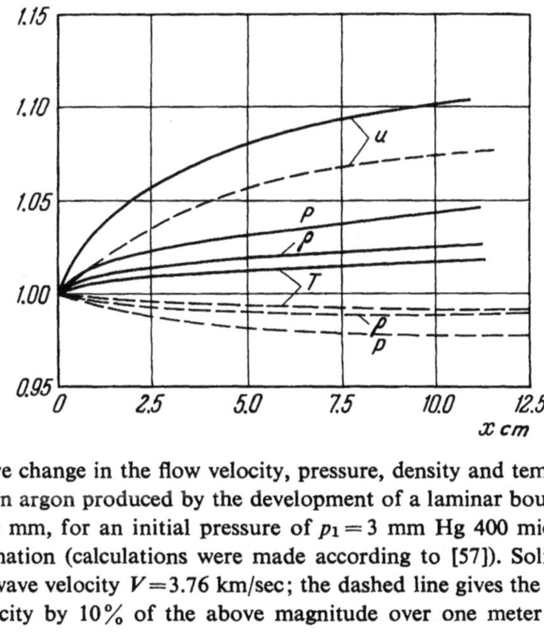

Fig. 31. Relative change in the flow velocity, pressure, density and temperature behind the shock front in argon produced by the development of a laminar boundary layer in a tube with $d=50$ mm, for an initial pressure of $p_1=3$ mm Hg 400 microseconds after shock-wave formation (calculations were made according to [57]). Solid line gives the constant shock-wave velocity $V=3.76$ km/sec; the dashed line gives the reduction in the shock-wave velocity by 10% of the above magnitude over one meter of tube length.

character of pressure, density and gas temperature changes behind the shock front in the presence of a boundary layer varies with changes in the form of the shock-wave velocity distribution in the tube; at the uniform propagation section and, apparently following very low deceleration rates, the above variables increase toward the end of the plug and when the deceleration rates are higher they are reduced. As in the case of Duff's model, the greatest gradients of the variables in the main body of the flow are observed near the wave front. The slow velocity (relative to the tube walls) increases perceptibly toward the end of the plug in all these cases. This quantity is most sensitive to the appearance of the boundary layer at the tube walls; hence the acceleration of the leading part of the interface was reliably established in a number of works. Changes in other quantities are less perceptible and at times do not surpass the accuracy of measurements. The results of Dem'yanov are not entirely correct, since he has made a number of simplifying assumptions in the process of solution. Thus, one of them is the assumption that the product $\varrho\eta$ (η is the coefficient of viscosity across the boundary layer is constant; experiments performed by Hartunian, *et al.* [360], show that this assumption is not valid for $\mathsf{M} > 6$. Dem'yanov has not considered the effect of intermixing of gases at the interface. Despite all this, his results can be used for estimating the changes in the flow variables in the main body of the flow behind the shock front in a tube in the presence of a laminar boundary layer. A detailed solution of this problem with a turbulent boundary layer is given by Spence and Woods.

Fayzullov [205] has attempted to clarify the observed temperature drop toward the end of the plug. He has postulated that each element of the plug retains that temperature to which the gas was heated by the passing wave front; since, during formation, the shock wave has propagated with a lower velocity, then the corresponding gas volumes contained in the end of the plug will have a lower temperature. It is apparent that this assumption is oversimplified. Fayzullov has not estimated changes in the distribution of flow variables in the plug produced by the passing through it of shock and compression waves, which overtake the leading shock wave in the formation and acceleration process. Such a model does not take into account deviations from one-dimensional flow and intermixing of gases in the region of contact, with the results that, in the majority of cases, about half of the gas volume heated by the shock wave is left behind the leading part of the interface.

The formulas obtained in the aforementioned works can be used if it is known whether the boundary layer is laminar or turbulent. The solution of this problem, as is usually the case in gasdynamics is related to a Reynolds number of the flow, which in this case is given by [448]

$$\mathsf{Re} = \frac{x}{\nu_2}\frac{(V - u_2)^2}{u_2}, \tag{6–2}$$

where ν_2 is the kinematic viscosity of the undisturbed flow, x is the distance from the shock front and u_2 is the flow velocity relative to the walls. For comparatively moderate shock-wave velocities, the particular Reynolds number Re^* for transition from laminar to a turbulent boundary layer, according to Hartunian, *et al.* [360], is $2-4\times10^5$, Martin [435] has obtained $1-3\times10^6$ for smooth and $3-4\times10^5$ for rough walls. For higher Mach numbers of the shock wave, the flow is stabilized, *i.e.*, the transitional Reynolds number Re^* increases appreciably (to $1-5\times10^7$ [360]) and the region in which the laminar boundary layer exists behind the shock front is increased. Hence for large Mach numbers the given values of Re* can in all cases be regarded as the lower limit of possible values. Values of the Reynolds number Re in the flow toward the end of the plug of maximum length (in argon and air) are presented in Fig. 32. It can be seen from this figure that in a tube with $d=5$ cm in argon with $M>4$ at the end of the plug $Re \approx 5\times10^{10}p_1^2$ and in air with $M=12$, $Re=1\times10^{12}p_1^2$, where p_1 is in atm. Thus, when $p_1=0.001$, 0.01 and 0.1 atm the Reynolds number in the tube do not exceed 5×10^4, 5×10^6 and 5×10^8 in argon and 1×10^6, 1×10^8, 1×10^{10} in air, respectively. Then for low initial pressures (here for $p_1 \leqslant 0.01$ atm) a laminar boundary layer will exist in the plug with a possibility of transition to a turbulent boundary layer in the cases of higher initial pressures. The majority of experiments concerned with the study of nonequilibrium phenomena was performed at relatively low initial pressures (when the nonequilibrium regions of the flow are appreciable and the effect of the boundary layer can be perceptible); under these conditions in the majority of cases a laminar boundary layer is observed in the flow behind the shock front. However, in each individual case, estimates must be made in order to make it possible to determine the character of the boundary layer as one of the major factors which effect the distribution of flow variables behind the shock front.

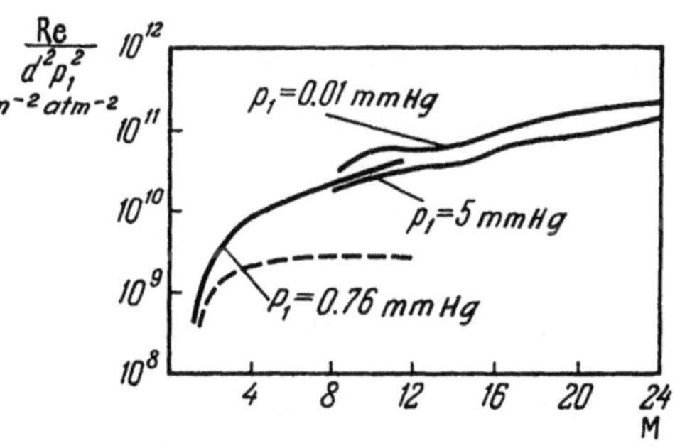

Fig. 32. Ratio of the Reynolds number at the interface to the product of the square of the tube diameter and the square of the initial pressure as a function of the Mach number of the shock wave. The solid lines pertain to air, while the dashed line is for argon [350]. 1) mm Hg.

Thus, the distribution of flow variables along the plug in a real flow in the shock tube is affected by the boundary layer (which retards the shock wave), as well as by waves traveling from the driver gas (in the process of shock-wave formation and acceleration). Usually these two factors are considered separately, *i.e.*, when analyzing the formation process it is assumed that there is no boundary layer, and when describing the boundary layer in the tube it is assumed that the shock wave has already formed. This makes it possible to expose the basic relationships governing each stage of the process of shock-wave propagation in a shock tube. However, actually both these factors act simultaneously and effect the flow variables behind the shock front. The section of uniform motion of a shock wave in a shock tube corresponds to a state in which the effect of the accelerating and decelerating factors on the shock-wave velocity is approximately the same. Unfortunately, a complete shock-tube theory, which would take into account the simultaneous effect of the above factors on the shock-wave velocity and on the distribution of the flow variables along the plug has not as yet been developed.

A certain assurance of successful study of nonequilibrium phenomena behind the shock front under these conditions is obtained by monitoring the shock-wave velocity distribution along the tube. It is best to study these phenomena at some distance from the beginning of the uniform flow section, when the nonequilibrium zone behind the wave front consists entirely of a gas through which the shock wave propagated with a constant speed (or with a moderate deceleration). It is desirable to simultaneously measure the velocity of the leading part of the interface in order to compare it with the shock-front velocity. These measurements make it possible to estimate the gradients of flow variables near the shock front, if, as was pointed out, use is made of the results of [57, 546, 450].

In any case, it is necessary to know the total length of the plug and be able to compare with it the length of the nonequilibrium zone behind the shock front. Results obtained by studying nonequilibrium at the end of the plug, where the boundary layers and the intermixing of gases at the interface exert a substantial effect, will be less reliable.

[2] Multidimensionality of the Flow

Frequently, it is assumed in the study of nonequilibrium phenomena that the flow behind the wave front is one-dimensional, *i.e.*, there are no gradients across the flow. The boundary layer at the tube walls disturbs the validity of this assumption. The extent to which this assumption is invalid can be determined by using relationships for the assumed boundary-layer thickness δ, inside of which the flow velocity comprises less than 99% of the velocity in the main body of the flow. For the laminar boundary layer behind the shock front, such relationships are discussed, for example, by Loytsyanskiy, Roshko, Hooker, Dem'yanov and others [101, 498, 379, 56]; according to [498, 379], we can assume that

$$\delta = \beta \sqrt{\frac{\nu_W x}{u_2}}, \qquad (6\text{–}3)$$

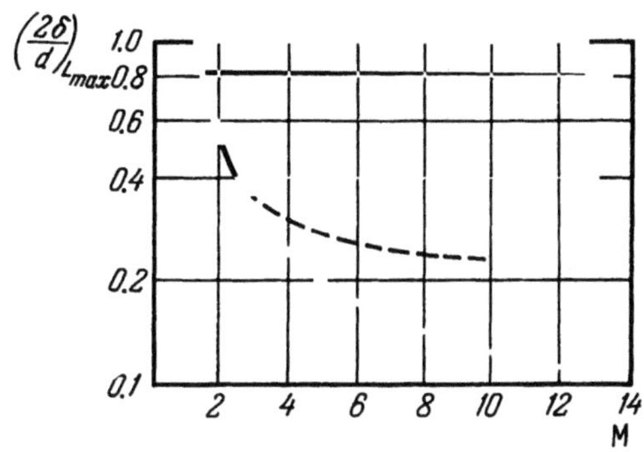

Fig. 33. Ratio of the assumed boundary-layer thickness to the shock-tube diameter at the end of a maximum length plug in argon (dashed line) and in air (solid line) [350] as a function of the Mach number.

where x is the distance from the shock front, ν_w is the kinematic viscosity at the tube wall and β is a numerical factor. Mirels [450] has found that the value of β decreases with an increase in the shock-wave velocity; for air with $p_1 \approx 0.0001$ atm, $\beta = 1.5$ for $\mathsf{M} = 5$ and $\beta \approx 1.1$ for $\mathsf{M} = 12$; for argon $\beta \approx 1.7$ for $\mathsf{M} = 5$ and $\beta = 1.3$ for $\mathsf{M} = 10$. For large Mach numbers these values are lower by only 10% than those obtained experimentally. Experimental results in xenon with $\mathsf{M} > 5$ and an initial pressure of $0.5 \leqslant p_1 \leqslant 4$ mm Hg also yield similar values of β [509].

It can be seen from Fig. 33 that, for sufficiently high Mach numbers, the boundary layer occupies a relatively small part of the tube cross section (even at the end of the plug).* However, despite the fact that the boundary layer is thin in comparison with the tube diameter, in the given case it contains the entire mass of the gas which has passed through the shock front, since the velocities of the front and of the leading part of the interface are then equal. This means that the flow in the end of the plug is not one-dimensional (see Fig. 29).

The boundary layer distorts the distribution of the flow variables in the plug in the case when the flow under study is probed by some beam across the tube axis with the results averaged over the entire cross section (for example, the interferometer method, the electron-beam method, some methods of absorption and X-ray spectroscopy, etc.). The appearance of a denser boundary layer with variable thickness can affect the measurements also near the wave front. The ratio of the gas mass Δm contained in the boundary layer to the mass m in the main body of the flow (for a probing beam

* The ratio $2\delta/d$ at the end of the maximum length plug is independent of the pressure, since $L_{max} \sim p_1$.

with a finite cross section) can be estimated from

$$\frac{\Delta m}{m} = \frac{2}{(d-2\delta)\,\varrho_2(x)} \int_0^\delta \varrho(y)\,\mathrm{d}y\,, \tag{6–4}$$

where $\varrho(y)$ is the density distribution across the boundary layer, $\varrho_2(x)$ is the density distribution along the main body of the flow. To make an estimate let us consider the case of a maximum length plug (see Sect. 5); then $\varrho_2(x)$ can be determined from Eq. (6–1) and the isentropic stagnation equation. The values of $\varrho(y)$ are found by solving the problem for the boundary layer with conditions at its external boundary which correspond to conditions in the flow behind the shock front. The ratio $\Delta m/m$ which is thus obtained, as a function of the distance from the front of a wave propagating with a velocity of 1.5 km/sec (laminar boundary layer, ratio of the temperatures of the main body of the flow and the wall is $T_2(T_w=4$ [101, 399]), is given in Fig. 34. In a tube with a moderate diameter ($d\sim 50$ mm) with a low initial pressure ($p_1\sim 10^{-3}$ atm) the mass of gas which is contained in the boundary layer can become appreciable even at small distances from the shock front. This must be remembered when considering results of study of nonequilibrium phenomena.

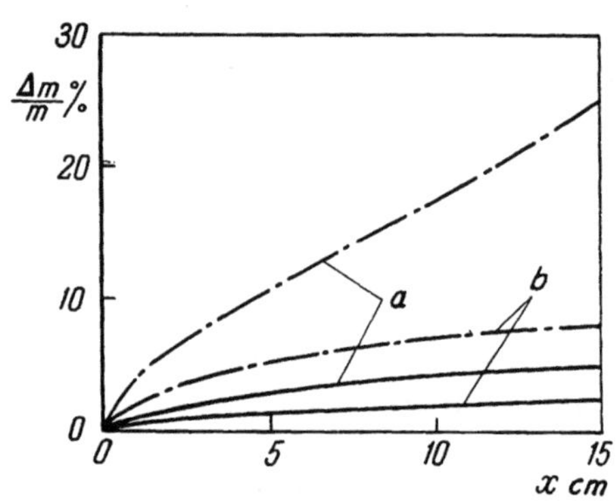

Fig. 34. Relative fraction of the mass of gas contained in the boundary layer in each section of a shock tube as a function of the distance from the shock front in air ($V=1.5$ km/sec) in a 50-mm diameter tube (a) and a tube with $d=100$ mm (b). Solid line denotes an initial pressure $p_1=1\times10^{-2}$ atm, the dotdash line pertains to $p_1=1\times10^{-3}$ atm.

As the temperature is increased, the dimensions of nonequilibrium zones behind the shock front are sometimes reduced to such an extent that the problem arises as to the manner in which the observed results can be affected by deviations from linearity in the immediate vicinity of the shock front. The cause of these deviations is the nonperpendicularity of the plane of the shock front to the tube walls (sloping of the shock front) and the curvature of the shock front. Let us first consider the problem of the shock-front curvature.

Lin and Fyfe [422] performing experiments in a 610-mm diameter tube with initial pressures of up to 0.02 mm Hg, have discovered that the observed (along the normal to the tube axis) shock front in air always has a finite "thickness," which increases with a reduction in the pressure and which is independent of the wave velocity in the range $12<\mathsf{M}<22$. They assume that this phenomenon should be related to the curving of the shock front due to the development of a boundary layer at the tube walls immediately behind the shock front. It was found that Δ, the observed shock front "thickness" is inversely proportional to the square root of the initial pressure, *i.e.*, $\Delta \sim p_1^{-\frac{1}{2}}$ in the range $p_1 = 0.02$–0.5 mm Hg. This relationship is written in the form $\Delta = 0.5\sqrt{l_1 R}$, where l_1 is the mean-free molecular path ahead of the shock front and R is the tube radius. From other experiments we wish to note the work by Daen and De-Boer [307]. They observed an increase in the thickness of the shock front in argon and helium at low Mach numbers. Direct measurements of the shock-front curvature (in argon) were performed by Duff and Joung [319], who have used piezoelectric transducers placed in the plane of the tube cross section to study the shape of the shock front in a tube 28.6 mm in diameter using initial pressures from 0.1 to 20 mm Hg. They have found that the shape of the shock front comes close to the spherical, with the bulge in the direction of motion; the magnitude of the deflection (apparent thickness) is given by $\Delta = 0.84 \cdot \sqrt{l_1 R}$ and is independent of the wave velocity in the range $1.8 \leqslant \mathsf{M} \leqslant 6.3$.

The solution of the problem of the shock-front shape (in the two-dimensional statement) which was undertaken by Hartunian [358], is based on the consideration of the interaction between the shock front and a laminar boundary layer which is developed at the tube walls. He has assumed that the flow behind the shock front and outside the boundary layer is potential. Relating the thickness of the flow displacement by the boundary layer with the velocity component normal to the tube wall and using a linearized relationship for an oblique shock, Hartunian has obtained an equation describing the shape of the shock front in the form $x = q\sqrt{y/p_1}$, where x and y are the space coordinates of the shock front along and across the tube, respectively (relative to the tube axis), q is a function which depends on the velocity, the kind of the gas and, in the presence of physicochemical transformations behind the shock front, also on the initial pressure. Thus the shock front has the shape of a paraboloid with its axis normal to the tube axis. Experiments have shown that, for large Mach numbers ($\mathsf{M}>8$) Δ is only weakly dependent on the shock-wave velocity. It is interesting to note that in a real gas the shock-front curvature is smaller than in an ideal gas. When a nonequilibrium zone is present, it should be expected that the magnitude of Δ will be contained inside the interval of values of Δ for an ideal and actual gas. Values of Δ for a shock wave in air, with consideration of the zone of nonequilibrium dissociation of oxygen were calculated by De-Boer [309]. He has refined Hartunian's theory by considering the shape of the shock front in tubes of different shape with a laminar or turbulent boundary layer at the walls, and on the assumption that the flow near the point where the shock front is in contact with the tube wall is viscous and compressible. A good agreement was thus obtained with experimental results (Fig. 35). De-Boer has shown that the driver gas has little effect on the shape of the shock front.

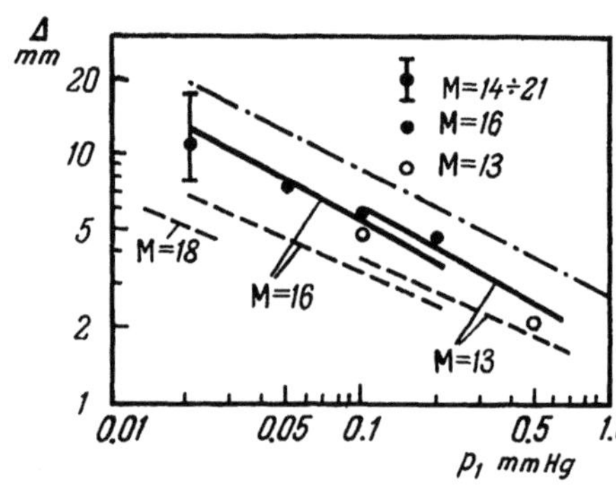

Fig. 35. Apparent thickness (deflection) of a shock front in air resulting from the curvature of its surface produced by the development of a boundary layer at the walls of a 610-mm diameter tube, as a function of p_1.

Solid line denotes De-Boer's calculations which took into account the nonequilibrium dissociation of the oxygen; the dashed line represents Hartunian's calculations for an equilibrium flow; the dash-dot line is the same as above for an ideal gas; ● and ○ denote values experimentally obtained by Lin and Fyfe.

Another source of errors in the study of the nonequilibrium region is the fact that the wave front slopes toward the tube axis at an angle other than 90°. This slope can be responsible for the random scattering of experimental points when measuring the apparent shock-front "thickness." The appearance of this slope can be due to the character of the diaphragm bursting, to the appearance of asymmetrical disturbances in the tube, of transverse waves, etc. In both small- [319] and large- [422] diameter tubes, the slope of the front is small in comparison with its curvature. Also, it can be measured by placing the piezoelectric transducers precisely along the perimeter of any given shock-tube cross section.

To reduce the effect of the boundary layer and of the curvature of the front on the measurements when using the electron beam, inspection holes are sometimes led into the tube; in this case only processes in the main body of the flow can be observed. The aerodynamic disturbance of the flow (which is thus produced) is reduced by specially placed plates [615].

It was noted already in the earliest shock-tube experiments that transverse waves, which form either in the process of shock-wave formation, or on interaction of the flow with the tube walls (effect of protrusions, voids, etc.), can serve as the source of inhomogeneity of the flow behind the shock front (see, for example, Fig. 43). Transverse waves can produce appreciable vibrations in the flow; Breadley [266] suggests that the interaction between the transverse waves and the gas flow can result in the appearance of periodic radiation pulsations following the propagation of the shock wave in the tube. Unfortunately, there are very little quantitative data on the possible effect of transverse waves on the behavior of the flow behind the shock front.

The most appreciable deviation from the one-dimensional picture of the flow takes place in the region behind the front of a shock wave which is reflected from the end wall of a shock tube, *i.e.*, at the point where the heated gas should be at rest. This deviation is produced primarily by the interaction between the reflected wave and the boundary layer which has existed in the flow behind the incident wave [431, 550, 512]. In this case a triple λ-shaped configuration of shock discontinuities forms near the wall and is clearly seen, as in Fig. 36. Strehlow and Cohen [550] have discovered that when the wave front moves after being reflected from the end wall the region in which the front branches out is first increased with a subsequent increase in the thickness of the boundary layer which remains behind the front of the reflected wave. Then the wave front is accelerated, the branching-out region ceases to increase and, finally, at a greater distance from the end

Fig. 36. Interferogram of the process of reflection of a shock wave from the end wall of a shock tube in carbon dioxide (M = 3.6, p_1 = 10 mm Hg) [66]. S = trace of the incident shock wave; R_0 = the trace of the center part of the front of the reflected shock wave, R_λ = trace of the sloping part of the reflected shock wave.

wall, the reflected wave moves with almost constant velocity. The branching-out region and the acceleration of the front are least perceptible for small Mach numbers of the incident wave and for gases with a high ratio of specific heats γ. Thus, in helium, no branching out or acceleration was observed, and only in some experiments with argon was the sloping of the front near the walls observed. The same results were obtained by Mark [431] and Toennies and Greene [572] up to $\mathsf{M}=10$. In diatomic gases (H_2, N_2, O_2, NO) a weak branching-out region appears at $\mathsf{M}>1.4$ and the front is accelerated at $\mathsf{M}>2$; at $\mathsf{M}>5$ the acceleration reaches 20%. In triatomic gases (CO_2, N_2O) the interaction of the reflected wave with the boundary layer is more pronounced, particularly at high Mach numbers. In these gases the front velocity is sometimes increased by 50%. Finally, in multiatomic gases (methane, propene) the branching-out region can fill out the entire tube cross section and result in the disappearance of the flat front. Hence it is most convenient to study nonequilibrium phenomena (in reflected waves) either in monatomic gases, or in gases which are added in small quantities to a monatomic gas. The experiments, however, have shown that even in a monatomic gas with a plane front and constant reflected wave velocity the measured velocity is lower than that theoretically calculated on the basis of the simplified shock-tube theory. This deviation increases with an increase in the Mach number of the incident wave. A similar reduction in the velocity of the reflected wave front amounting to 7–10% less than the theoretically calculated value was observed by Bazhenova, *et al.* [15], in nitrogen with $p_1=12.7$ mm Hg and $\mathsf{M}=3$–8. This cannot be explained by heat losses due to heat transfer to the shock-tube walls following the reflection.

The most important characteristic for kinetic studies is the gas temperature; hence, the deviations discussed above should be expressed in terms of the deviation of the temperature from the theoretical values. It is pointed out by a number of authors [550, 572, 29] that the reduction in the velocity of the reflected wave front in argon can result in a substantial temperature reduction in comparison with the theoretical value (up to 15–20%). On the other hand, pressure measurements in a reflected shock wave in monatomic gases [452, 265, 415] point to only moderate deviations of the actual from the theoretical temperature immediately behind the wave front (within the limits of several tens of degrees). This conclusion is also obtained on the basis of pressure measurements in air [357]. An indirect temperature measurement performed by Kudryatsev, *et al* [91], has shown that the measured temperature behind a shock front in argon is lower than that theoretically calculated by 7% (Fig. 37); these results, however, are not final.

The most important feature from the viewpoint of evaluating the feasibility of studying nonequilibrium phenomena in reflected shock wave is the

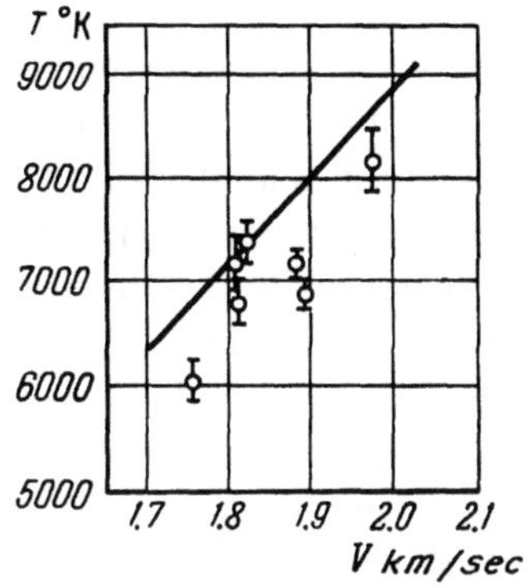

Fig. 37. Measured temperatures in a reflected shock wave in argon as a function of the velocity V of the incident shock wave. The solid line represents the theoretically calculated temperatures.

fact that the flow variables behind the front of such a wave do not remain constant. A detailed study performed by Zaytsev, *et al.* [66], using an interferometer has shown that at a distance of 3 mm from the end wall the density of nitrogen through which the front of a reflected wave has passed increases continuously, exceeding (at $M \geqslant 4$) the theoretical value (Fig. 38). A similar result was obtained by Gardiner and Kistiakowsky [325] who have used xenon; first, the actual density is equal to the theoretical, and then it increases; at $M = 3.9$, 350 microseconds [after passing] this excess comprises 4%. The pressure of the gas also increases [542, 265] and also exceeds the theoretical value; but it is the gas temperature which undergoes the most appreciable change [91]. Within 80 microseconds of the time when the

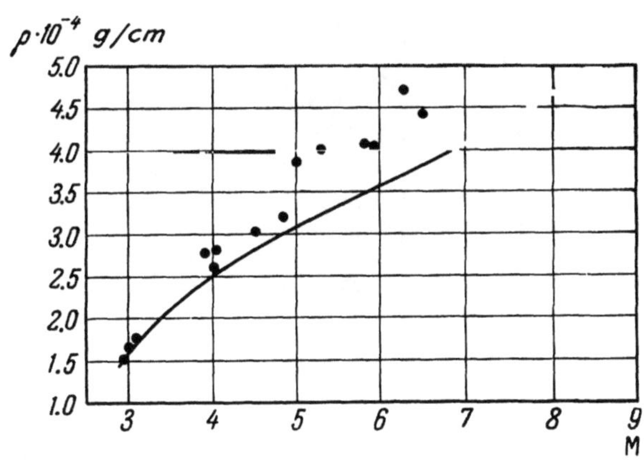

Fig. 38. Density of gas behind a reflected shock front in nitrogen at a distance of 3 mm from the end wall of the shock tube as a function of the Mach number of the incident wave for an initial pressure of $p_1 = 10$ mm Hg. The dots denote experimental data, while the solid line gives the theoretically calculated densities.

reflected wave front passes in argon, the temperature is reduced by 15% and then, approximately 150 microseconds later it again increases. Also the temperature behind the front of a reflected shock wave does not remain constant in other gases (air, oxygen, nitrogen and carbon dioxide). The increase in the branching-out region (with a reduction in γ) results in a smooth increase in the temperature behind the reflected wave front. Average temperatures (for a time up to 40 microseconds) at a distance of 10 mm from the end in these gases, however, come close to those theoretically calculated (Fig. 39) despite appreciable scattering; then the temperature drops. Experiments performed by Zaytsev, *et al.* [66], verify the fact that, for $M=6$ and $p_1=10$ mm Hg in the intermediate vicinity of the end wall (at a distance less than one cm) within the experimental error ($\pm 5\%$) the interaction between the reflected wave and the boundary layer does not affect the gas density. It can be postulated that the subsequently observed density and pressure rise in this region can be due to the arrival of disturbances from the inhomogeneous branching-out region. After the reflected front has moved away through a distance of more than 5–10 cm, the state of the gas in this volume during the time from 40 to 200 microseconds after reflection remains homogeneous. The gas-temperature fluctuations in this region, noted by Kudryatsev, *et al.* [91], also do not exceed limits of density measurement accuracy obtained by Zaytsev and other authors. This makes it possible to be optimistic in estimat-

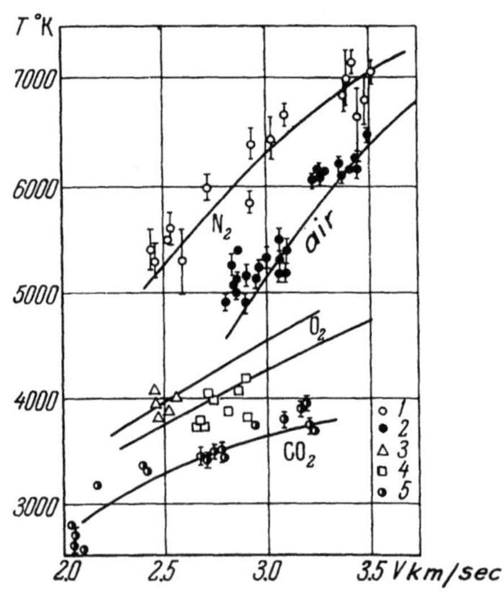

Fig. 39. Measured temperatures in a reflected shock wave in air, nitrogen, oxygen and carbon dioxide as a function of the velocity of the incident shock wave. 1) Nitrogen, $p_1=12.7$ mm Hg; 2) air, $p_1=10$ mm Hg; 3) oxygen, $p_1=30$ mm Hg; 4) oxygen, $p_1=10$ mm Hg; 5) carbon dioxide, $p_1=12.7$ mm Hg; the solid line denotes the corresponding theoretical values.

ing the feasibility of studying nonequilibrium phenomena in reflected waves. However, the complexity of phenomena which take place in a shock tube following the reflection of waves is responsible to the relative scarcity of results concerning the kinetics of physicochemical transformations obtained in studying reflected waves compared with those obtained with incident waves. A certain degree of success was obtained by using only the method of "freezing" with subsequent chemical analysis of a gas sample, as well as in a number of other cases (see Sect. 14). When setting up experiments for the study of nonequilibrium phenomena in reflected shock waves, the homogeneity of the test gas volume should be carefully investigated.

[3] Heat Transfer by Radiation

In strong shock waves the flow variables which characterize the state of the gas in the plug can change appreciably due to heat transfer by radiation. In fact, experiments by Petschek, *et al.*, in argon with $M = 13$–17 have shown that the gas behind the wave front luminesces intensely and that then the flow variables do not remain constant [475]. Their calculations have shown that the observed reduction in the radiation intensity can be ascribed to cooling due to radiative heat transfer. This effect is appreciable only at sufficiently high temperatures, when the gas is dissociated and ionized. Now we shall point out the character of changes in thermodynamic variables in the flow behind the shock front for this case.*

The radiative removal of energy means that the flow is no longer adiabatic; the equation for the energy balance of unit mass (or volume) of gas must be supplemented by terms which take into account the spontaneous and induced radiation of this gas volume and the absorption of the light energy by molecules contained in this volume. Adding to this equation the equation of state, the equation of chemical and ionization equilibrium (assuming that the gas is in equilibrium) and relationships which follow from the law of conservation of mass and momentum fluxes, it is possible to obtain distributions of the temperature, pressure, degree of ionization and other variables behind the shock front. This problem is solved simplest for one-dimensional, steady-state flow behind the shock front in a monatomic gas (for example, in argon [476]).

It is natural that the radiative heat transfer results in a reduction in the gas temperature as one moves away farther from the front; the degree of ionization is also decreased. The effect of the radiative heat transfer will be most appreciable for low initial pressure in small diameter tubes.

* A detailed analysis of the effect of radiation on the shock-wave structure can be found, for example [71] (see also [476, 178, 166a]).

To obtain a quantitative estimate of changes in temperature, pressure and other variables one, first of all, must know the emissive power (or emissivity) of the gas under the test conditions.

Petschek, *et al.* [475], have taken into account only the radiation in the continuous spectrum, but in a wide frequency range; in later works [476, 22, 23], this range was reduced. Yakubov and Sevast'yanenko [178] have shown that continuous radiation is not the only process responsible for the cooling of argon. Thus, for example, it was found that at $T \sim 12{,}000\,°\mathrm{K}$ and $p_2 = 4.25$ atm, about one-half of the radiated energy is devolved upon the bright-line spectrum. Figure 40 shows the temperature distribution behind

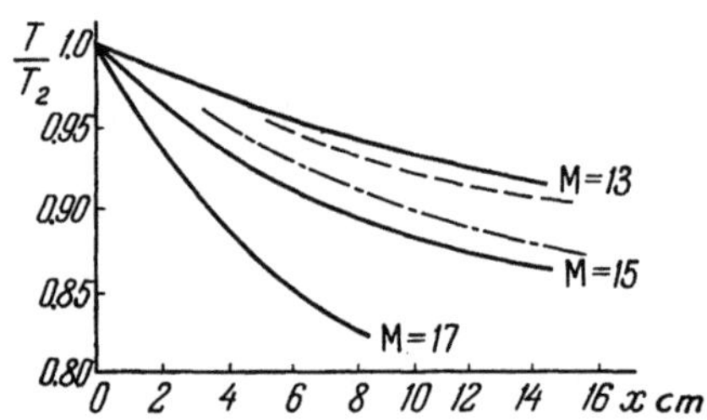

Fig. 40. Relative reduction of the temperature as a function of the distance from the shock front in argon as a result of radiative heat transfer. Solid line denotes calculations due to Yakubov and Sevast'yanenko for $M = 13$, 15 and 17; calculations by Pomerantz [476] for $M = 17$ are denoted by a dashed line when absorption was taken into account and by the dot-dash line when absorption was not considered. The initial pressure was 10 mm Hg.

the shock front in argon (in a 4-cm diameter tube) with $M = 13$–17 and $p_1 =$ 10 mm Hg, obtained by Yakubov and Sevast'yanenko. They have not considered the absorption of light by the radiating gas. For comparison we have drawn on the same graph the theoretically calculated temperatures from [476] for $M = 17$ and $p_1 = 10$ mm Hg, with and without consideration of the absorption. These calculations did not include the radiation in the bright-line spectrum which has resulted in minimizing the contribution of the radiative heat transfer compared with calculations of Yakubov and Sevast'yanenko. Under these conditions ($M = 17$, $p_1 = 10$ mm Hg) even when the absorption is taken into account (as in [476]), an appreciable reduction was observed in the degree of ionization of argon (almost two-fold at a distance of 10 cm). Other variables change less perceptibly (the temperature drops by 7%, and the pressure increases by 2.5% at the 10-cm distance).

As the shock-wave velocity increases, the radiative cooling becomes increasingly more perceptible, and the rate of change of all the quantities behind the shock increases. The presence in the gas of easily ionized admixtures appreciably increases the effect of radiative heat transfer. The contribution of radiative heat transfer is particularly great in very strong shock waves [236, 408]. This must be included in calculations when analyzing phenomena in strong shock waves.

Thus, dissipation processes which accompany the flow of gas heated by a shock wave in a shock tube (boundary layers, heat transfer, etc.), should to a given extent affect the flow variables behind the wave front. A number of investigators have attempted to clarify the actual extent of this effect. Temperature, density and pressure measurements have shown that under conditions when it is possible to expect local thermodynamic equilibrium, the flow variables behind the front of an *incident* wave become close to their theoretical values. This shows that shock tubes can be used extensively for obtaining a gas with specified properties and makes it possible to a certain extent to take a more optimistic view of the aforementioned difficulties. In summing up the consideration of conditions under which shock waves propagate in real shock tubes it must be emphasized that the study of non-equilibrium phenomena in shock waves requires, in each individual case, careful analysis of all the factors affecting the flow variables in the shock wave.

[7] AUXILIARY MEASUREMENTS OF FLOW VARIABLES IN SHOCK TUBES

As was pointed out, in order to study the structure of shock waves it is required to measure certain flow variables along the plug. Sometimes it is ufficient to measure one such variable (for example, distribution of the molecular concentration and the density, etc.). However, among all the variables which are measured in these experiments, there are some that must be known in any kind of experiment. These include the rate of propagation of the shock wave, initial pressure and initial composition of the gas; sometimes the initial gas temperature must also be known. The purity of the test gases is also of great significance when studying relaxation processes. Hence, before we pass on to the study of experimental methods for studying relaxation processes in shock waves (which is done in the next chapter), we must consider the methods of measuring the velocities of shock-wave propagation in the shock tube and of determining the initial state of the test gas.

[1] Measuring the Shock-wave Velocity

The main quantity which characterizes the state of the gas behind the shock front is the rate of propagation. There exist two methods for measuring the shock-wave velocity: 1) the measurement of the average wave velocity at any section of a tube; and 2) continuous recording of the shock-wave propagation. These will now be considered briefly.*

Methods most extensively used are those which involve the measurement of time intervals Δt during which the shock front passes a given distance (base) Δx, separating any two recording devices which record the time of shock-wave passage. Here it is assumed that the shock-wave velocity varies at this base so insignificantly that the rate of propagation can be regarded as constant. To determine the accuracy of this assumption, the velocity is sometimes measured over several bases located at short distances from each other. These measurements of shock-wave velocity are the simplest. The main elements of these arrangements are signal pickups for receiving signals corresponding to the time of shock-wave passage, and electronic devices which record these signals. The main requirements of these sensors reduce to the condition that the element which is subjected to the shock wave have a low inertia, and that the leading front of the pulse thus obtained be sufficiently steep and have an amplitude ensuring reliable recording. In a number of cases it is necessary to ascertain that the leading front of the signal coincides with the time of the passing of the shock wave. It is desirable that the pickups which are used be of the same design and equally placed with respect to the tube and its inside surface.

From among the great number of sensors which are used for registering the time of the passing of the shock front, we wish to point out the following.

1. *Ionization pickups* operate on the principle of closing the spark gap as a result of partial ionization of the gas (or admixtures it contains) across the shock-wave front. The pulse with a steep leading front is obtained in this case by a low-capacitance discharge. A specimen of an oscillogram with two pulses supplied by ionization pickups, obtained by measuring the velocity of a shock wave in oxygen is presented in Fig. 41b. These pickups are used successfully in those cases when the gas temperature behind the wave front exceeds two thousand degrees. It should be remembered that at very high temperatures a breakdown is possible in inert gases which is accompanied by the flow of current through the pickup even before the shock wave reaches the pickup due to photoelectric emission from the tube walls which is produced by the intense ultraviolet radiation from the heated plug. In this case the voltage applied to the pickup should not exceed several volts. In

* A more detailed presentation of methods of measuring the shock-wave propagation including diagrams, sketches of recording devices and specimens of signals, can be found in works by Gaydon and Hurle [328], Breadley [266], and Greene and Toennies [347], as well as in the collection "Shock Tubes" [203]. These books contain an extensive bibliography of works on the subject. The oscillograms presented in Fig. 41 were obtained by the authors.

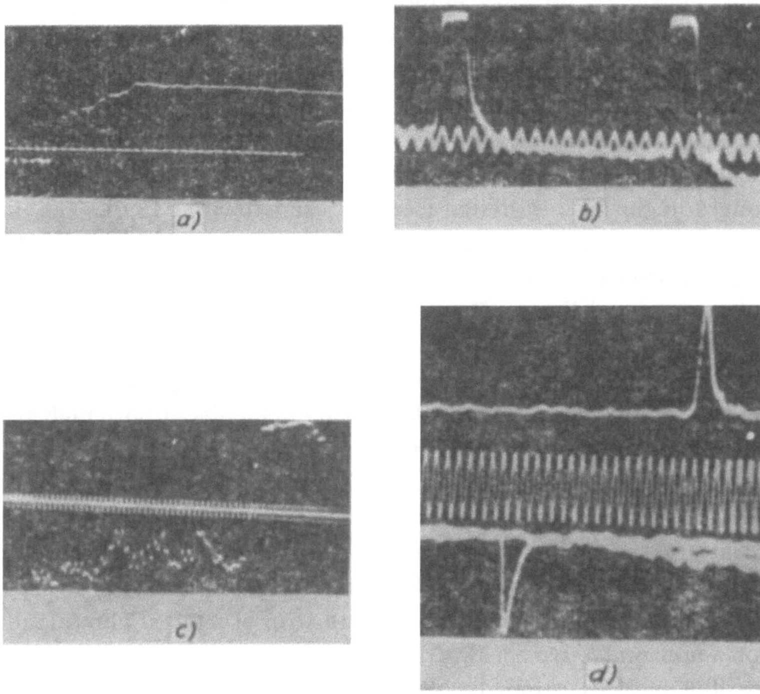

Fig. 41. Specimens of oscillograms of signals obtained when measuring shock-wave velocities by different methods. The sinusoidal curve provides time calibration. a) Photomultipliers which record the absorption of light by the gas (see Sect. 11); b) ionization pickups; c) piezoelectric pickups; d) photomultipliers in combination with the Tepler shlieren system.

gases which absorb this short-wave radiation (oxygen, carbon dioxide, etc.), the ionization pickups operate more reliably. It is entirely possible that some role in the charge formation in the ionization pickup is played by the appearance of charges produced by the rubbing of the gas at the tube surface.

2. *Piezoelectric pickups* utilize the piezoelectric properties of quartz, ceramics, barium titanate or other piezoelectric elements. The receiving planes of the piezoelectric element are held tightly between the pickup housing from one side and a plate contained in an insulating insert from the other. This ensures electrical and mechanical contact between the piezoelectric element and the pickup housing. When a shock wave passes near the end wall of the pickup the gas pressure acts on the end wall and is transmitted to the piezoelectric element. When measuring shock-wave velocities, it is important to ensure that only the leading front of the pressure pulse be properly reproduced. The width of the receiving plane of the piezoelectric

element is not significant, since in the given case it is sufficient that the shock wave act only on a small part of the piezoelectric element. Vibrations at the natural vibrational frequency of the piezoelectric element are superimposed on the signal, produced by the passing shock wave, which is registered off the pickup. This is of no significance in velocity measurements. When measuring the gas pressure the effect of the natural vibrations of the pickup is eliminated by changes in the pickup design (see, for example [65, 189]). The piezoelectric pickup is insulated against the vibrations transmitted by the shock-tube shell, which carries disturbances produced by the bursting diaphragm, by rubber rings and liners. The piezoelectric elements are installed in the pickups such that the signals following one another upon the passing of a shock wave near the pickup are of opposite polarities (Fig. 41c). These pickups can be used for measuring the shock-wave velocities at both high as well as relatively low initial pressures (up to 1 mm Hg).

3. *Thermal or film pickups* are essentially thermal resistances installed in the side wall of the shock tube. The thermal effect of the gas heated by the shock wave on the wall varies the current in a circuit in which the pickup is connected. To reduce the inertia effect, the thermal pickups are very thin (fractions of a micron) and consist of a thin strip of a metal film (platinum, gold) on a nonconductive backing (best of all glass or quartz). The resistance of such film pickups comprises several ohms. For the pickup to operate properly it is sufficient to have a heating effect of several tens of degrees, which is ensured even with low wave velocities. Hence the film pickups are used in such experiments where the gas temperature behind the wave front is moderate (less than 1000–2000°K). Together with this, these pickups operate successfully even at higher temperatures. However, at $T > 8000$°K their effectiveness drops due to the effect of photoelectric emission under the action of radiation.

4. *Optical arrangements* make it possible to use changes in the refraction index across the wave front or the reflection of light from the front surface. In combination with a photomultiplier this makes it possible to record the time of passage of even weak waves (for a sufficiently high initial pressure). The sensitivity of recording the changes in the refraction index can be appreciably increased by using Tepler's shlieren system with a long-focus optical arrangement. A specimen of an oscillogram with such pulses obtained by a combination of Tepler's instrument with a photoelectronic multiplier and an oscillograph is given in Fig. 41d. The use of this method for high-velocity shock waves is made difficult by the intensive radiation of the gas (sometimes some improvement is obtained by using light filters and a double knife edge ahead of the photoelectronic multiplier). Among the shortcomings of these arrangements is also the need of using sufficiently strong sources, the pres-

ence of scattered light and also the sensitivity of the setup to vibrations; these setups must be rigidly connected with the tube.

5. In stronger shock waves use is sometimes made of *recording the radiation* of the gas by photomultipliers placed behind slits, which are placed such that light beams perpendicular to the tube axis are isolated. Despite the simplicity of this method, its use is limited by the fact that in very weak waves the gas does not luminesce, and in stronger waves it is only the easily excited admixtures (sodium, calcium, etc.), which luminesce; in addition, the start of perceptible glowing in this case can lag behind the shock front at an undetermined distance. Finally, in very strong waves it is possible to get false signals produced by diffusion of the emission and excitation of the gas ahead of the wave front. Here the recorded signal ahead of the wave can also have a sufficiently steep front, which is due to the "wave" of excitation of gas particles as a result of diffusion of the radiation.

6. When working with gases which have a perceptible absorptivity in the accessible spectral regions the time of the passing of the shock wave can be registered by the changes in the *absorptive power* across the front. As an example, we can cite changes in the rate of propagation of a shock wave in iodine vapor by the absorption of light in the visible spectral region (Fig. 41a); the system for measuring and recording the absorptive power which is used in this case is described in Sect. 11. When studying shock waves with very low initial pressures (for example, in air with $p_1 = 0.02$–0.2 mm Hg) it may be very successful to record the change in the absorption of radiation by oxygen in the vacuum ultraviolet spectral region. Some care in using this method must be exercised in the case when the excitation of molecules which are involved in the observed absorption is delayed.

7. The *thermoanemometer* (heated thin wire stretched across the flow) is a good indicator of the time of passing of the shock wave, since the conditions of heat transfer between the wire and the medium surrounding it change perceptibly upon the passing of the front. In other cases the use of the thermoanemometer is limited by its mechanical strength.

8. Among other methods of recording the passage of the wave front we should note the use of a *beam of positive ions,* the intensity of which changes upon the passing of the shock wave. For this purpose, an ion beam which is obtained in a discharge tube is directed along the normal to the shock-tube axis. However, this method is not used extensively.

The time intervals needed by the wave to pass between the pickups are recorded by various time-measuring devices, such as, cathode oscillographs with horizontal and circular scanning, scanning of a stationary beam by a photographic camera with a rotating drum (photoregistering device) and also oscillographs with multiline scanning. In this case it is possible to observe the shape of the pickup pulse directly on the screen, which appreciably increases the reliability and accuracy of the method. When recording pulses by the various methods, it is necessary to properly synchronize the oscillograph scannings and to check for the possible relative shift of the scans. In a number of cases use is made of thyratron arrangments, which start up chronographic time-measuring devices after receiving a signal from the first pickup and shut them off upon receiving a signal from the second pickup. A combination of pulse series generator and a counting device is sometimes used here for time measurement. However, although this system is convenient, it must then be ascertained that the leading pulse front is sufficiently steep and that the scattering of the actuation time of the thyratron systems does not contribute greatly to the measurement error.

The accuracy of measuring the shock-wave velocity by the base methods is determined primarily by the time-measurement accuracy. The use of oscillograph methods to record the reduction in the time-measurement error is limited by the fact that the oscillograph beam has a finite thickness. The use of oscillographs with long and high-velocity scanning is very useful. When using counting devices to ensure sufficient measurement accuracy, the resolving power of counting a large number of pulses during short time periods must be high.

Another method of velocity measurement is continuous recording of the shock-wave propagation. For this purpose a narrow inspection hole is placed along the tube axis and the image of the tube is projected on a rapidly moving film (photographic velocity recorder). Sometimes the motion of the film is replaced by the motion of the image which is obtained by a rotating mirror. The position of the shock wave is determined either (in the case of sufficiently strong waves) by the radiation of gas in the shock wave,* or (in

Fig. 42. Specimens of photographic scanning of the shock-wave propagation in argon. a) Radiation in the incident wave (M = 20); b) shlieren photograph of the flow in the incident wave S and of the radiation of gas in the reflected wave R (M = 10), K is the trace of the interface.

* It should be remembered here that the luminescence region may not coincide with the wave front.

weaker waves) by shlieren photographs of the flow which are obtained by shadow photography or by Tepler's systems. The rate of propagation of shock waves by the continuous method is determined from the relationship $V=(v_f \arctan \alpha)/m$, where v_f is the velocity of the film, α is the angle between the direction of motion of the film and the image of the shock wave and m is the scale-reduction factor of the image. The photographs which are thus obtained yield an x–t diagram of the propagation of not only the wave front, but also (in a number of cases) of the leading part of the interface, of the reflected wave, etc. (Fig. 42). This is the known advantage of continuous measurement methods over base methods. A shortcoming of these methods is the considerable complexity. A detailed analysis of errors and selection of the optimal operational regime when using continuous recording of the shock-wave propagation can be found in [17].

Fig. 43. A sequence of photo frames of the propagation of a shock wave in air with $M=2.4$ and $p_1=0.2$ atm (at a distance of about $60d$ from the diaphragm) [175]. F is the shock front, K is the interface. Transverse waves can be seen in the plug.

Sometimes high-speed photography is used for velocity measurements. In this case the inspection holes in the tube are enlarged and several photographs are obtained which follow one another in sequence at fixed time intervals. This method also makes it possible to obtain a general picture of the flow in the tube (Fig. 43). The accuracy of velocity measurements in this case is determined by the glow duration of the light pulse and errors in measuring the frame speed. Among other methods of continuous recording of shock-wave velocity, we can note the radio-frequency method, when a sufficiently strong shock wave acts as a reflector of radio waves which propagate toward the shock wave. Measuring the phase shift between the incident and reflected radio waves makes it possible to follow the front velocity.

When working with reflected shock waves it is important to know the shock-wave velocity at the approach to the end wall of the tube. If the tube is designed so that directed measurement (let us say, by continuous methods) of the velocity at the end wall is impossible, then it is necessary to measure the velocity distribution of the incident wave at the approach to the end and then to extrapolate. This procedure does not always give good agreement between the calculated values of the flow variables and the actual values of these variables behind the reflected shock but can aid in reducing the error.

From the results of a large number of experiments in shock tubes, it is seen that it is relatively easy to measure the velocity of shock waves with an accuracy of up to 1–2%, however, it is difficult to obtain higher accuracies.

[2] Measuring the Initial Pressure and Temperature

Another important characteristic of the initial state of a gas is the pressure in the low-pressure chamber before the arrival of the shock wave. In most experiments, the initial pressure comprises several mm Hg; cases of higher pressures (up to atmospheric [234]) or lower pressures (up to 2×10^{-2} mm Hg [424]) are encountered less frequently.

The modern physical equipment makes it possible to measure these low pressures with sufficient accuracy. Among the methods and devices used we can point out static manometers which use the pressure effect (membrane, U-shaped, compression manometers), thermoelectric devices which are based on the change in the thermal conductivity of the gas with a change in the pressure (Pirani gages, thermocouples), ionization vacuum meters which measure the ionic current produced by various agents, gas discharge tubes which use the physical characteristics of high-voltage dischargers, etc. [2, 36, 89, 31a]. Some of these instruments are used as indicators which show the extent of the preliminary evacuation of the tube before it is filled with the test gas. The most frequently used instruments for this purpose are thermocouple and ionization vacuum meters which overlap a pressure range from 1×10^{-7} to 1 mm Hg; the accuracy which is obtained (about 15%) is entirely sufficient for estimating the quality of the preliminary evacuation and airtightness of the shock tube. Stricter requirements are placed on manometers which measure the initial pressure of the test gas. When working with nonagressive gases, most frequent use is made of U-shaped mercury and oil-filled manometers; here a sufficient degree of accuracy (0.5–1.0% and better) can be achieved by measuring the levels by a cathetometer. For precision measurements the manometer readings are sometimes corrected to a constant temperature level and a correction is introduced for the capillary depression [6, 375]. Replacing the mercury by oil (for example, diffusion oil) increases the sensitivity by approximately a factor of 15, and the use of an inclined calibrated tube increases it by an additional factor of 5–10. In the case of aggressive gases it is convenient to use membrane vacuum meters with special diaphragms; the readings of these instruments at pressures of several mm Hg are not always accurate and require calibration by a more accurate instrument. Regular calibration of readings by comparison with the McLeod manometer makes it possible to use successfully, alphatrons for shock-tube investigations with a vacuum of up to 0.02 mm Hg with an error not exceeding 0.5% [424, 401a]. To obtain low pressures in the tube with a sufficient accuracy, a measuring volume which is many times smaller than that of the low-pressure chamber is sometimes used. The pressure is measured when the gas is contained in the measuring volume under a higher pressure. Then the gas flows into the tube chamber. Thus, an appreciable increase in the sensitivity is obtained; this method is both simple and expedient. Sometimes the characteristic of the test gas makes it possible to use special methods for measuring the initial pressure. When working with gases which absorb light intensely, it is possible to determine the absolute concentration by measuring the absorption power by optical methods and recalculate it in terms of the pressure; this method was used by Generalov *et al* (see Sect. 11) [48], to measure the initial pressure of iodine vapor. Tumakayev and Lazovskaya [202] have measured the concentration of mercury atoms before the start of the experiment by using Rozhdestvenskiy's hooks method (see Sect. 10).

Sometimes, when working with shock tubes, the initial temperature of the gas is not measured at all and is assumed to be close to the room temperature. However, certain studies in gases with a low saturated vapor pressure (iodine, mercury, cesium, etc.), require that the shock tube or a part of it be preheated [48, 59, 361]; in this case the initial temperature must be measured and the uniformity of heating the tube and the gas contained in it must be checked. This is done most conveniently by thermocouples. Sometimes the shock-tube temperature is kept constant even when temperatures close to room temperature are used (for example, in the work by Hooker and Millikan [380] with $T_1 = 300 \pm 1\,°K$).

[3] Preparing the Test Gas

In the study of nonequilibrium phenomena, special care should be given to the purity of the gas in the shock tube. In fact, considerable data show that the use of gases with unknown admixture content without proper purification can sometimes result in erroneous conclusions with respect to the rate at which the nonequilibrium processes in the gas take place. Occasionally (particularly at low temperatures), the difference in relaxation times thus obtained can be of several orders of magnitude [87, 384, 428]. At higher temperatures the errors become less perceptible (the rate at which the processes take place in the main gas increases), however, even this must be checked quantitatively. We illustrate this by the following example. It followed from the known results due to Blackman [254], who has measured τ_{vib}, the vibrational relaxation time for O_2 at $760° \leqslant T \leqslant 3000\,°K$, that the Landau-Teller theory of vibrational relaxation is not completely correct; Blackman's data were used not only in fluid-flow calculations (see [426, 74, 613]), but also for estimating the intensity constants of the intermolecular interaction potential for the reaction O_2–O_2 [441]. Blackman has used chemically purified oxygen which has contained up to 0.025% of water vapor; the low-pressure chamber of the shock tube was prevacuumed only to a pressure of 0.5 mm Hg (for a working pressure of about 5 mm Hg). To ascertain the reliability of these data and to resolve the problem of the effect of admixtures on the vibrational relaxation time in a wider temperature range, Generalov [42] has performed more careful measurements. The oxygen was multiply fractionated and all the admixtures were removed from it (up to a content appreciably less than 0.01%, except for argon which comprised 0.8%). The shock tube was prevacuumed to a pressure of 0.02 mm Hg and then was "washed" by filling it a number of times to a pressure of 50 mm Hg and removing the test gas to a pressure of 0.02 mm Hg. The argon admixture, which was specially added, has shown that the amount of argon (0.8%) which is contained in purified oxygen cannot have an appreciable effect on τ_{vib}. Experiments in purified oxygen have shown conclusively that Blackman's results at $T < 2000\,°K$ were in error; at $T = 1200\,°K$ the vibrational relaxation time for O_2 molecules is actually higher by a factor of two than that obtained by Blackman. The temperature dependence of τ_{vib} obtained by Generalov is satisfactorily described by the Landau-Teller theory. Similar results were subsequently obtained by White and Millikan [599]. The effect of admixtures was studied by adding to the oxygen small amounts of water, ethyl alcohol, carbon dioxide and nitric acid vapor. In almost all cases, Generalov has discovered a perceptible reduction in τ_{vib} at $T<2500$–$3000\,^{\circ}$K and an increase at higher T compared with the values of τ_{vib} for pure oxygen (Fig. 44); the most appreciable effect from among the aforementioned admixtures was exerted by the alcohol vapor. Let us note, by the way, that the same amount of admixtures also affects the rate of oxygen-molecule disintegration. This example shows the necessity of checking the composition and carefully purifying the test gas and the shock tube.* Thus, the investigator is forced to exercise care with the surface finish of the inside walls of the tube and of the supply communications. In conjunction with this the study of nonequilibrium phenomena is frequently performed in stainless steel tubes or in tubes with special corrosion-resistant coatings. Glass and quartz tubes are also sometimes used, the former, however, may break as a result of the experiment [401a, 137]. A necessary condition for obtaining reliable results is regular mechanical removal of dust and solid particles from the tube, satisfactory airtightness and vacuuming of the low-pressure chamber and 'washing" the tube by the test gas. Sometimes this procedure can be replaced by continuous "blowing-down" of the low-pressure chamber by the test gas under working pressure up to the time of arrival of the shock wave, when an increase in the pressure in the chamber closes the automatic gas inlet and outlet valves (see, for example [475]). To improve the quality of the vacuuming, Hooker and Millikan [380] have heated the tube by superheated water at $160\,^{\circ}$C and have reduced the pressure in the low-pressure chamber to 2×10^{-6} mm Hg; the in-leakage rate due to improper hermetization was 1×10^{-6} mm Hg per minute. Haught [361] had the tube cleaned, heated and vacuumed continuously for five days; the experiment was performed only once a week.

Study of nonequilibrium phenomena in gas mixtures necessitates particular consideration of the composition of these mixtures. Usually, the composition of a mixture is determined by measuring the partial pressure of its components. Providing for homogeneous intermixing of these components is a more complicated problem. Several hours, and sometimes longer, are needed for obtaining a sufficiently homogeneous mixture. Forced circulation is used to accelerate the intermixing and reducing the effect of barodiffusion (which is particularly appreciable when working with components with perceptibly different masses). The obtaining of a homogeneous

* Among other examples we can note the effect of easily ionized admixtures (sodium, etc.), on the kinetics of the ionization products in air, argon, etc. (see also [343, 321, 322, 287, 288, 360a]).

mixture of gases at atmospheric pressure in a tube about five meters in the Knight and Venable experiment [407] required 1 to 2 hours of intensive circulation; the homogeneity was checked by a mass spectroscope and by checking the reproducibility of results. Mixing of BrCN with argon required 24 hours [469]. When making up the mixtures it must be remembered that the gases used may be adsorbed by the mixing chamber and shock-tube walls; for example, when working with water vapor the concentration was reduced by surface adsorption by a factor of two to three [590]. The necessary level of partial pressure of water vapor in this case was established by continuously blowing the mixture through the low-pressure chamber of the tube in quantities equal to 30–40 volumes of the tube proper. After the adsorption equilibrium has been reached in the tube, the mixture, as was shown by analysis, was homogeneous and its composition was constant.

In certain cases, solid admixtures (aerosols) must be placed in the tube

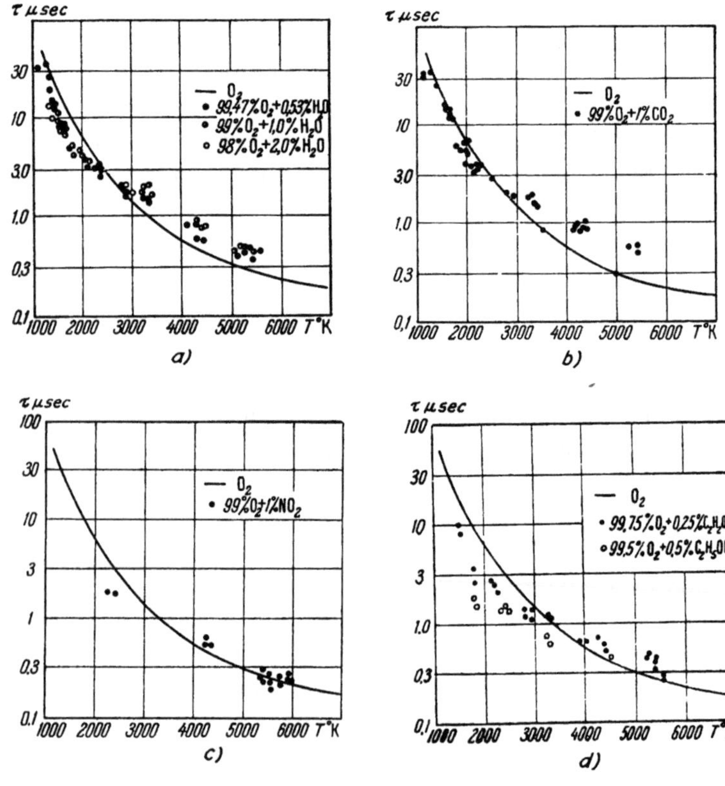

Fig. 44. Temperature dependence of the vibrational relaxation time in oxygen. a) when water vapor has been added to the oxygen; b) when carbon dioxide has been added; c) when nitric acid has been added; and d) when alcohol vapor have been added. The solid line denotes the time of vibrational relaxation in pure oxygen.

Clouston and co-workers [298] have placed sodium salts in the flow of gas which was supplied to the low-pressure chamber and heated them; heating of salts applied directly to a platinum spiral was found to be more convenient. Basu [242] has created an aerosol in a vertical shock tube by placing a powder together with a small volume of gas at a high pressure in a section of the tube above the test section. The concentration of the admixture as a function of time in the given tube cross section was calculated on the basis of the weight of the powder collected with time at the bottom of the tube which was weighed by a sensitive scale.

The test gas is purified and its composition checked by various physico-chemical methods, which have been widely descirbed in the literature (see, for example [63]). Sometimes the purification is performed most simply by multiple fractionation using cooling and purification or chromatographic purification methods. Among methods of composition analysis when working with the shock tube, the most extensive use is made of mass spectrograph, chromatographic, spectral and other methods.

3

Experimental Methods of Study of Nonequilibrium Phenomena in Shock Waves

[8] GENERAL REQUIREMENTS OF THE RECORDING APPARATUS

The experimental study of phenomena in shock waves is based on the use and further development of processes which take place at a rapid rate in gases under high-temperature conditions. Before we consider specific examples for the measurement of any given quantities, let us discuss certain general aspects of experimental methods which are characteristic of the majority of schemes for the study of nonequilibrium phenomena in shock waves.

As will soon be seen, the predominant place among these shemes is occupied by optical methods which make it possible to obtain a high time resolution without exerting a perceptible effect on the object under study. Some success was achieved also by the use of ultrahigh-frequency radio waves, x-rays, electron beams, etc. The signals which are thus obtained are recorded either photographically (high-speed filming), or by electronic schemes with an oscillograph recording system; in the latter case the emission detector is almost always a photoelectric multiplier.

Two versions are possible when using the photographic recording method: instantaneous photographing of the object upon illumination by a very short duration high-intensity source (obtaining shadow and interferometer photographs of the flow pattern or instantaneous photographs of the absorption distribution in a gas, etc.), or the photographing of the image of the object on a rapidly moving photographic film (for example, continuous photography of the position of interference lines in a given tube cross section, continuous photography of the emission spectrum of a gas, etc.).

Instantaneous photographs yield the spatial distribution of gas characteristics in the region behind the wave front; in this case it is important to have a good resolution in the plane of the image. If it is assumed that the optical system has been adjusted precisely, *i.e.*, that an ideally clear image of the object under study has been obtained on the photographic film, then as is known, the major role in obtaining high-quality results will be played by the resolving power and aureole formation of the photographic material being used, which are due to the scattering of light in the emulsion layer [77, 58]. The resolving power of negative films ordinarily used comprises about seventy lines per mm, however, this resolution can be obtained only with great difficulty due to aberrations. The measurement of the intensity of the recorded emission encounters difficulties in photometering the negative when making photographs of extremely short duration, when the relation between the blackening intensity and the quantity of illumination (the characteristic curve of the emulsion) may be nonlinear. Hence some method must be used in each experiment to construct a control characteristic curve of the emulsion used. These difficulties are all substantially magnified when studying the spectral distribution of the emission intensity (problem of heterochromic photometering [167]) when filming the spectrum image in time. In this case it is difficult to use an optical wedge or an ordinary stagewise reducer which is employed in constructing the characteristic curve of the emulsion.

Photoelectric multipliers are more universal emission receivers; the major characteristics, connecting diagrams and use of photoelectric multipliers are given in a number of books [221, 122, 114]. One of the most important characteristics of photoelectric multipliers is the dependence of the anode current on the intensity of the light flux which flows to the photocathode. Within certain limits (practically for fluxes from 10^{-13} to 10^{-4} lumens) one may assume that this relationship is linear; here the maximum anode current which is drawn off the photoelectric multiplier can be as high as 25 mA [122]. For a number of studies importance is acquired by the spectral characteristic of the photoelectric multiplier, which is determined by the spectral sensitivity of the photocathode and by the transmittance of the photomultiplier shell walls. The available photomultipliers which are used in spectrophotometric work make it possible to study a wide range of wavelengths from $\lambda = 1500$ A to $\lambda = 12{,}000$ A. The use of converters makes it possible to widen these boundaries considerably (recording of the ultraviolet edge radiation, of x-rays, corpuscular flows).

The main source of difficulties and errors when working with photomultipliers are the so-called "noises" which are produced by a large number of factors, for example, statistical fluctuations of the quantum emission of primary electrons from the photocathode, fluctuation in the thermoelectron population, fluctuations in the elementary magnitude of the secondary emis-

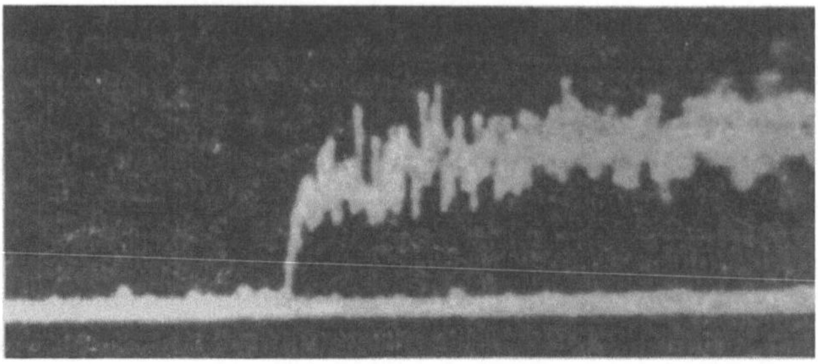

Fig. 45. Specimen of an oscillogram of the dependence of the radiation intensity on the distance behind the shock front in oxygen using a photomultiplier with a high noise level.

sion factor of the dynodes and in the scattering of its values for different dynodes, scattering in the passing times of electrons, etc. To this should also be added the thermal fluctuations in the load resistance, shock effect of the circuit and anode current of the first amplifier tube and other external interferences. A specimen of an oscillogram (oxygen radiating in a shock wave) with a high noise level is presented in Fig. 45. The root-mean-square of the fluctuation current as a result of the shrot effect is $\overline{\Delta I_{sh}^2} = 2ei\Delta f m^2 (1+B)$, where i is the primary photocurrent, m is a multiplying factor ($I = im$), Δf is the frequency range of the recording apparatus, $(1+B)$ is a coefficient which takes into account the increase in the noise level upon multiplication and which is usually equal to 1.5–3.0, e is the charge of an electron [221]. The root-mean-square error thus produced in measuring the photomultiplier signal is equal to (assuming that $I = im = \varphi\Phi m$, where φ is the photocathode sensitivity and Φ is the magnitude of the light flux which is incident on the photocathode)

$$\beta = \sqrt{\frac{\overline{\Delta I_{sh}^2}}{I^2}} = \sqrt{\frac{2e\Delta f\,(1+B)}{\varphi\Phi}}. \tag{8–1}$$

Thus the noise level is proportional to the square root of the light intensity, and the recorded signal is directly proportional to this quantity. Hence the signal/noise ratio, which is equal to $(I/\sqrt{\overline{I_{sh}^2}}) \equiv (1/\beta)$, increases with an increase in the light intensity. A reduction in the frequency range also reduces β, however, the reduction in Δf impairs the resolving power in time.

Among other errors which arise when working with photomultipliers we must note inaccuracies which are due to insufficiently stabilized [electrical power] supply to the photomultiplier. This is the more important in the case when calibrating measurements are performed before the experiment.

A further amplification of the signal coming from the photomultiplier (as well as of the signal obtained from other types of receivers, for example, transforming detectors of ultrahigh-frequency oscillations, ionization probes, thermoanemometric and film-type thermal transducers, etc.), requires the use of amplifiers with a sufficiently high linear amplitude characteristic, since the oscillograph recording uses a beam of finite thickness. The ratio of the thickness of the oscillograph beam to the amplitude of the useful signal observed on the screen with a good signal/noise ratio can make a basic contribution to the measuring error. Clear focusing of the beam and the selection of the best cathode-ray tube can help somewhat; however, in any case it is desirable to have a large-amplitude useful signal (up to 60–80 mm) on the oscillograph screen.

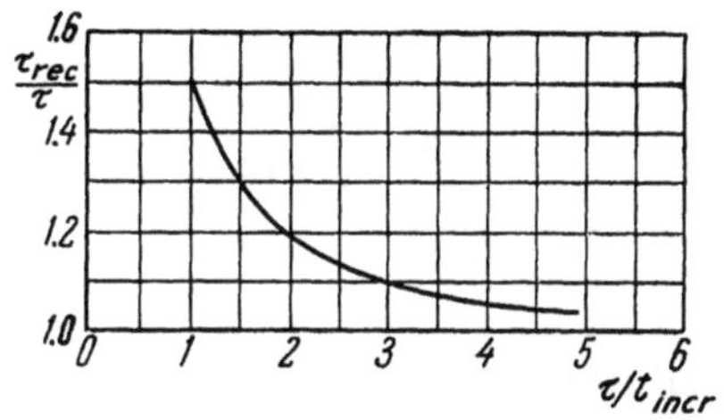

Fig. 46. Effect of frequency characteristics of electronic amplification arrangements on the duration of the recorded signal of exponential shape [124].

τ is the actual relaxation time of the signal (at the input to the arrangement), t_{incr} is the time during which the front increases for a stepped signal [see (8.2)] and τ_{rec} is the recorded relaxation time.

Among the most important factors which distort the observed signals are frequency distortions, which are most perceptible in recording very short signals. It is known [124] that when the voltage at the amplifier input, the resistances being taken off, increases in steps, the time duration t_{incr} of the increase in the front of the pulse at the amplifier output is determined by

$$t_{\text{incr}} = \frac{0.35}{\Delta f}. \tag{8–2}$$

Thus, for example, for an arrangement with Δf equal to 14 Mc/s, $t_{\text{incr}} = 0.025$ μsec. Assuming that the shape of the signal when studying relaxation processes behind a shock wave approaches the exponential, it is not too difficult to determine the systematic error due to frequency distortions by using the graph given in Fig. 46, which is taken from [124]. Knowing the ratio of the actual relaxation time τ of the phenomenon under study to t_{incr}, it is possible to use this graph to determine the systematic error in measuring τ. It turns out that in the case when the measured τ is equal to the relaxation time of the measuring arrangement, then the pulse for a signal exponentially varying with time which is obtained on the oscillograph screen will be by 50% longer than actual (for example, 0.038 μsec instead of 0.025 μsec). As τ becomes larger, this error will be less perceptible.

When measuring the density, concentration, or other flow variables behind the shock front, the probing radiation in the majority of cases is directed across the tube axis. This beam is not infinitesimally thin; usually it is limited by slits in the inspection section of the shock tube. The fact that the slit has a finite width is responsible for the fact that the flux which passes the inspection apertures at each time instant is averaged over the entire slit width. If the width of such a slit is s, and the shock-wave velocity is V, then the receiver at any given time instant records a signal

$$J = \frac{V}{s} \int_{t-\frac{s}{2V}}^{t+\frac{s}{2V}} I \, dt, \tag{8–3}$$

where I is the actual (undistorted) signal. For an exponential signal $I = I_0 \exp[-t/\tau]$, we get

$$J = \frac{2\tau I_0 V}{s} e^{-\frac{t}{\tau}} \operatorname{sh} \frac{s}{2V\tau}. \tag{8–4}$$

Expanding in a series we arrive at

$$J = I_0 e^{-\frac{t}{\tau}} \left(1 + \frac{s^2}{24V^2\tau^2} + \cdots\right). \tag{8–5}$$

Thus the distortion effect of the slit on the exponential signal will be insignificant if

$$\frac{s^2}{24V^2\tau^2} \ll 1. \tag{8–6}$$

If we assume that the permissible systematic error in the recorded value of the signal cannot exceed five percent, then, setting $[(s/V\tau)^2/24] \approx 0.05$, it is possible to obtain the maximum permissible (for the given τ and V) value of the slit width. Thus, when $\tau = 0.1$ μsec and $V = 4.3$ km/sec, $s_{max} = 0.47$ mm. As was pointed out in the preceding chapter, a perceptible error in measuring very short relaxation times can be introduced by the curvature of the shock front, which is due to the interaction of the shock wave with the boundary layer.

We wish to note another important circumstance which is related to the specifics of investigating processes in moving gases and which also affects the resolving power of the study methods used. In the study of changes in

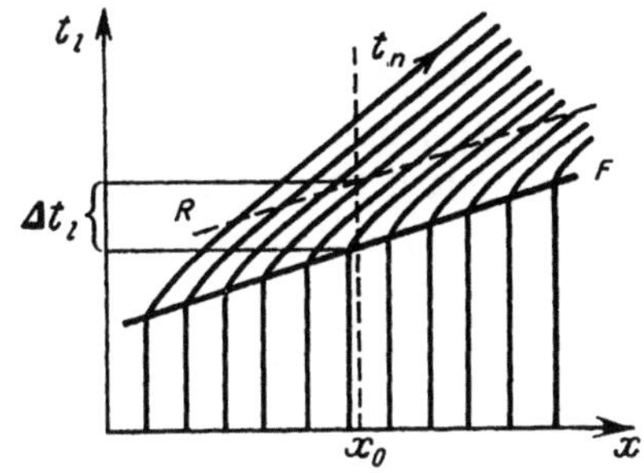

Fig. 47. A t, x diagram of the path of gas particles in the nonequilibrium region behind a shock front, which illustrates the relationship between the time t_l in the laboratory coordinate system and the natural time t_n during which a particle travels in the flow. F is the trace of the shock front; R is the region in which equilibrium is established, x_0 is the position of the observation point (in the shock tube), for which the time of passing of the nonequilibrium zone comprises t_l. Here, different gas particles appear in the line of sight of the recording instrument.

the state of a gas behind the front of a passing shock wave in any fixed cross section of the shock tube, it should be remembered that the instrument does not follow the evolution of processes in any one given physically infinitesimal gas volume heated by the shock wave, but rather records changes in state of various volumes which move with the flow. This motion, in the laboratory coordinate system which is tied to the shock-tube walls, is related in the direction of motion of the wave front (Fig. 47). However, we are interested in the development of processes on a time scale tied to some one given gas volume. The time measured from the oscillograph screen (or, briefly, the time t_l) is uniquely related to the time of development of processes in each physically infinitesimal gas volume (or, briefly, to the natural time t_n). At some time t_0, let a shock front pass through a fixed gas volume with the coordinate x_1, coinciding, for example, with the position of the inspection hole in the test section of the tube. After a time dt_l the shock front will pass through a distance $dx = V dt_l$. On the other hand, some infinitesimal gas volume will traverse this distance during a time $dt_n = dx/v_2$, where v_2 is the velocity of the volume under consideration relative to the shock front. From this

$$dt_l = \frac{v_2}{V}\, dt_n = \frac{\varrho_1}{\varrho_2}\, dt_n. \tag{8–7}$$

For finite time intervals

$$t_n = \int_0^{t_l} \frac{\varrho_2}{\varrho_1}\, dt_l, \tag{8–8}$$

where the origin of the laboratory time measurement is usually made to coincide with the instant of the passing of the shock front. For flows with constant density $t_n = (\varrho_2/\varrho_1)t_l$, from which it follows that for sufficiently strong waves, t_n greatly exceeds t_l. Consequently, also the resolving power, referred to the natural time, will be lower than the resolving power of the apparatus in the laboratory time by the same amount. This limitation is removed when studying phenomena in reflected waves, when the gas is practically stationary.*

When relaxation processes take place behind the shock front the density will not be constant; hence, to make a transition from t_l to t_n, the integral of Eq. (8–8) must be evaluated.**

[9] CERTAIN RELATIONSHIPS FOR THE FLOW OF NONEQUILIBRIUM GAS

Measurements performed using shock tubes results in obtaining characteristics such as the excitation times, dissociation and ionization rates, etc. Realizing these values from distribution curves obtained in experiments is a complex problem and, as a rule, the calculations are performed with simplifying assumptions to the effect that the gas flow is one-dimensional and that the shock wave moves with a constant velocity for a sufficiently long time. Hence, the phenomenon under study may be regarded as steady-state. The question of the validity of these assumptions was considered in the preceding chapter.

Let us now note the main relationships which are needed for relating the measured variables and the kinematic characteristics. For the sake of concretness, we shall restrict ourselves to the consideration of chemical reactions in a gas behind a shock front (including dissociation) and of excitation of vibrational degrees of freedom. For other processes these relationships can be obtained similarly.

To describe the processes behind the front of a one-dimensional steady-state shock wave, we now write three groups of relationships.

1. *The laws of conservation of the mass, momentum and energy fluxes*, assuming that the effect of viscosity, thermal conductivity, and diffusion on the flow variables in the volume behind the shock front

$$\varrho_1 V = \varrho_2 v_2 , \tag{9–1}$$

* However, in the reflected wave the gas pressure increases appreciably, which results in accelerating all the processes in the gas.

** Subsequently, we shall try not to use these subscripts.

$$p_1 + \varrho_1 V^2 = p_2 + \varrho_2 v_2^2, \tag{9-2}$$

$$h_1 + \frac{V^2}{2} = h_2 + \frac{v_2^2}{2}, \tag{9-3}$$

where h is the enthalpy per unit mass.

2. *The equation of state and Dalton's law*, on the assumption that the gas can be considered as an ideal mixture of perfect gases:

$$p = \frac{\varrho RT}{\mu} \quad \text{or} \quad p = nkT, \tag{9-4}$$

$$p = \sum p_i \quad \text{or} \quad \sum \xi_i = 1. \tag{9-5}$$

If the gas is made up of different atoms, then these equations must be supplemented by others which reflect the fact that the ratio of the atomic population of one kind to that of the other is constant.

3. *Kinetic equations*, which describe the relaxation processes behind the shock front. The number of particles of the ith component of the gas per unit volume in a coordinate system moving with a gas element which has a velocity v changes (if we disregard diffusion processes) due to two reasons: 1) the changes in the gas density; and 2) chemical transformations in the gas:

$$\frac{dn_i}{dt} \equiv \frac{\partial n_i}{\partial t} + v\frac{\partial n_i}{\partial x} = -n_i\frac{\partial v}{\partial x} + \left(\frac{dn_i}{dt}\right)_x$$

or

$$\frac{\partial n_i}{\partial t} = -\frac{\partial}{\partial x}(n_i v) + \left(\frac{dn_i}{dt}\right)_x. \tag{9-6}$$

In a steady flow $(\partial n_i/\partial t)=0$, and

$$\frac{\partial}{\partial x} n_i v = \left(\frac{dn_i}{dt}\right)_x \tag{9-7}$$

The rate of formation of the ith component (dn_i/dt), as is usual in chemical kinetics, is determined by the reaction mechanism. Thus, for example, for thermal dissociation of oxygen molecules

$$O_2 + M \to O + O + M, \tag{9-8}$$

$$\left(\frac{dn_{O_2}}{dt}\right)_x = -k_d n_{O_2} n + k_{rec} n_O^2 n, \tag{9-9}$$

where k_d and k_{rec} are the dissociation and recombination rate constants, respectively.

The change in the vibrational energy store E_{vib} per unit volume in the process of molecular vibration excitation for the case when only one component undergoes relaxation is described by the relaxation Equation (2–1)

$$\left(\frac{\mathrm{d}E_{\text{vib}}}{\mathrm{d}t}\right)_{\text{rel}} = \frac{1}{\tau}\left[E_{0\,\text{vib}} - E_{\text{vib}}\right]. \tag{9–10}$$

Here the subscript "rel" denotes the fact that the change in E_{vib} is determined by relaxation, while the subscript "0" denotes the equilibrium value. The variation in E_{vib} in the flow may also be described by a relationship similar to Eq. (9–6) [setting $(\partial/\partial t)=0$]:

$$v\frac{\partial E_{\text{vib}}}{\partial x} = -E_{\text{vib}}\frac{\partial v}{\partial x} + \left(\frac{\mathrm{d}E_{\text{vib}}}{\mathrm{d}t}\right)_{\text{rel}} \tag{9–11}$$

or

$$\frac{\partial}{\partial x}E_{\text{vib}}v = \left(\frac{\mathrm{d}E_{\text{vib}}}{\mathrm{d}t}\right)_{\text{rel}}. \tag{9–12}$$

From Eqs. (9–7) and (9–12), we immediately obtain the following for changes in n_i and E_{vib} in terms of the natural time

$$\frac{\mathrm{d}}{\mathrm{d}t}n_i v = v\left(\frac{\mathrm{d}n_i}{\mathrm{d}t}\right)_{x}, \tag{9–13}$$

$$\frac{\mathrm{d}}{\mathrm{d}t}E_{\text{vib}}v = v\left(\frac{\mathrm{d}E_{\text{vib}}}{\mathrm{d}t}\right)_{\text{rel}}. \tag{9–14}$$

The form of kinetic equations used in describing the process of ionization and the excitation of electronic levels is similar to that of Eqs. (9–13) and (9–14).*

To establish a relationship between the kinetic characteristics (τ, k, etc.) and the measured density, temperature and concentration distributions behind the wave front, it is necessary first of all to make an assumption with respect to the mechanism of the process, *i.e.*, to select the basic elementary reactions. Naturally, it is impossible to fully take into account all the possible processes in a gas. Using published data and comparing the conclusions which follow from assuming a given mechanism with the experimental results, it is usually possible to select the basic, most effective means of transformations in the system. As an example, let us consider the dissociation of oxygen molecules. In this case, in addition to the direct disintegration of O_2 it is possible to have disintegration with intermediate ozone formation:

$$O_2 + M \rightarrow O + O + M; \tag{9–15}$$

$$O_2 + O_2 \rightarrow O_3 + O; \tag{9–16}$$

$$O_3 + M \rightarrow O_2 + O + M. \tag{9–17}$$

* In very strong shock waves in this case (as in the consideration of photochemical processes) the radiative transfer must be taken into account. Very few specific results are available in this field (see Sect. 20).

Using previously determined dissociation rates for O_3, it was possible to show by comparing calculated results with experimental data that the contribution of reactions involving the formation of ozone is insignificant; for this reason, only the direct disintegration of O_2 is considered in the subsequent analysis [202].

Henceforth, the problem can be solved in two ways. On the one hand, it is possible, on the basis of various rate constants (or relaxation times), to seek the density, temperature and concentration distributions behind the wave front by solving Eqs. (9–1)—(9–5), (9–13) and (9–14). Due to the fact that the system of equations which gives the solution is complicated, the solution can be obtained only by the use of computers (if no additional simplifying assumptions are made). After a number of transformations of the right-hand side, differential equations such as Eq. (9–13) are easily solved by the Runge-Kutta method. This is precisely the method by which Rink [491], having measured the density distribution behind the shock front, selected the dissociation rate constant for the O_2 molecules. From among other examples of this kind we should note the work by Wray and Teare [616], who have measured the rate of dissociation of nitric oxide behind the wave front. Despite the effectiveness of the use of computers for analysis of experimental results, the selection method is convenient only in the case of simple systems with a moderate number of unknown parameters.

A second possibility is to determine the kinetic characteristics directly from the experimentally obtained distribution of the flow parameters behind the wave front. Without making any additional assumptions, in this case it is possible to solve Eqs. (9–13) and (9–14) by assuming that τ or k are unknown.

For steady-state one-dimensional flow behind the shock front, on the basis of Eq. (9–13) and making use of Eq. (9–1) we get

$$\frac{1}{v}\frac{\mathrm{d}}{\mathrm{d}t}(n_i v) = \frac{\mathrm{d}n_i}{\mathrm{d}t} - \frac{n_i}{\varrho}\frac{\mathrm{d}\varrho}{\mathrm{d}t} = \left(\frac{\mathrm{d}n_i}{\mathrm{d}t}\right)_{\mathrm{x}}. \tag{9–18}$$

The rate of change in the number of the ith gas component per unit volume can be represented in the form

$$\frac{\mathrm{d}n_i}{\mathrm{d}t} = n\frac{\mathrm{d}\xi_i}{\mathrm{d}t} + \frac{n_i}{\varrho}\frac{\mathrm{d}\varrho}{\mathrm{d}t} - \frac{n_i}{\mu}\frac{\mathrm{d}\mu}{\mathrm{d}t}. \tag{9–19}$$

Substituting the above equation into Eq. (9–18) and using Eq. (4–2) we get

$$\frac{\mathrm{d}\xi_i}{\mathrm{d}t} - \frac{\xi_i}{\mu}\sum \mu_i \frac{\mathrm{d}\xi_i}{\mathrm{d}t} = W_i, \tag{9–20}$$

where

$$W_i = \frac{1}{n}\left(\frac{\mathrm{d}n_i}{\mathrm{d}t}\right)_{\mathrm{x}}. \tag{9–21}$$

The form of W_i, as was pointed out, depends on the assumed reaction mechanism. Thus for the reaction $O_2+M \rightarrow O+O+M$, where M can be O_2 or O, in the beginning of the dissociation zone, when it is possible to disregard the recombination of oxygen atoms

$$W_{O_2} = -\frac{p}{kT}\left[k_d(O_2, O_2)\,\xi_{O_2}^2 + k_d(O_2, O)\,\xi_{O_2}\xi_O\right], \qquad (9\text{–}22)$$

where $k_d(O_2, O_2)$ and $k_d(O_2, O)$ are the dissociation rate constants for oxygen molecules in the O_2–O_2 and O_2–O collisions, respectively. Equation (9–20) is simplified if we remember that $\xi_{O_2}+\xi_O=1$ (shock wave in pure oxygen). Then

$$\frac{d\xi_{O_2}}{dt} = (1 + \xi_{O_2})\,W_{O_2}\,. \qquad (9\text{–}23)$$

Knowing the magnitude of one of the dissociation rate constants (for example, $k_d(O_2, O)$ which is determined in the study of O_2+Ar mixtures [282]), it is possible to determine $k_d(O_2, O_2)$ from experiments in pure O_2 by measuring the O_2 concentration distribution [105].

In the case of dissociation of carbon dioxide molecules behind the shock front in pure CO_2, the mechanism of reactions is, apparently, more complicated; it is possible to assume that at $T<6000\,°K$ this dissociation takes place in the manner

$$CO_2 + CO_2 \rightleftarrows CO + O + CO_2\,, \qquad (9\text{–}24)$$

$$CO_2 + O \rightleftarrows CO + O_2\,, \qquad (9\text{–}25)$$

$$O_2 + M \rightleftarrows O + O + M\,. \qquad (9\text{–}26)$$

It can then easily be found from the above that behind the shock front

$$\frac{d\xi_{CO_2}}{dt} = (1 + \xi_{CO_2})\,W_{CO_2} + \xi_{CO_2}W_{O_2}\,, \qquad (9\text{–}27)$$

$$\frac{d\xi_{O_2}}{dt} = (1 + \xi_{O_2})\,W_{O_2} + \xi_{O_2}W_{CO_2}\,, \qquad (9\text{–}28)$$

where W_{CO_2} and W_{O_2}, according to Eq. (9–21), are determined by the rates of the dissociation of the CO_2 and O_2 molecules, referred to the total particle number per unit volume. The complete solution of the problem of relations between reaction rate constants and the distributions of the CO_2 and O_2 concentration behind the shock front is performed most expediently using computers [383] (see Sect. 11). If, however, we restrict ourselves only to the first of the three above reactions (without taking recombination into account), then

$$\frac{d\xi_{CO_2}}{dt} = -(1 + \xi_{CO_2})\,k_d(CO_2, CO_2)\,\xi_{CO_2}^2\,\frac{p}{kT}\,. \qquad (9\text{–}29)$$

Also here, consequently, it is necessary to measure the concentration distribution for CO_2 in order to determine the reaction rate constant.

The temperatures for all the above cases can be determined by using the law of conservation of energy, and by considering the energy expended by breaking up the bound in the dissociating molecules and on the basis of the experimentally obtained distribution of the gas component concentration. A number of specific examples is presented in Sect. 11 concerning the description of methods of concentration measurement using spectral absorption analysis.

The form of the kinetic equation demonstrated by Eqs. (9–22) and (9–23) which describe the process of molecular dissociation behind the wave front is not the only one possible. In a number of works concerned with the study of the dissociation rate of molecules, use is made of other expressions for particle concentration and, according to this, the kinetic equations are observed in a different form. For convenience in comparison, these kinetic equations are presented in Table 4 for the reaction $O_2+O_2 \rightarrow 2O+O_2$ in pure oxygen. All of these equations can easily be converted into one another; the value of the reaction rate constant k_d in all of them is the same and is determined from Eq. (9–9).

Thus, turning to consideration of the process of vibration excitation, we wish to note that Eq. (9–14) when used for describing the vibrational relaxation behind the wave front, should be referred to a unit mass of the gas. Then the change in the vibrational energy of a unit gas mass $\varepsilon_{vib}=E_{vib}/\varrho$ behind the shock front, according to Eqs. (9–14) and (9–1), will be described by the same relaxation equation

$$\frac{d\varepsilon_{vib}}{dt}=\frac{1}{\tau}\left[\varepsilon_{0\,vib}-\varepsilon_{vib}\right]. \tag{9–30}$$

The determination of the relaxation time thus requires measuring the vibrational energy as a factor of time after the passing of the shock front. Sometimes the total vibrational energy content per unit mass of a gas can be estimated by measuring the number of individual vibrational levels (see Sect. 11). The temperatures and, consequently, ε_{0vib} can be obtained on the basis of the law of conservation of energy (Sect. 11). Excitation of molecular vibrations in a mixture of several gases undergoing relaxation is considered in Sect. 22 as applied to air.

It is frequently assumed in processing experimental data (as well as in calculations) that the pressure in the nonequilibrium region remains constant; according to what was said in Sect. 3, this assumption is valid in the majority of cases.

The determination of kinetic characteristics is substantially simplified

Table **4.** Different Forms of the Kinetic Equation for Dissociation Reaction $O_2+O_2 \rightarrow O+O+O_2$

Determination of particle concentration	Concentration expressed in terms of the number of particles per cm³	Relation between the concentration and the mole fraction	Form of kinetic equation
Molar fraction	$\xi_{O_2} = \dfrac{n_{O_2}}{n_O + n_{O_2}}$	1	$\dfrac{d\xi_{O_2}}{dt} = -(1+\xi_{O_2})\,k_d \dfrac{p}{kT}\xi_{O_2}^2$
Degree of dissociation	$a = \dfrac{n_O}{2\left(n_{O_2} + \dfrac{n_O}{2}\right)}$	$a = \dfrac{1-\xi_{O_2}}{1-\xi_{O_2}}$	$\dfrac{da}{dt} = \dfrac{\varrho \mathscr{N}}{\mu_{O_2}}(1-a)^2\,k_d$
Ratio of the particle number of the given type per cm³ to the equivalent molecular population	$\gamma_{O_2} = \dfrac{n_{O_2}}{n_{eq}}$	$\gamma_{O_2} = \dfrac{\mu}{\mu}\xi_{O_2}$	$\dfrac{d\gamma_{O_2}}{dt} = -\,k_d\gamma_{O_2}^2\dfrac{\varrho}{\varrho_1}n_1$

n_1 is the number of particles per cm³ ahead of the shock front, ϱ is the density of the gas behind the shock front, n_{eq} is the equivalent molecular number (number of molecules per cm³) behind the shock front on the assumption that no chemical processes take place, $n_{eq} = n_1(\varrho/\varrho_1)$, μ is the average molecular weight and $\mathscr{N}$ is Avogadro's number.

when the test gas is a small admixture to an inert diluting agent. Examples of this are experiments in mixtures of oxygen, nitric oxide, water vapor and carbonhydrates with argon [278, 244, 616, 81]. In this case not only the pressure, but also the temperature and density of the gas behind the shock front change only slightly. The use of these mixtures for experiments in reflected waves is also preferable since, as was noted in Sect. 6, the deviations from one-dimensionality are least of all manifested in monatomic gases.

[10] MEASURING THE GAS DENSITY

The gas density is one of the major parameters which characterize the state of a fluid. It is the study of the gas density distribution in a shock wave which is responsible for the main results in investigating the excitation of molecular rotations, vibrations and dissociation.

[1] Study of the Reflection of Light from a Shock Front

One of the tasks in the study of shock-wave structure is measuring the density distribution in the wave front proper. These methods were worked out by Horning, Cowan, Greene and Andersen [303, 381, 344, 346, 233, 234, 427]. The method they have used for obtaining the density distributions in the shock front consists in measuring the reflecting power of the front. A diagram of the installation set up by these authors is given in Fig. 48. A high-intensity parallel beam of light was incident at angle θ on the forward surface of the shock front at instant the latter passed the observation holes. After being reflected, the light entered a photomultiplier, the signal from which was fed to an oscillograph. The experimental value of the reflection

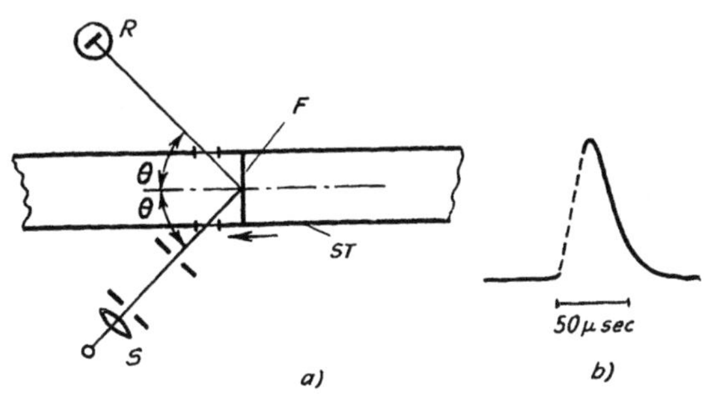

Fig. 48. Scheme for measuring the density distribution in a shock front on the basis of the light reflection coefficient (a) and a specimen of the signal which is obtained (b). *ST* is the shock tube, *F* is the shock front; *S* is the light source, *R* is the receiver of the reflected signal, θ is the angle of incidence of the light ray on the surface of the front.

coefficient was compared with the theoretical, which was calculated on various assumptions with respect the density distribution in the shock front, and that density distribution was selected which was in best agreement with experimental data. The successful use of the above scheme requires that a number of conditions be satisfied. The light source used should be of sufficient intensity and stability and the light being used should be sufficiently monochromatic. Stopping down the working light beam has ensured the isolation of a central, uncurved section of the shock-front surface, and also reduction of the interference level, the latter being produced by scattering of light from the shock-tube walls. Strict requirements are placed on the sensitivity of the radiation receiver, since the reflection coefficient is a quantity of the order of 10^{-5}–10^{-7}.

The signals recorded on the oscillograph screen have the shape shown in Fig. 48b. The duration of the pulse is determined by the time during which the shock front passes the observation holes. The main quantity which is measured is the amplitude of the observed signal, which, after calibration, serves as a basis for obtaining the reflection coefficient.

Using Fresnel's formulas, it is possible to obtain the following relationship [303] for the reflection coefficient R of light reflected from a multilayer flat interface (on the condition that $R \ll 1$)

$$R = \frac{1 + \tan^4 \theta}{4} |F(y)|^2 , \quad F(y) = \int_{-\infty}^{\infty} \frac{dn}{dx} e^{-2\pi i y x} \, dx , \tag{10–1}$$

where x is the distance along the shock-tube axis,* n is the refraction index, $y = 2 \cos\theta / \lambda$, and λ is the wavelength of the radiation used. The density is related to the refraction index by the Gladstone-Dale relationship

$$\frac{n - 1}{\varrho} = K , \tag{10–2}$$

where K is the Gladstone-Dale constant, which depends on the nature of the gas, its composition and λ. Thus it is possible to specify various density distributions (and, consequently, various refraction indices) in the shock front and then to select that which gives the best agreement with the measured data (Fig. 49). It was found that a relationship such as

$$\frac{\varrho - \varrho_1}{\varrho_2 - \varrho_1} = \frac{1}{1 + \exp(-4x/L)} , \tag{10–3}$$

* The coordinate origin is then placed in the point of inflection of curves which characterize the density distribution in the shock front (see Fig. 49).

where L is the effective shock-front thickness, which is defined by

$$L = \frac{\Delta \varrho}{\left(\dfrac{d\varrho}{dx}\right)_{max}}, \tag{10–4}$$

where $\Delta\varrho = \varrho_2 - \varrho_1$, is such a distribution. Using Eq. (10–2) for the distribution (10–3) it is possible to evaluate the integral in Eq. (10–1). According to [211], in this case

$$R = \frac{1 + \tan^4\theta}{4}(\Delta n)^2 \left[\frac{\pi^2 L \dfrac{\cos\theta}{\lambda}}{\mathrm{sh}\left(\pi^2 L \dfrac{\cos\theta}{\lambda}\right)} \right]^2, \tag{10–5}$$

where Δn is the difference in the refraction indices to both sides of the shock front.

This is how the values of L were obtained for a large number of gases [234, 427] (Ar, N_2, O_2, CO, CO_2, N_2O, CH_4, Cl_2, HCl, etc.). Despite great experimental difficulties, the accuracy in measuring L was about 15%, which is quite satisfactory. Here, the majority of experiments was performed with weak waves ($M \sim 1.1$–1.5) at a high initial pressure (near atmospheric). It was found that the shock-front thickness comprises only several (or several tens) of mean-molecular-free paths. It was discovered that the measured thickness of the front in argon and nitrogen is inversely proportional to the pressure in the range $2.5 \leqslant p_1 \leqslant 11$ atm. Obviously, this relationship will also be satisfied at lower pressures.

In the limit, as $y \to 0$, $F(y) \to (\Delta n)^2$. According to Eq. (10–1), in this case

$$R = \frac{1 + \tan^4\theta}{4}(\Delta n)^2, \tag{10–6}$$

i.e., R is independent of the density distribution across the front. Using Eq. (10–2), it is easy to relate the value of Δn with the density difference to both sides of the front $\Delta\varrho$. It is easiest to let y go to zero by letting θ approach 90°.* Performing measurements in a sliding beam, we will find the following for the ratio of the refraction indices of the test gas R_x and, let us say, of argon R_{Ar}

$$\frac{R_x}{R_{Ar}} = \frac{\left[\left(\dfrac{\Delta\varrho}{\varrho_1}\right)(n-1)\right]_x^2}{\left[\left(\dfrac{\Delta\varrho}{\varrho_1}\right)(n-1)\right]_{Ar}^2}. \tag{10–7}$$

* Then, however, the angle θ should differ so much from 90°, that the condition $R \ll 1$ should remain valid.

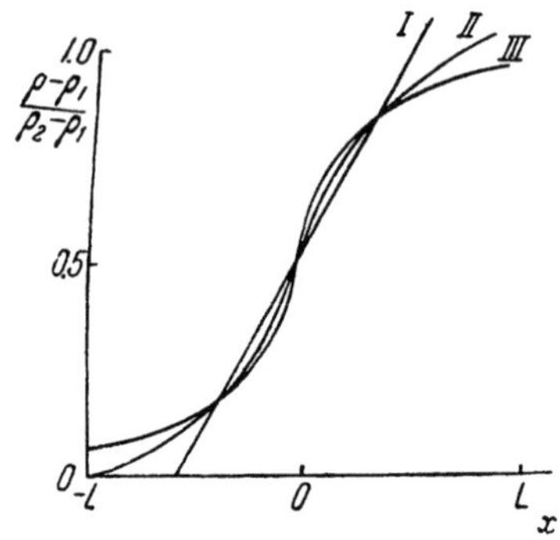

Fig. 49. Different density distributions in a front which are used for comparison with the experimentally obtained refraction index R. The dependence of $(\varrho - \varrho_1)/(\varrho_2 - \varrho_1)$ on x is determined by

$$\text{I} - \begin{cases} 0 & x \leqslant -\dfrac{L}{2}; \\ \dfrac{\left(x + \dfrac{L}{2}\right)}{L} & -\dfrac{L}{2} \leqslant x \leqslant \dfrac{L}{2}; \\ 1 & x \geqslant \dfrac{L}{2}; \end{cases}$$

$$\text{II} - \frac{1}{1 + \exp\left(-\dfrac{4x}{L}\right)};$$

$$\text{III} - \frac{1}{2} + \left[\frac{\left(\text{arc}\,\tau\,\dfrac{\pi x}{L}\right)}{\pi}\right].$$

The density directly behind the shock front in polyatomic gases is determined by the degree of completness of the rotational relaxation of molecules. Hence, these measurements make it possible to pronounce an opinion on the rotational relaxation time. For this the measured densities directly behind the shock front are compared with calculated densities obtained upon assuming the absence or, conversely, complete excitation of molecular rotations. Thus, for experiments in nitrogen, such values of $\Delta\varrho/\varrho_1$ immediately behind the wave front were obtained which point to the fact that when the wave-front thickness is about 50 mean-free paths, the rotational relaxation ends already in the wave front. The agreement between the measured and calculated values of ϱ_2/ϱ_1 makes it possible to estimate here the upper limit of the rotational relaxation times. With an increase in the Mach number and reduction in the number of collisions, not enough time is available for establishing an equilibrium in the front, which is shown by the measured density value. In the case of very large τ_{rot}, for example, for hydrogen, the above method makes it possible to estimate only the lower limit of the possible values of τ_{rot}; thus excitation of rotations in hydrogen requires at least 150 collisions.

The above method for shock-wave study makes it possible to study the development of processes in a gas in cases when the number of collisions required for them to occur does not exceed two hundred. At the same time it is by far not universal; the researcher who would attempt to use for the study of the structure the front of a strong shock wave accompanied by radiation of the gas, will be faced with serious difficulties.

[2] The Tepler Shlieren Scheme

Extensive use is made of shlieren systems in the study of gas-flow phenomena. As we know, the term shlieren is used to define an optical inhomogeneity which occupies a relatively small volume. The operating principle of a shlieren system consists in separating a beam of rays which undergo refraction in the gas volume under study. Study of changes in the refraction index makes it possible sometimes to determine the density distribution in the flow behind the wave front.

A detailed substantiation of the method used in measuring the refraction index distribution by using a large number of various shlieren schemes is given in a series of works by Schardin [521, 522]. The greatest success in fluid-flow studies was achieved by Tepler's shlieren scheme (Fig. 50). According to Schardin, the illumination E^* of a point in the field of the image

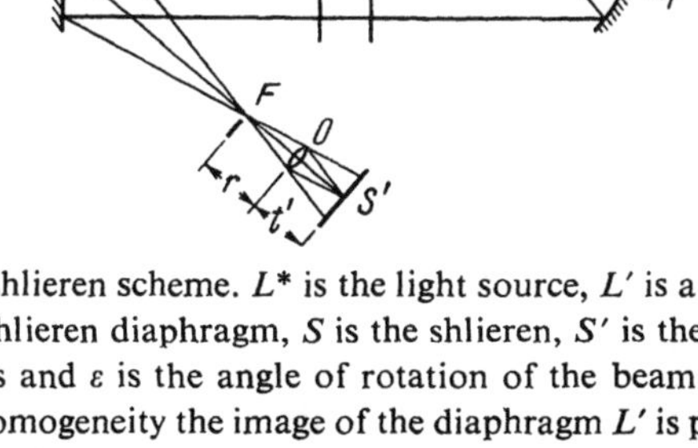

Fig. 50. The Tepler shlieren scheme. L^* is the light source, L' is a diaphragm, K_1 and K_2 are mirrors, F is the shlieren diaphragm, S is the shlieren, S' is the image of the shlieren, O is the objective lens and ε is the angle of rotation of the beam in the shlieren. In the absence of optical inhomogeneity the image of the diaphragm L' is partially covered by the shlieren diaphragm F. The appearance of the inhomogeneity S causes the beam to rotate through an angle ε and the image of the diaphragm L' is displaced, which produces a change in the illumination.

of a subject (in our case in the field of the shock-tube duct) depends on the luminous intensity B of the source, light losses in the system, (the coefficient of losses η), on the focal distance f', the distance r from the shlieren diaphragm to the objective lens O and the distance t' from the objective lens to the plane of the image, on the height b' of the light source image in the plane of the shlieren diaphragm, and also on the aperture of the optical system ω^*. The dependence of E^* on the distance a' from the edge of the shlieren diaphragm to the boundary of the light source image is the determining relationship in using the Tepler shlieren scheme. Thus,

$$E^* = \eta B\left(\frac{f' + r}{t'}\right)^2 \frac{a'b'}{f'^2} \cos^4 \omega^*. \tag{10–8}$$

Usually ω^* is so small that $\cos^4 \omega^* \approx 1$; frequently, also $r \ll f'$. Hence,

$$E^* = \eta B \frac{a'b'}{(t')^2}. \tag{10–9}$$

A shift in the light source image in the plane of the shlieren diaphragm in the case when the refraction index has a gradient in the direction perpendicular to the edge of the diaphragm (assuming that it has the form of a straight edge), produces changes in the values of a'. Then for E^*

$$E_s^* = E^* + \Delta E^* = \eta B \frac{b'}{(t')^2}(a' + \Delta a'). \tag{10–10}$$

The magnitude of $\Delta a'$ depends on the angle of deflection of the light beam in the shlieren, *i.e.*, $\Delta a' = \varepsilon f'$. From this

$$\varepsilon = \frac{(t')^2}{\eta B b' f'} E_s^* - \frac{a'}{f'}, \tag{10–11}$$

i.e., the angle of deflection ε in any point of the optical inhomogeneity under study is, with an accuracy of a constant component, directly proportional to the illuminance of the image of the corresponding point. On the other hand, for uniform flow along the x-axis

$$\varepsilon = \frac{1}{n}\int_{y_1}^{y_2} \left(\frac{\partial n}{\partial x}\right)_y dy, \tag{10–12}$$

where y is the distance along the beam directed across the flow. Disregarding the dependence of $\partial n/\partial x$ on y, we get

$$\varepsilon = \frac{d}{n}\frac{\mathrm{d}n}{\mathrm{d}x}, \tag{10–13}$$

where d is the width of the flow in the test section of the tube. Thus, the angle of beam rotation in the shlieren is proportional to the gradient of the refraction index in a direction perpendicular to the direction of the light beam. For experiments under conditions when the Gladstone-Dale relationship (10–3) is satisfied, we get (due to the fact that n is practically equal to unity)

$$E_s^* = \eta B \frac{b'}{(t')^2}\left[a' + f'dK\frac{\mathrm{d}\varrho}{\mathrm{d}x}\right], \tag{10–14}$$

where K is the Gladstone-Dale constant.

For large density gradients (which can take place behind a shock front) the magnitude of ε can become so large that the entire image of the light source will be found outside the shlieren diaphragm; then any further change in the density will not affect the change in the illuminance E_s. To eliminate this saturation effect, the shlieren diaphragm is placed at a certain angle β to the light source image so that the image should not entirely go past the edge of the diaphragm [610].

Two methods exist for recording the light beam which passes the shlieren diaphragm; one of them being photographic and the other photoelectrical. The use of a short-duration pulsating light source makes it possible to obtain a photograph of the distribution of the density gradient in the region behind the shock front. This was the manner in which Glick and Wurster [338] have measured the length of the nonequilibrium dissociation zone behind a shock front in oxygen. One of the main difficulties encountered in the use of this method is calibration of the blackening on the film, since the direct use of Eq. (10–14) does not yield sufficiently accurate results. The experimentally obtained density gradients were calibrated by Glick and Wurster by comparing the observed results with interferometric data. Schardin recommends that calibration be performed by using plano-convex lenses with a large radius of curvature (or wedges), which makes it possible to obtain the specified value of the calibrating angle ε_0 which corresponds to some value of $\mathrm{d}\varrho/\mathrm{d}x$.

Photoelectric recording methods, two versions of which exist, are used much more extensively. In the scheme with a narrow slit (differential scheme), the image of the test section of the shock tube is obtained in the S' plane (Fig. 50), which is precisely the location of a slit which isolates the minimum possible (along the tube axis) image region; a photomultiplier is placed behind this slit. Then the signal is fed to an amplifier and oscillograph [37]. When the slit (with area σ) is uniformly illuminated, and AC amplifiers are used, the signal $I(t)$ which appears on the oscillograph screen is determined by

$$I(t) = C\frac{\pi\sigma Bdb'f'K}{(t')^2}, \tag{10–15}$$

where C is the amplification factor. When a shock wave passes the slots in the test section, curves are produced on the oscillograph screen, a sample of which is shown in Fig. 51a. In the scheme with a wide slit (integral scheme), a wide slit, *i.e.*, an aperture with edges perpendicular to the direction of flow, is installed in the observation section of the tube. If ξ is the coordinate along which the distance from the edge of the tube is measured along the tube axis, and s is the slit width, then integrating Eq. (10–15) over the entire length of the slit, we obtain the following for a signal which is obtained on the oscillograph screen at the time t (all the remaining constants are lumped under C'):

$$I(t) = C' \int_0^s \frac{\partial}{\partial \xi} [\varrho(\xi, t)]\, d\xi = C' [\varrho(s, t) - \varrho(0, t)]. \tag{10–16}$$

Thus, in this case the signal on the oscillograph screen is proportional to the density differences at the edges of the slit. The use of this method is successful in the case when the gas density at one or the other edge is a known quantity. This is obtained most simply in the case when the width of the list is greater than the dimensions of the nonequilibrium region of density increase near the shock front. Then the oscillograph curve has the shape shown in Fig. 51b [487, 488].

The signals which are obtained can be calibrated, as before, by an optical wedge with a specified ε_0. However, calibration is most frequently performed by changing the value of a', *i.e.*, the shlieren diaphragm is displaced through a specified distance and a light modulator is used to obtain a calibrating signal on the oscillograph screen. This scheme is used to determine the density distribution behind the shock front, which makes it possible to calculate the dimensions of the nonequilibrium zone. The above method was used by a number of investigators [487, 611] to obtain the vibrational relaxation time behind a shock front in carbon dioxide.

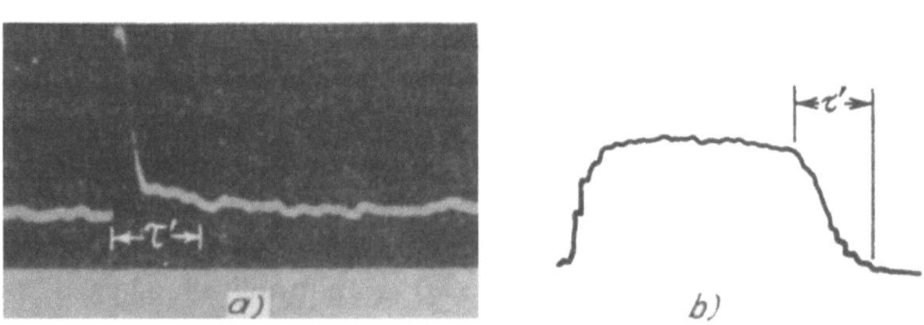

Fig. 51. Samples of oscillograph curves obtained by shlieren schemes with photoelectric recording using the narrow slit method (a) [nitrogen] and the wide slit method (b) [carbon dioxide]; τ' is the nonequilibrium region.

The limit of the accuracy in determining density distributions by the Tepler shlieren scheme is determined by the effect of a number of factors, among which are the diffraction of light in parallel beams, the optical inhomogeneity under study and diffraction at the shlieren diaphragm. The first type of diffraction results in "washing out" the image of each point to dimensions $u'=(\lambda f')/l$, where λ is the wavelength, and l is the size of the optical inhomogeneity. According to Schardin, the minimum discernible gradient of the refraction index in this case is found to be equal

$$\left(\frac{\mathrm{d}n}{\mathrm{d}x}\right)_{\min} \approx \frac{\alpha\lambda}{2ld} \quad \text{or} \quad \left(\frac{\mathrm{d}\varrho}{\mathrm{d}x}\right)_{\min} \approx \frac{\alpha\lambda}{2Kld}, \tag{10–17}$$

where α is the fraction of the useful part of the signal in the entire signal, $\alpha=(\varepsilon f')/a'$. Remembering that in the presence of nonequilibrium phenomena behind the front of a plane shock wave the gas pressure does not vary significantly, and also assuming that the gas composition is unknown (μ=const), it is possible to reduce this estimate to the most graphically presentable magnitude of the discernible gradient of the gas temperature:

$$\left(\frac{\mathrm{d}T}{\mathrm{d}x}\right)_{\min} = \frac{RT^2}{\mu p}\cdot\frac{\alpha\lambda}{2Kld}. \tag{10–18}$$

Thus, for example, for carbon dioxide at $T\sim2000\,^{\circ}\mathrm{K}$ and $p\sim1$ atm, with $\alpha\sim0.5$ and $\lambda=0.5$ microns we will get (for a tube 50 mm in diameter) $(\mathrm{d}T/\mathrm{d}x)_{\min}\approx(80/l)$ degrees/cm.

The effect of the second type, *i.e.*, diffraction, at the edge of the shlieren diaphragm is the more discernible the smaller a'. In this case the "degree of washing out" in the plane of the image is equal to $u''=(\lambda t')/a'$. This may turn out to be of substance when recording by the photographic method and when using photomultipliers in conjunction with a narrow slit (differential method); in the latter case, the smallest value of a' is determined by the relation between u'' and the slit width s in front of the photomultiplier.

Other error-contributing factors are inaccurate calibration, fluctuation in the illumination intensity, unstable photomultiplier amplification, photomultiplier noise, the finite thickness of the oscillograph beam and other usual shortcomings of photoelectric recording methods. Here let us point out the limitations which manifest themselves when measuring the density gradient by the differential shlieren method as a result of photomultiplier noise. Using Eqs. (8–1) we will find that the root-mean-square error produced by photomultiplier noise will comprise 100% (the noise level is equal to the useful signal) for density gradients

$$\left(\frac{\mathrm{d}\varrho}{\mathrm{d}x}\right)_{\mathrm{th}} \approx \frac{t'}{f'Kd}\sqrt{\frac{2e\Delta fa'}{\varphi\eta\sigma Bb'}}. \tag{10–19}$$

Thus in order to increase the signal-to-noise ratio (reduce the threshold value $(d\varrho/dx)_{th}$), it is necessary to use the most intense light sources, long-focus instruments and to adjust the instrument so that a' should not be too large.

[3] The Interferometer Method

Numerous measurements of density distribution behind a shock front in the nonequilibrium region were performed by the use of the interferometer. The interferometer most extensively used in gas flow studies is the Mach-Zehnder device [211, 252, 255, 256] (Fig. 52). The optical system L_1 creates a parallel beam of monochromatic light with the same phase (in the plane R). Then the semitransparent plate P_1 splits the beam in half; mirrors K_1 and K_2 direct the light to the second semitransparent plate P_2. The system consisting of L_2 and L_3 focuses the light and constructs the image of a certain cross section of the shock tube G in the plane G'. When the optical path traversed by both beams after passing P_2 is the same, no interference is observed in the G' plane. To create interference patterns the plate P_2 is rotated through a moderate angle ε, which produces in the P plane a path difference which is proportional to ε [211]. The angle of rotation of P_2 is such that it makes it possible to observe the interference at any point of beam overlapping, which makes the Mach-Zehnder scheme quite convenient. Upon obtaining the interference picture it is possible to determine on the basis of the shift in the observed interference fringes δ (upon the appearance of the difference $\Delta n \cdot d$ in the optical paths between the interferometer "shoulders") changes in the refraction index in the gas volume under study $\Delta n = \lambda\delta/d$ or, using Eq. (10–2), to measure the appearing density difference $\Delta\varrho = (\lambda\delta/Kd)$. This relationship

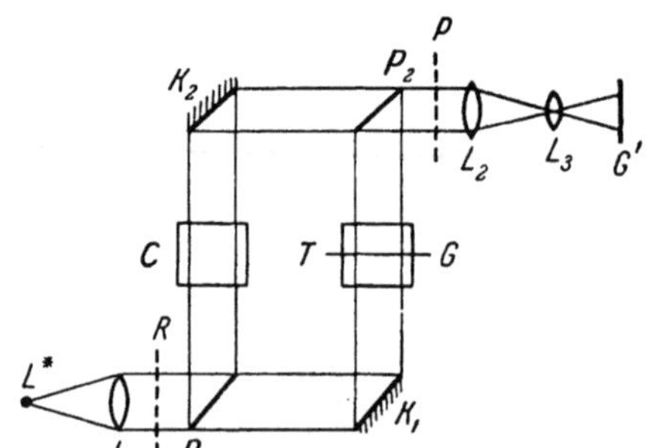

Fig. 52. Schematic diagram of the Mach-Zehnder interferometer. L^* is the light source, L_1, L_2 and L_3 are optical systems for obtaining parallel light beams and images, K_1, K_2 are mirrors, P_1 and P_2 are semitransparent plates, T is the cross section of the test section of the shock tube, and C are compensating plates.

can be written in the form

$$\frac{\varrho}{\varrho_1} = 1 + \frac{C}{p_1}\delta, \tag{10–20}$$

where $C = (RT_1\lambda)/(\mu Kd)$ is a constant which depends on the kind of gas and the wavelength. Calibration and determination of C are most frequently performed by using the relationship between the fringe shift and the change in the initial pressure in the shock tube (for a constant initial temperature), since in this case $C = \Delta p_1/\delta_1$.

In practice, the utilization of the above arrangement requires the use of high quality optical instruments; thus, the optical holes in the shock tube and other components of the arrangement should be plane-parallel with an error of less than one-quarter of the wavelength. The instrument and the shock tube should be installed so as to ensure high vibration resistance and sufficiently constant temperature of the different parts of the instrument.

The interferometer picture which is obtained is recorded by instantaneous photography [254, 437, 438, 67], and filming [275, 170, 67]. In the first case, using a very short-duration spark light source (about 0.1–0.3 microseconds), a single photograph of the position of the interference fringes in the plane of the test section of the tube is obtained, while in the second, a dependence of the position of fringes in certain flow cross section on the time is obtained. An interesting refinement and development of these methods was performed by Zaytsev, Shatilov, *et al.* [67]. To obtain a series of frames of the interference picture they have used a multiframe electron-optical converter.

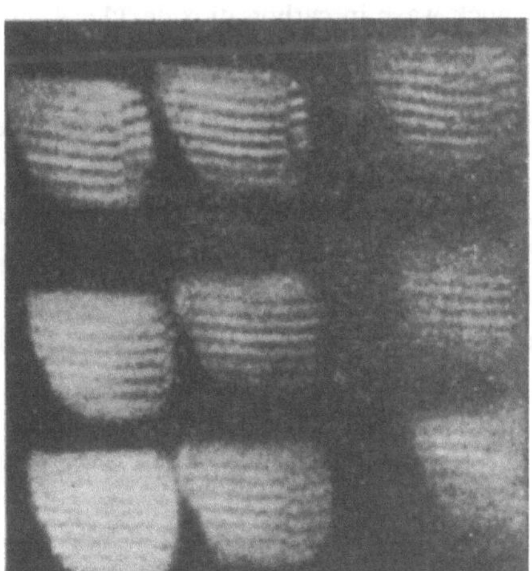

Fig. 53. Specimens of interferograms of the propagation of a shock wave in nitrogen, which were recorded by an electron-optical converter. The frequency was 10^5 frames per second and the exposure time is 0.5 microseconds [67].

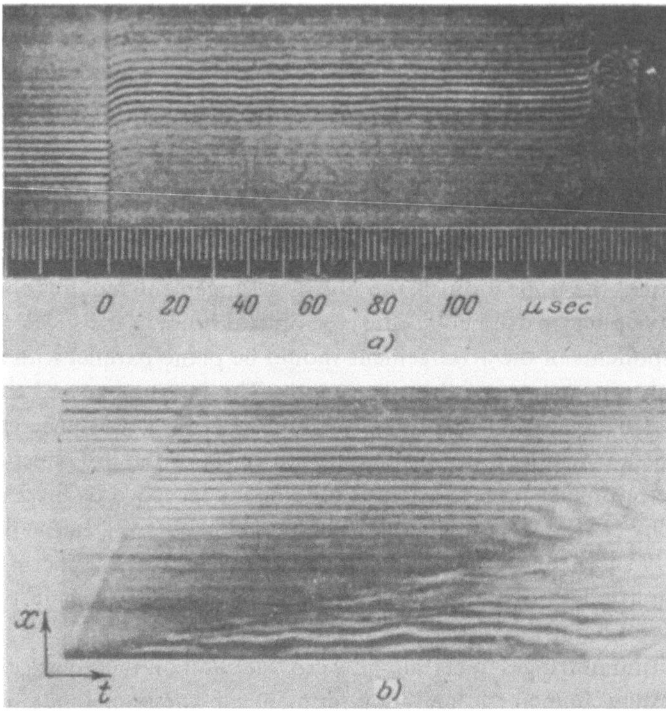

Fig. 54. Samples of interferograms of the density distribution in a shock wave which were obtained in [66, 67]; a) by photographic scanning with a vertical slit (incident shock wave in nitrogen, $M = 8.7$, $p_1 = 30$ mm Hg) and b) by the method of scanning with an inclined slit (reflected shock wave in carbon dioxide, $M = 4$, $p_1 = 10$ mm Hg, $\alpha = 45°$).

Nine consecutive frames appeared on the converter screen (Fig. 53); the picture-taking rate could be 1×10^5, 2.5×10^5 and 5×10^5 frames/sec with an exposure time of 1, 0.5 and 0.2 microseconds. Photographs of even higher quality are obtained by scanning with a moving photographic film using a slit perpendicular to the flow (Fig. 54a) as well as a slit inclined at a angle α to the velocity vector of the shock wave (Fig. 54b). In this case, in addition to the density distribution in the gas volume, Zaytsev and others were able to obtain (with an accuracy of up to 5–10%) the velocities of a plane front of density disturbance (shock front) using the relationship

$$V = \left| \frac{v_f}{m} \cdot \frac{\tan \beta}{\tan \alpha - \tan \beta} \right|, \tag{10–21}$$

where v_f is the film velocity, β is the angle between the direction of motion of the film and the image of the trace of the front, and m is the image scale reduction factor. The scanning method above makes it possible to measure continuously the density in a fixed flow cross section with a spatial resolution of 1–1.5 mm and a time resolution of 2–3 microseconds. This method was used by Zaytsev and others [67] to obtain a density distribution behind a shock front which points to the development of excitation of molecular vibrations in nitrogen (Fig. 55). The excitation and phenomena and the disintegration of molecules of oxygen, nitrogen, carbon monoxide and of other gases were also studied by interferometer methods in [254, 437, 438, 275, 553] and others.

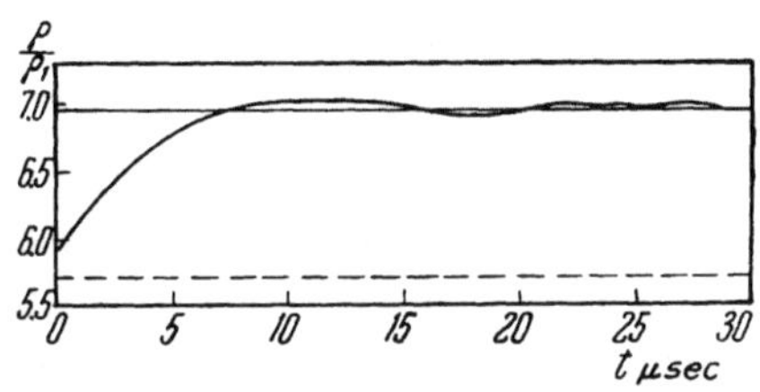

Fig. 55. Time dependence of the density ratio after the passing of a shock front in nitrogen ($M = 8.7$, $p_1 = 30$ mm Hg), obtained by an interferometer. The solid horizontal line denotes the equilibrium value of the density, while the dashed line gives the density without taking into account vibrational excitation.

The fringe shift which takes place in a shock front is measured (which is needed for absolute measurements) either by nonmonochromatic light with fixing of the shift of the zero fringe [67], or using an inclined beam with supplementary light source (see [254, 256]).

The accuracy in measuring the density distribution behind the shock front when using an interferometer is determined by ordinary factors which are due to shortcomings of photographic recording methods (granularity of the film, insufficient image contrast, etc.), and also by a number of other factors, among which are distortion of the unexcited fringes due to the insufficiently high quality of optical instruments and of their adjustment, the fact that the shock-front plane and the optical axis are at an angle, washing out of the fringes due to the fact that the illumination time is finite (in the case of instantaneous photography) or insufficiently high film velocity (on continuous filming), etc. Most authors assume that the fringe shift δ can be measured with an accuracy of $\Delta\delta = 0.04$–0.1 of the fringe width. The relative error in the density determination which is thus produced is given by

$$\frac{\Delta\varrho}{\varrho} = \frac{C}{\frac{\varrho}{\varrho_1} p_1} \Delta\delta . \qquad (10\text{–}22)$$

This error increases with a reduction in the initial pressure. Thus, for example, when $\Delta\delta \approx 0.05$ for oxygen with sufficiently high wave velocities ($\varrho/\varrho_1 \approx 6$) in a tube with $d=5$ cm and $\lambda=5000$ A the relative error in measuring the density distribution in the gas behind the front of an incident wave is approximately $(\Delta\varrho/\varrho) \approx (38/p_1)\%$, where p_1 is given in mm Hg. A reduction in the pressure decreases the total fringe shift, which limits the applicability of the interferometer method of density measurement at low initial pressures. The total error in the determination of the absolute magnitude of the density across the shock front when using photography with white light was analyzed in [67]; for experiments in nitrogen with $p_1=10$ mm Hg it comprises 7–8% for the incident wave and 4–5% for a reflected wave and increases with a reduction in the wave velocity.

The radiation of the test gas at high temperatures makes the use of the interferometer as well as of the Tepler shlieren scheme difficult. The use of various procedures, *i.e.*, utilization of light filters, isolating parallel beams by diaphragms, etc., makes it sometimes possible to reduce the radiation effect.

The error due to the temperature dependence of K is insignificant, since as was shown by experiments in air and in CO_2, the gradient $(\partial n/\partial T)_\varrho$ in these gases is quite low. The most significant limitation of density measurements on the basis of the refraction index (interferometer) or of its gradient (Tepler's shlieren scheme) is the dependence of the Gladstone-Dale constant (or of the specific refraction $r=\frac{2}{3}K$, which is due to the polarizability of atoms and molecules) on the gas composition in the case of dissociation, ionization and other transformations which take place in gases at high temperatures. Alpher and White [228] assume that for an ideal gas mixture it is possible to postulate that the specific refraction is additive, *i.e.*, to assume as valid the relationship

$$n - 1 = \sum K_i \varrho_i , \tag{10–23}$$

where ϱ_1 is the partial density of components which have a specific refraction $r_i=\frac{2}{3}K_i$. Thus, to measure the density of gas mixtures it is necessary to know the specific refractions of the components and the mixture composition; this makes this density determination method not too effective in the given case. The problem simplifies when the magnitudes of K_i are close to one another which is the case of molecular and atomic oxygen, when $(K_{O_2}/K_O)=1.06$. In the case of nitrogen this difference is higher $(K_{N_2}/K_N)=0.79$, K_i for nitric oxide exceeds the values observed for air [228]. Due to the closeness of K_O and K_{O_2} it is possible to assume that the density measurements by the shlieren and interferometer methods in air are quite reliable up to temperatures of $\sim 5000\,°K$ (on the assumption that the specific refraction is constant).

For experimental determination of the specific refraction of various gases at high temperatures it is possible to use the refraction index in that region behind the shock front where the gas composition (and the specific refraction of all the components with the exception of the sought constituent) are known [228, 229, 598]. In the case of equilibrium the gas composition is most frequently calculated on the basis of the measured initial conditions and the wave velocity; here one must be sure that the fluid in the region under study is actually at equilibrium.

An appreciable difference in the refraction of the electron and ordinary gases aid in the successful use of interferometric methods in measuring the electron concentration distribution behind the shock front [229] (see Sect. 13).

Of great interest in the study of nonequilibrium excitation of the electronic levels in atoms is the use of the method of "hooks" due to Rozhdestvennyy for study of the distribution of the number of excited atoms in a shock wave, which was undertaken by Tumakayev, Dunayev, Shukhtin and Lazovskaya [59, 202]. As is known, this method of hooks [171] is based on the study of the anomalous dispersion near absorption lines by spectral expansion of the interference fringes picture. The hooks in the fringes appear when an additional plane-parallel plate is introduced into the light beam. Then the atomic number density N_k at the level k which corresponds to the obtained line will depend on the distance Δ_k between the apexes of the hooks which are situated to both sides of the spectral line.

$$N_{\text{vib}} = \frac{\pi m c^2 k_{\text{vib}}}{e^2 \lambda_{\text{vib}}^3 l_{\text{vib}} f_{\text{vib}}} \Delta_{\text{vib}}^2, \tag{10–24}$$

where e and m are the charge and mass of an electron, λ_k is the wavelength of the line under consideration, l_k is the height of the test gas column, f_k is the oscillator power for the pertinent optical transition and k_k is a constant which depends on the thickness and on the refraction index of the plate inserted into the interferometer.

Tumakayev and Lazovskaya [202] have studied the dependence of the number of normal and excited atoms of mercury at the levels $6s^1S_0$ and $6p^3P_{0,1,2}$ on the distance in the flow ahead of the shock wave and behind its front. To obtain Rozhdestvennyy's hooks near a number of mercury lines they have used a combination of an interferometer with a spectrograph with a plane diffraction grating (Fig. 56). The interference pattern was photographed in three sections of the spectrum (near 2537 A, 4047 A–4358 A and 5461 A) using a pulsating light source with a flashing duration of 1–2 microseconds. The spatial resolving power of this installation was determined by the flash duration and the velocity of the flow under study (the interference bands here were not localized at infinity) and comprised about two and four mm for a shock-wave propagation velocity of 1 and 2 km/sec, respectively. Samples of interferograms thus obtained for $\mathsf{M} \sim 9$ are presented in Fig. 57. Only one interferogram, corresponding to a given distance from the shock front, was obtained in each experiment. An appreciable number of excited atoms ahead of the shock wave in mercury vapor was discovered. Thus, for $\mathsf{M} \sim 10$, the atomic number density on the $6p^3P_1$ level ahead of the shock front was about 1×10^{13} cm^{-3}, *i.e.*, about 1% of the maximum num-

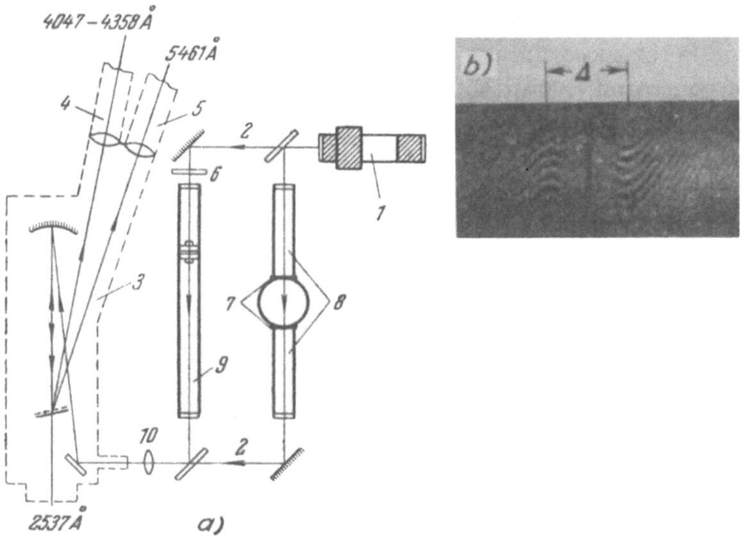

Fig. 56. a) Optical arrangement of the installation described in [59] used in the study of anomalous dispersion by Rozhdestvennyy's method of hooks in mercury vapor behind a shock front. 1) Pulsating light source; 2) Rozhdestvennyy's interferometer; 3) spectrograph with a diffraction grating; 4) and 5) additional photographic cameras for recording the 4047–4358 A and 5461 A mercury lines; 6) is a planeparallel plate for producing the hooks; 7) is the inspection section of the shock tube (a cross-sectional view); 8) are evacuated cylindrical tubes; 9) is a compensator; 10) is an objective lens. b) Is a sample of an interferogram near the 2537 A line in mercury vapor, which fill the shock tube before performance of the experiment (Δ is the distance between the hook apexes).

ber of mercury atoms at this level behind the shock front. Tumakayev and Lazovskaya have discovered that the population of the electronic levels of mercury atoms behind the shock front does not take place immediately, but rather after a certain time. The scattering (up to 15%) which was observed by the authors is due to errors in measuring $N_k f_k$ (about 5–6%) and of the rate of propagation of the shock wave (1–2%), inaccuracy in determining the location of the flow region being photographed relative to the front and, finally, by the effect of the unchecked (within the limits of one percent) air admixture.

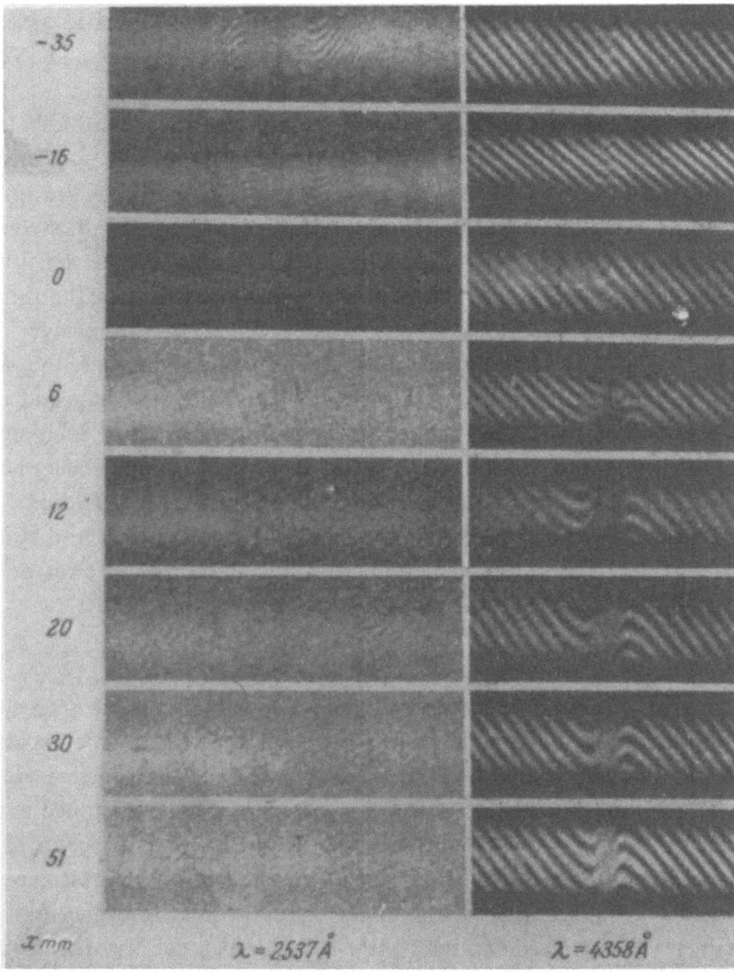

Fig. 57. Samples of interferograms with Rozhdestvennyy's hooks which characterize changes in the dispersion near the 2537 A resonance line (the first column) and in the line of the visible triplet for mercury 4358 A (second column), as a function of the distance x from the shock front in mercury vapor for $M = 9$. The negative values of x are for the region ahead of the shock front.

[4] The Electron Beam Method

As the initial pressure is made smaller, the sensitivity of the shlieren and interferometer methods of density measurement drops. Hence other methods, including that of measuring the density distribution in the shock tube by the scattering of an electron beam passed through the test gas perpendicular to the tube axis, have been developed for the study of shock-wave structures at low pressures. Some results of density measurements in rarefied gases under stationary conditions and in steady-state flows are described in [524, 528, 386].

In flows with energies of the order of ten and several tens of keV the electrons which collide with the molecules and atoms of the gas undergo a primarily elastic scattering. Let us consider the distribution of the amount n_y of electrons along a beam which is emitted by a source with a diameter F_s and which is incident on a receiver with a diameter F_{rec}, when the gas density is constant (Fig. 58). Let $F_{rec} \geqslant F_s$; then at small distances, *i.e.*, in the immediate vicinity of the source, all the electrons are received by the receiver and the observed signal is steady. Then, as the distance is increased, the signal is incident at an angle as a result of a single electron scattering and the fact that they move away from that region of the beam which is incident at the receiver. At a certain distance y_d the incidence of the signal becomes decelerated, *i.e.*, as a result of multiple scattering in the entire gas volume, a part of the electrons which earlier left the beam now return to it and is incident on the receiver. The subsequent reduction in the number of electrons in the beam is an exponentional function of the distance [524, 528]:

$$n_y = n_{y_d} e^{-\alpha(y-y_d)}; \qquad (10\text{–}25)$$

where α is an attenuation coefficient which depends on the electron energy in the beam and very little on the atomic weight of the gas which scatters the electrons; the magnitude of α is practically proportional to the gas density ϱ. For electron energies of 16.5 keV and $p=30$ mm Hg, the distance y_d in air is about 4 cm, *i.e.*, it is comparable with the shock-tube diameter. Thus, in quantitative measurements of density using an electron beam it is necessary to consider the possibility of the effect of the geometric dimensions on the electron scattering effects. Under these conditions a decisive role is assumed by statistical calibration of the readings of the installation used. For this purpose the shock tube is filled by the test gas under different pressures and the readings of the receiver are recorded.

The use of the electron beam for measuring the gas density in a shock tube started with the works by Venable, Kaplan, Ballard and others [585, 239, 317]. The experimental arrangement is simple: an electron beam produced by an electron gun is passed through the test section of a shock tube and is received by a scintillator of a photomultiplier. Here we shall describe in more detail the installation described in [33]. The electron gun with an accelerating potential of 30 keV is assembled on an electron-beam tube projector; the beam focusing and deflection system uses deflection plates of the type employed in television tubes. The electrons enter the low-pressure chamber through a hole with a diameter of about 1 mm, which is covered by

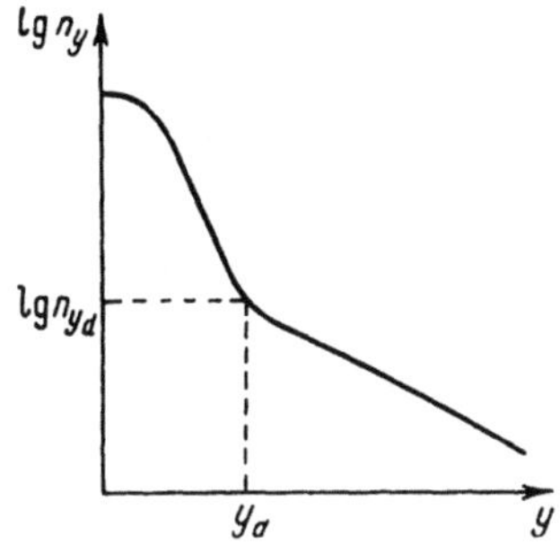

Fig. 58. Distribution of the logarithm of the number of electrons n_y along a beam with a source and receiver with finite dimensions. An equilibrium between the singly-scattered electrons and with electrons which are returned to the beam as a result of multiple scattering is established at the distance y_d.

an aluminum foil several microns thick. This arrangement is preferable to that used by Duff [317], which consisted of an open inspection hole (with a 20-micron diameter hole in platinum foil), which distorts the readings of the gas influx from the low-pressure chamber into the electron gun volume at the time of shock-wave passage. The photomultiplier was screened from the optical radiation of the gas behind the shock front by coating the scintillator by a thin layer of aluminum. The signal from the photomulplier is fed to an oscillograph. Specimens of oscillograms obtained with this installation are shown in Fig. 59. The error in measuring the absolute density magnitude due to current fluctuations and random errors on transition from calibration to measurements did not exceed 9–10%; the relative density values were measured with a smaller error. The investigation of density distribution in the plug at different distances from the diaphragm performed by Busygin and Tumakayev [33] has produced interesting information on the shock-wave formation.

Appreciable difficulties in density measurement by the electronic beam are encountered in the gas when an attempt is made to determine the density

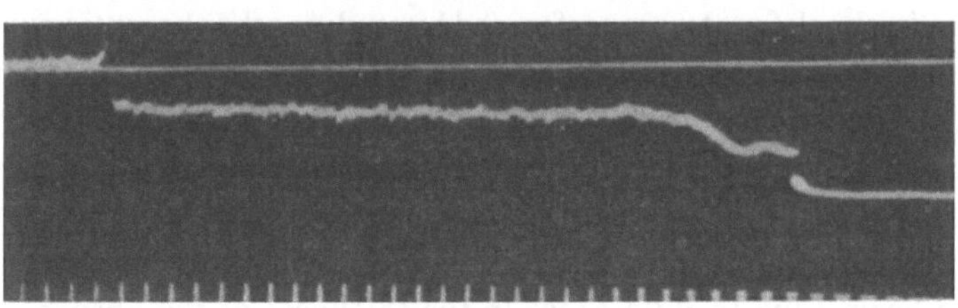

Fig. 59. Oscillogram of the change in intensity of an electron beam in the flow behind a shock front in argon with M ~ 3 [33]. The straight line is the signal level without scattering of electrons in the gas behind the shock front. At the time of front arrival the signal drops sharply; the appearance of a contact region in the test section results in a further smooth reduction in the signal. The sinusoidal line gives the time marks.

distribution in polyatomic gases under molecular disintegration and conditions of new compound formation, since it is difficult to determine the cross section for scattering of electrons by a complex molecular mixture by simple calibration. The cross sections for scattering of electrons by the individual components of such a mixture using a beam energy of several tens of keV have not as yet been reliably determined; it is also difficult to take into account changes in the state of the gas. All this should be remembered when setting up experiments for the study of the state of a polyatomic gas behind the front of a strong shock wave using an electron beam.

Studies on neutral argon with a high time resolution (fractions of a microsecond) made it possible to obtain some information about the wave-front structure up to $M=10$ [585, 239, 320, 277a]. Study of the density distribution, plug length and other characteristics of the shock wave with argon at low initial pressure (up to 0.5 mm Hg) with a resolution of about 3 microseconds has exposed particular features of processes in a shock tube under a low pressure [317].

Ivanov [75] has used the electron beam to study the density distribution in the front of as tanding shock wave which forms in front of a body placed in a supersonic flow of rarefied air. He noted a perceptible increase in the shock-front thickness compared with the value predicted by Mott-Smith [454].

To measure the gas density behind a shock front in a tube, McChesney [440] has suggested the use of electrons which are a product of beta decomposition. It has been calculated that to ensure an accuracy (standard deviation) of not worse than 5% during 2 microseconds under usual shock-tube conditions, it is necessary to have a very high activity source (about 100 millicuries).

[5] The Use of X-Ray Radiation

A number of successful experiments dealing with gas-density determination behind a shock front were performed by studying the absorption with a soft x-ray radiation. The change in the intensity P of a beam of monochromatic x-rays upon passing through a layer d of a substance is governed by $P=P_0 \exp(-\mu_m \varrho d)$, where μ_m is the attentuation coefficient which determines the reduction in the intensity of the primary beam by a unit mass of the absorbing medium [85]. The above relationship cannot be used for a highly diverging beam (due to differences in the thickness of the absorbing layer) and for wide beams, when a certain part of the scattered and characteristic radiation is added to the primary beam. The value of μ_m is independent of the physical state of the fluid; for complex chemical compounds of the type of $X_a Y_b Z_c$, the total molecular attenuation coefficient μ_m is obtained additively from atomic attentuation coefficients*

$$\mu_m = a\mu_X + b\mu_Y + c\mu_Z. \tag{10–26}$$

This is also valid for a mixture of gases in which chemical transformations take place, which gives the x-ray methods of gas-density measurement great advantages over the interferometer, Tepler and the electron beam methods.

The attenuation of a beam of x-rays is determined by photoelectric absorption when the incident energy is transformed into the kinetic energy of the displaced photoelectron and into the potential energy of the excited atom as well as by scattering. The scattering coefficient is usually much smaller than the absorption coefficient; for $\lambda > 0.5$ A and elements heavier than iron they can be disregarded. Some idea about the character of the passing of x-ray radiation through a gas is given by Fig. 60. It can be seen from the figure that light gases exhibit absorption only in the spectral region with larger wavelengths. When working with nonmonochromatic radiation it is possible to refer to the effective attenuation coefficient, which depends on the spectral composition of the incident radiation. In this case the values of the effective attenuation coefficient are obtained most reliably by static calibration with the same source which is used for measurement. When working with shock tubes, it is customary to use nonstandard x-ray tubes with tungsten, copper and chromium targets; the accelerating potential is 20–30 kV, which makes it possible to obtain a continuous spectrum (in the direction perpendicular to the cathode beam) with a maximum of $\lambda \approx 16{,}130/U$, where U is the accelerating potential in volts, alongside the characteristic spectra [85, 211]. Knight and Venable [407] have described a demountable

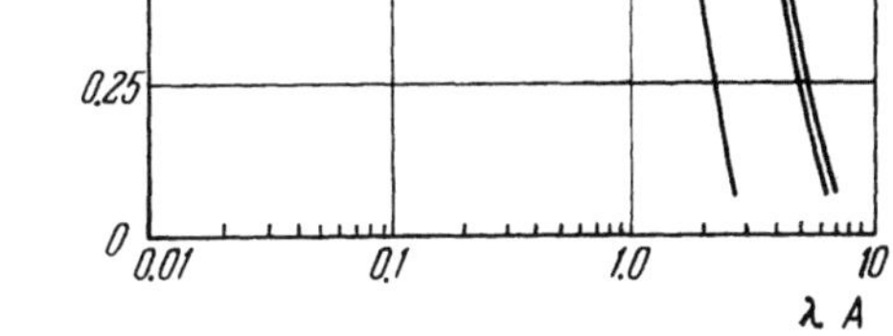

Fig. 60. Dependence of the transmittance P/P₀ of x-ray radiation by various gases on the wavelength at a pressure of 10 atm, temperature 6000°K and gas column length of 5 cm.

* The change in the physicochemical state of a substance has a perceptible effect on the fine structure of x-ray absorption spectra [35]; these effects are not considered here.

tube which uses radiation pulses with an accelerating potential of 26–28 kV and a current of 0.15–0.2 A; the outlet aperture of the tube they used was covered by beryllium foil. The tube was installed directly in the test section of the shock tube. The radiation was received by a phosphorescent converter installed on the photomultiplier; the resolving time was about 1 μsec. Knight and Venable have studied the absorption in a heavy inert gas and have come to the conclusion that this method can be used successfully to measure the density distribution behind a shock front in xenon and krypton. Subsequently, other investigators (Rink, Chesik and others [291, 491, 492, 493]) have successfully used the x-ray absorption method to measure the rate of disintegration of O_2, H_2 and D_2 molecules, by adding to these gases a heavy inert diluting gas, since the absorption of x-rays in the above gases is insignificant for the selected wavelengths. The use of radiation with a higher wavelength for the study of absorption in the test gas proper is made difficult by the need to use a very high intensity x-ray beam, etc. Despite the use of a heavy diluting gas, which comprised up to 30% of the mixture composition, it was still possible to obtain interesting data on the rate of disintegration of hydrogen, deuterium and oxygen molecules.

[11] ABSORPTION METHODS OF MOLECULAR CONCENTRATION MEASUREMENT

The composition of a gas in the nonequilibrium region behind a shock front can be studied by spectral absorption analysis methods. In the study of shock waves these methods are used to analyze processes in polyatomic gases which have intensive absorption spectra in the visible, close or far ultraviolet spectral region; several investigators have studied the absorption spectra in the infrared region, however, the number of these investigations is small [308]. A summary of a number of works concerned with the study of nonequilibrium phenomena in shock waves by spectral absorption analysis methods in di- and triatomic gases is given in Table 5. Instructions and

Table 5. Works Concerned With the Study of Relaxation Processes in Di- and Triatomic Gases by Absorption Spectroscopy Methods

Light-absorbing gas	Wavelength, A	Temperature region, °K	Phenomenon studied	References
Oxygen	2245	1200–10,000	Vibrational relaxation, dissociation in a pure gas and in mixtures with other gases	[41–44] [104] [105, 109]

Table 5. *Continued*

Light-absorbing gas	Wavelength, A	Temperature region, °K	Phenomenon studied	References
	2272	3000–3500	Dissociation	[102]
	1470	1200–8000	Vibrational relaxation and dissociation in a mixture with argon	[276] [277, 278]
	1470	5000–18,000	Simultaneous excitation of vibrations and disintegration in mixtures with argon	[615]
Chlorine	3400–4200	1550–2650	Dissociation in mixtures with argon	[373] [389]
Bromine	4390	1010–2225	Dissociation in the pure gas and in mixtures with argon	[463]
	4360–4870	1400–2700	Same as above	[271]
	4360	1400–1990	Same	[374]
Iodine	4660, 5050, 6600	760–5600	Dissociation in the pure gas	[48]
Iodine	4360, 4900	1060–1860	Dissociation in a mixture with argon, nitrogen and other gases	[269, 270]
Nitric oxide	2360	450–1300	Excitation of vibrations in the pure gas and in mixtures with argon	[494]
	2465	3000–4300	Decomposition in a mixture with argon	[324]
	1270	1500–8000	Excitation of vibrations in decomposition in the pure gas and in mixtures with argon	[616, 614]
Hydroxyl OH	3094	2400–3200	Decomposition of water vapor	[244]
CN	3883, 4216	1700–2500	Decomposition of the cyanogen gas	[608]
C_2N_2	2188		Same	[608]
H_2	1148, 4	2950–5300	Dissociation in mixtures with argon	[468]
NO_2	4358	300	Decomposition of N_2O_4	[285]
	4360	600–1038	Decomposition of NO_2Cl	[374]
	4500, 4360, 5460	1350–2400	Decomposition of NO_2 in a mixture with argon	
NO_2, NO_3	3660, 4050, 4360, 5460, 6520	450–1200	Decomposition of N_2O_5 in a mixture with argon	[525]
CO_2	2380, 3000	1300–6300	Decomposition of CO_2	[49, 111]
SO	2548–2622	3000–4000	Decomposition of SO_2 in a mixture with argon	[329]

references to other works in which the absorption of light by diatomic gases was studied can be found in [192].

[1] Dependence of the Absorption on the Molecular Concentration

As is known, the spectral absorption analysis is based on two main propositions: 1) the light energy losses dI in a homogeneous medium are proportional to the thickness of the infinitesimally thin layer dx and are independent of the energy of the incident light (the Bouguer-Lambert law, substantiated by Vavilov's experiments); and 2) for monochromatic radiation dI is proportional to the number of absorbing molecules (atoms) in a layer dx (Baire's law). Thus, if n is the number density of particles of the absorbing fluid, then for the intensity of monochromatic radiation $I_{\nu'}$ with the wave number ν'

$$\frac{\mathrm{d}I_{\nu'}}{I_{\nu'}} = -\sigma_{\nu'} n \, \mathrm{d}x . \tag{11–1}$$

After integrating for a finite thickness of the layer d

$$I_{\nu' d} = I_{\nu' 0} e^{-\sigma_{\nu'} n d} , \tag{11–2}$$

where $I_{\nu' 0}$ is the intensity of the incident radiation, $\sigma_{\nu'}$ is the absorption cross section per one molecule in the volume. The product $\sigma_{\nu'} n$ is frequently called the absorption coefficient per unit length of optical path (or simply, the absorption coefficient) and is denoted by $k_{\nu'}$. The absorption of light by some mth line which occupies some interval $\Delta \nu'_m$ along the spectrum can then be described by the ratio of $W_{\nu' m}$ of the absorbed fraction of intensity to the intensity of the incident radiation in the same interval. Here it is assumed that $I_{\nu' 0}$ is constant. Then, according to Eq. (11–2),*

$$W_{\nu' m} = \int_0^\infty \frac{I_{\nu' 0} - I_{\nu' d}}{I_{\nu' 0}} d\nu' + \int_0^\infty (1 - e^{-k_{\nu'} d}) \, d\nu' . \tag{11–3}$$

The quantity $W_{\nu' m}$ is called the total absorption of the line (or sometimes also the equivalent line width) [30, 125, 204, 216]**.

We now use the Ladenburg-Millikan relationship

$$\int k_{\nu' m} d\nu' = \frac{\pi e^2}{m_e c^2} N_k f_e , \tag{11–4}$$

* In these calculations it is usually assumed arbitrarily that the spectrum consists of one line only; then $\Delta \nu'_m \to \infty$ and the limits of integration move out to infinity.

** If the distribution of parameters over the spectrum is specified by the frequency ν (see, for example, Sect. 12), then the total absorption of a line is $W_{\nu m} = c W_{\nu' m}$, where c is the speed of light.

where e and m_e are the charge and mass of an electron, N_k is the molecular number density of the starting level, f_e is the oscillator force for the optical transition under study. Then in the limiting case of very weak absorption ($k_{\nu'}d \ll 1$) we will get that $W_{\nu'm}$ is proportional to N_k, *i.e.*, for weak lines the absorptivity is independent of the form of the line and is proportional to the molecular number density N_k at the starting level. In the opposite case of very strong (reabsorbed) lines it is found [30, 125, 204] that $W_{\nu'm} \sim \sqrt{N_k f_e \Delta_{\nu' int} d}$, where $\Delta_{\nu' int}$ is the half-width of lines which is determined by the molecular interaction. Thus for very strong lines the absorptivity is proportional to the square root of the molecular number density at the starting level. In the intermediate cases the value of $W_{\nu'm}$ can be obtained only by numerical integration. Tables of values of $W_{\nu'm}$ in a wide range of values of $k_0 d$ (k_0 is the maximum value of $k_{\nu'}$ in the line when only the Doppler widening occurs) and

$$a = \frac{\sqrt{\ln 2}\Delta_{\nu' int}}{\Delta \nu'_D}$$

(ratio of the half-width determined by the molecular interaction to the Doppler half-width $\Delta \nu'_D$) can be found, for example, in [191], and a graphical representation is contained in [204, 216] and in a number of other works. The value of k_0 is determined by the probability of the optical transition and, for example, in the particular case of Schumann-Runge bands of the O_2 molecule for the equilibrium state of the gas it is equal to:

$$k_0 = \frac{2}{\Delta \nu'_D}\sqrt{\frac{\ln 2}{\pi}}\frac{\pi e^2}{m_e c^2} n f_e \frac{\nu'}{\nu'_{\max}} q S_{J''} \frac{e^{-[G_0(v'') + B_{v''}J''(J''+1)]\frac{hc}{kT}}}{\frac{1}{3}Q(T)}, \tag{11-5}$$

where v'' and J'' are the numbers of the vibrational and rotational levels, q is the Frank-Condon factor (the probability of changes in the vibrational state), $S_{J''} = J'' + 1$ for the R branches and $S_{J''} = J''$ for the P branches, $Q(T)$ is the total statistical sum, $B_{v''}$ is the rotational constant for the v'' level; $G_0(v'')$ is the height of the v'' vibrational level over the zero level, and $\nu'_{\max}$ is wave number of the absorption maximum in the system. The value of f_e is determined by the square of the matrix element of the electron dipole transition moment

$$f_e = \frac{8\pi^2 m_e c}{3h} R_e^2 G \nu'_{\max}, \tag{11-6}$$

where G is the frequency of degeneration of the final state.

In practice, the spectral instrument with an apparatus function $\psi(\nu')$ generates a certain frequency interval $\Delta \nu'$, which usually contains a large number of lines. In this case A, the absorption power recorded by the instrument is

$$A = \frac{\int\limits_{\Delta\nu'} \psi(\nu')\,(1 - e^{-k_{\nu'} d})\,d\nu'}{\int\limits_{\Delta\nu'} \psi(\nu')\,d\nu'}. \qquad (11\text{–}7)$$

A peculiar feature of molecular spectra of gases at high temperatures is the presence of a very large number of lines, a part of which may overlap. The temperature increase and the populating of excited vibrational, rotational and electronic levels which accompanies it results in such an increase in the number of lines that the dependence of $k_{\nu'}$ on ν' in a certain range $\Delta\nu'$ becomes very complex; and an exact analytical determination of $k_{\nu'}$ becomes practically impossible. Some success in analyzing the absorption in this case can be obtained by using some model of a random distribution of lines in the spectrum, taking into account their overlapping [88]. When lines overlap and the bands do not have sharply defined edges, the spectrum will approach continuum; in this case $k_{\nu'} = \text{const}$ in the interval $\Delta\nu'$ and Eq. (11–7) reduces to

$$A = 1 - e^{-kd}. \qquad (11\text{–}8)$$

In a number of cases it is possible to study the absorption directly in the continuous spectrum which is formed by the bound-free transitions from the basic or the excited state. It is obvious that for the small intervals $\Delta\nu'$ (when it may be assumed that $k_{\nu'} = \text{const}$) Eq. (11–8) is also valid in the continuous spectrum. The absorption coefficient k which is contained in Eq. (11–8) is, according to Baire's law, proportional to the molecular number density n. When there is equilibrium with respect to all degrees of molecular freedom, it is convenient to represent k in the form of the product $k = \sigma n$, where σ is the absorption coefficient per molecule in a unit volume, which depends on the temperature; $\sigma = \sigma_0 \varphi(T)$. The value of φ depends on the distribution of molecules among the levels under study; if this is a Boltzmann distribution, then φ is a Boltzmann multiplier which depends (for the vibrational degrees of freedom) on the vibrational temperature T_{vib}. In the particular case when no equilibrium exists between the translational and vibrational degrees of freedom, T_{vib} is not equal to T. The value of σ_0 is determined by the probability of an optical transition which depends, in particular, on the magnitude of the square of the matrix element of the electron dipole moment R_e^2 for the frequency range under consideration. At present, it is impossible to determine theoretically the values of R_e^2 for molecular systems even for diatomic molecules with any degree of reliability; the efforts of a large number of investigators are directed toward the experimental determination of the above quantity for various systems of molecular spectra. Hence the most natural way for using spectral methods for the study of nonequilibrium phenomena in shock waves has become the direct measurement of absorption coefficients in shock-tube experiments in order to determine the absolute values of k as well as the dependence of k on the temperature and on the molecular number density.

[2] The Ultraviolet Spectral Region

Specifically, let us consider the methods used in the study of vibrational excitation and dissociation of oxygen and carbon dioxide molecules which were developed by Losev and Generalov [102, 41], etc. It is well known that oxygen heated to a high temperature has an appreciable absorption capacity in the quartz ultraviolet spectral region as a result of absorption in the bands of the Schumann-Runge system [103, 109, 110]. Heated carbon dioxide absorbs light in this spectral region to even a greater extent [49]*. The absorption of ultraviolet radiation was recorded by the following optical arrangement (Fig. 61). The source of ultraviolet radiation with a continuous spectrum was a quartz arc-discharge lamp, filled with xenon at a high pressure. Using a quartz lens, the light was directed perpendicular to the shock-tube axis and was passed through the observation section of the tube; the transverse dimensions of the light beam were limited by diaphragms to 0.2–0.3 mm by width and 15 mm by height. A second quartz lens

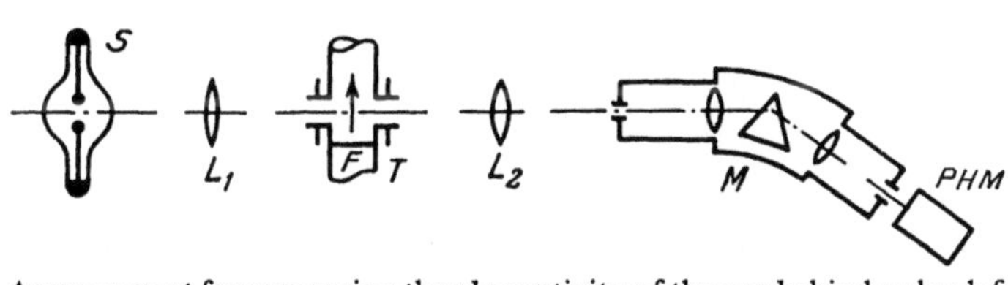

Fig. 61. Arrangement for measuring the absorptivity of the gas behind a shock front with photoelectric recording. S is the light source; L_1 and L_2 are lenses, M is a monochromator; PHM is a photomultiplier; T is the shock tube and F is the shock front.

has projected the output diaphragm at the inspection section of the tube at the inlet slit of the quartz monochromator with a 60-degree Corneau prism; the dispersion of the instrument for $\lambda = 3000$ A was about 30 A/mm. Behind the output slit of the spectral instrument there was installed a photomultiplier with an aperture closed with ultraviolet-transmitting glass which made it possible to work in the spectral region of $\lambda \geqslant 2200$ A. The signal from the load resistance was fed through a cathode repeater to an oscillograph. To

* The possibility of appearance of an appreciable absorption of light by the carbon dioxide in the quartz ultraviolet spectral region was predicted by Gaydon [39].

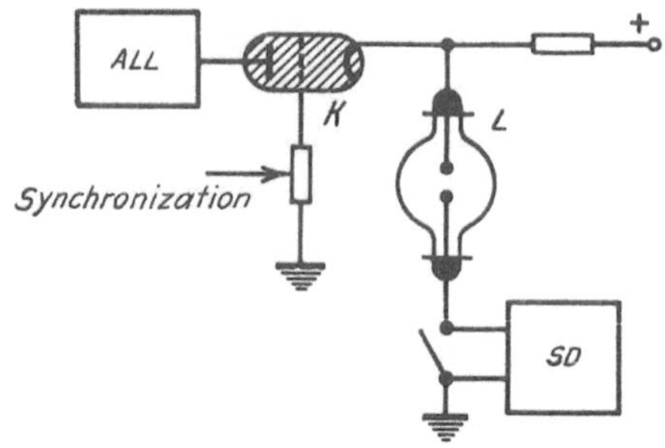

Fig. 62. Block diagram of the supply circuit of an arc-discharge light source L with an additional current pulse at the instant when the shock wave passes the inspection section of a shock tube. SD is a starting device for igniting the lamp; ALL is an artificially produced long line for the formation of an additional current pulse and K is a thyratron key.

increase the signal-to-noise ratio, at the time when the shock wave passes the inspection holes an additional, practically rectangular high-current pulse (up to 500 A) was supplied to the arc-discharge lamp. For this they have employed an arrangement shown in Fig. 62; its main elements being a light source consisting of an artificially produced long line, synchronizing pulse thyratrons and a starting device for igniting the lamp in the steady-state regime. The transmission band for frequencies of the above arrangement has extended to 14 Mc/s which, according to the presentation in Sect. 8, has provided a resolving power in time of about

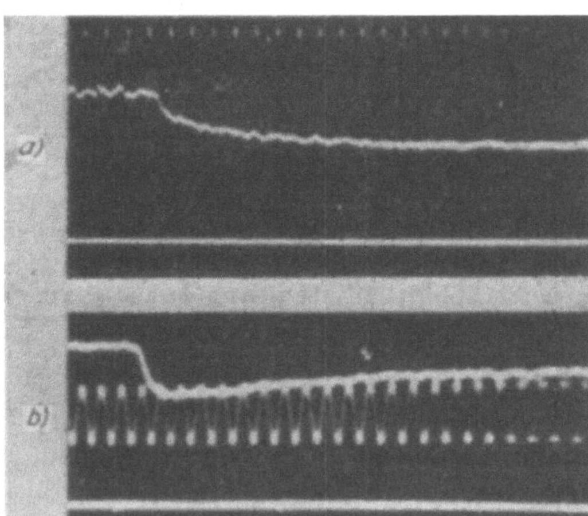

Fig. 63. Specimens of oscillograms of intensity changes in ultraviolet probing radiation ($\lambda = 2245$ A) behind a shock front propagating in oxygen. a) With the velocity $V = 2.1$ km/sec with $p_1 = 1.5$ mm Hg; b) with the velocity $V = 3.13$ km/sec with $p_1 = 1.5$ mm Hg. The time marks are made each microsecond. At the bottom is a zero line obtained when no signal is present.

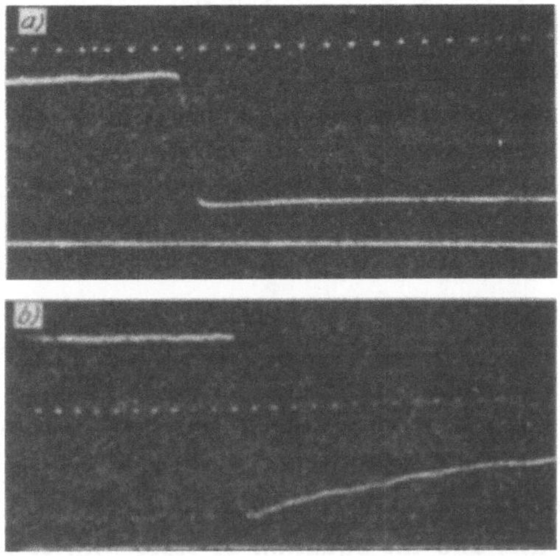

Fig. 64. Specimens of oscillograms of intensity changes in ultraviolet probing radiation ($\lambda = 2380$ A) in carbon dioxide. a) $V = 2.65$ km/sec, $p_1 = 7.5$ mm Hg; b) $V = 3.05$ km/sec, $p_1 = 5$ mm Hg. Time marks denote each second. At the bottom is the zero line.

0.025 μsec. When measuring small values of the absorptivity use was made of an oscillograph with a linear amplitude characteristic for signals which occupy 80 mm on the screen.

Samples of oscillograms of the absorptivity distribution behind a shock front in O_2 and CO_2 thus obtained are presented in Figs. 63 and 64. After the shock wave appears in the observation section of the tube, the signal begins to fade, *i.e.*, the absorption starts; in oxygen for comparatively moderate wave velocities, this reduction in the signal takes place sufficiently smoothly (Fig. 63a). When the shock-wave velocities are higher, after a more rapid reaching of the absorption maximum, the absorption is smoothly decreased (the signal becomes stronger). In oxygen this is followed by the establishment of some constant absorption level; in carbon dioxide the absorption reduction did not end until the end of the plug. Finally, after the arrival of the contact region, the signal level has returned to the initial value (this section is not seen on the oscillograms which are given here). Checking experiments have shown that the radiation of the test gases due to the high intensity of the translucent pulse (when the supply potential of the photomultiplier is low) does not make any perceptible contribution to the observed signal.

In the range of velocities behind the shock front under study one should expect an intense excitation of vibrations and of dissociation of O_2 and CO_2 molecules. In the wavelength interval under consideration ($\lambda = 2245$ A for O_2,

$\lambda = 2380$ A and $\lambda = 3000$ A for CO_2) the cold gas does not absorb light; in this spectral region only vibrational excited molecules can absorb light. Hence it is natural to relate the observed absorption increase to the process of vibrational excitation and the reduction in absorption to the dissociation of vibrationally excited molecules.

To determine the dependence of the absorptivity A on the temperature and on molecular number density, it is necessary to measure these three quantities simultaneously. The values of A were found from the oscillograms as the relative intensity reduction after the shock-wave arrival; the values of T and n had to be determined separately. Temperature measurement by the pyrometric method (see the following section) can give the desired result only for points sufficiently removed from the shock front; the measurement of n by an independent method is difficult even far from the front. Hence, in determining the temperature and the molecular number density, it is necessary to use the conservation laws and the equation of state with certain assumptions regarding the character of processes in the shock wave. Thus, one may assume that the constant absorption cross section (in oxygen) corresponds to the equilibrium state; this is indicated by measurements of the gas temperature and density under similar conditions [205, 327, 254, 437]. The next assumption states that the relaxation of vibrational energy in the maximum absorption region has basically ended and that appreciable dissociation has not as yet started. This breaking up of the region behind a shock front into individual zones of vibrational excitation and of dissociation is, as was noted in Chapter 1, somewhat arbitrary and is not always in accord with reality, since actually it should be expected that both these processes overlap in this region. Calculations performed in [108, 282, 316] and subsequent experiments show that in oxygen with a wave velocity of less than 4 km/sec this overlapping is insignificant. The basis for this assertion is the large difference in the relaxation times for vibrational energy and the characteristic dissociation time for oxygen at temperatures below 7000°–8000°K [108]. Even a greater difference between the rate of vibrational excitation (at least of the excitation of deformation and symmetric valent vibrations) and the dissociation rate apparently takes place in carbon dioxide, which is indicated by the measured time of vibrational relaxation in CO_2 in a number of works [544, 387, 611] and the form of oscillograms of the absorptivity distribution behind a shock front in carbon dioxide which were obtained. This is the viewpoint postulated by Gaydon and Hurle [327], who have measured the temperature distribution behind the shock front in carbon dioxide. They assume that even antisymmetric valent vibrations of the CO_2 molecule succeed in being excited before the dissociation starts. However, no direct proof of these assertions exists.

The measurements made and the assumptions make it possible to determine the dependence of A on T and n for O_2 and CO_2 in a wide range of values of arguments ($1200° \leqslant T \leqslant 7000$ °K and $1.3 \cdot 10^{17} \leqslant n \leqslant 2 \cdot 10^{18}$ cm^{-3} for oxygen and $1300° \leqslant T \leqslant 6300$ °K and $2.2 \cdot 10^{17} \leqslant n \leqslant 3.2 \cdot 10^{18}$ cm^{-3} for carbon dioxide). It was discovered that in both oxygen and carbon dioxide the absorption obeys Baire's law (11.8) $A = 1 - \exp[-\sigma(T)nd]$; the values of $\sigma(T)$ thus obtained are given in Fig. 65. In oxygen a contribution to the

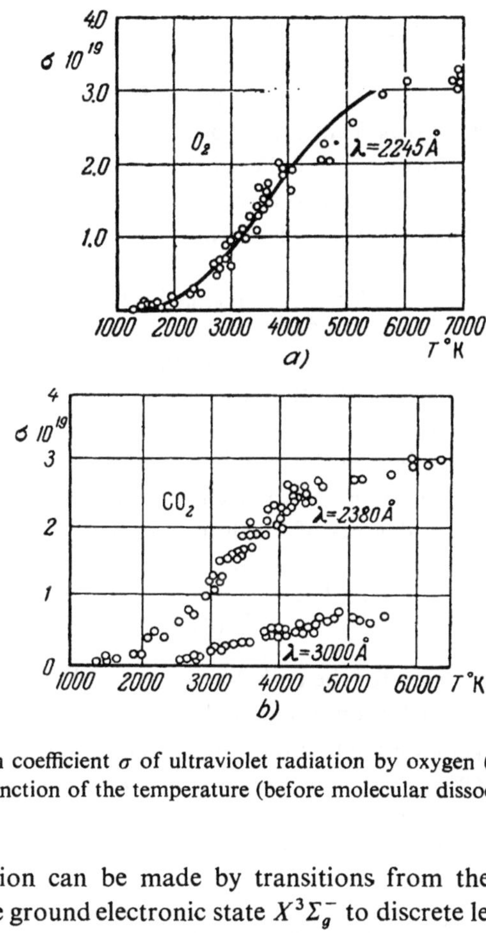

Fig. 65. Absorption coefficient σ of ultraviolet radiation by oxygen (a) and carbon dioxide (b) as a function of the temperature (before molecular dissociation starts).

observed absorption can be made by transitions from the excited vibrational levels of the ground electronic state $X^3\Sigma_g^-$ to discrete levels of the state $B^3\Sigma_u^-$ (band spectrum) or to a continuum of energies the maximum of which exceeds the dissociation limit from the $B^3\Sigma_u^-$ state (continuous spectrum). The main contribution to the absorption in O_2 with $\lambda = 2245$ A is given by transitions from the fourth and fifth vibrational levels of the $X^3\Sigma_g^-$ state; the calculated value of $\sigma(T)$ is shown in Fig. 65 by a solid line. Identification [of levels] in carbon dioxide is more difficult, one may only put forward a hypothesis that an extremely intensive continuum which is determined by the transitions $1\pi_g \rightarrow 3\sigma_u^1$ increases in width with an increase in the temperature.*

[3] Determining the Relaxation Time and Dissociation Rate

Knowing the temperature dependence of the absorptivity, it is possible to determine the times of vibrational relaxation and the dissociation rates by using the results of measurements of the distribution of absorption in the gas behind the shock front and Eqs. (9–1)—(9–5), (9–13) and (9–14) given in Sect. 9.

First, let us consider the problem of determining the vibrational relaxation time in oxygen. It follows from Eq. (9–30) that to determine τ_{vib} it is necessary to know the distribution of the vibrational energy of a unit gas mass ε_{vib} and the translational temperature T behind the shock front. We write the enthalpy of a mass of gas in the form

$$h = \frac{7}{2}\frac{RT}{\mu} + \varepsilon_{\text{vib}}(T_{\text{vib}}). \qquad (11\text{–}9)$$

It is here assumed that in the process of approaching equilibrium, the distribution of molecules by vibrational levels remains a Boltzmann distribution with a temperature T_{vib} approaching T. This assumption may be regarded as valid when the transfer of vibrational photons between molecules takes place at a higher rate than the energy transfer between the translational and vibrational degrees of freedom (see. Sect. 17). Then

$$\varepsilon_{\text{vib}}(T_{\text{vib}}) = \frac{R}{\mu}\frac{\theta'}{\left(e^{\frac{\theta'}{T_{\text{vib}}}} - 1\right)}. \qquad (11\text{–}10)$$

To the above expression for the enthalpy we can add the term $\varepsilon_e(T_{\text{vib}})$ which accounts for the contribution of energy by excited electronic levels which is a function of T_{vib}, *i.e.*, assuming that the populating of the lower electronic levels of O_2 proceeds at the same rate as the vibrational excitation. Since the contribution of ε_e to the enthalpy up to temperatures of $T < 7000\,^\circ$K is insignificant, the error which this produces is minor.

By simple transformations we obtained from Eqs. (9–1)—(9–4) and (11–9) a relationship between the vibrational temperature of the gas and the density ratio:

$$\left(1 - \frac{\varrho_1}{\varrho_2}\right)\left[\frac{V^2}{2} + \frac{7}{2}\frac{RT_1}{\mu} - 3V^2\frac{\varrho_1}{\varrho_2}\right] = \varepsilon_{\text{vib}}(T_{\text{vib}}). \qquad (11\text{–}11)$$

* It was established in special experiments that neither the admixtures (which comprised less than 0.01 % in O_2), nor the products of dissociation (O, CO, etc.), makes an appreciable contribution to the observed absorption.

We now write the expression for the absorptivity obtained in the preceding chapter in the form

$$A = 1 - e^{-\sigma_0 \varphi(T_{\mathrm{vib}}) \frac{\varrho_2}{\varrho_1} \frac{p_1 d}{k T_1}} \equiv 1 - e^{-a \varphi(T_{\mathrm{vib}}) \frac{\varrho_2}{\varrho_1}} \tag{11–12}$$

where a is a quantity, constant for a given experiment, which depends on its initial conditions. To establish the relationship between T_{vib}, ϱ_2/ϱ_1 and A, we must expand the expression for the Boltzmann multiplier $\varphi(T_{\mathrm{vib}})$. This relationship is quite complicated, since it must take into account the contributions of many lines which belong to different bands. However, it is possible with sufficient accuracy to select such an expression for $\varphi(T_{\mathrm{vib}})$ which would be in satisfactory agreement with the values of σ obtained in the same experiments (Fig. 65). As such an expression we can use the relationship for the population of the vth level of an harmonic oscillator

$$\varphi(T_{\mathrm{vib}}) = \left(1 - e^{-\frac{\theta'}{T_{\mathrm{vib}}}}\right)^{-1} e^{-\frac{v\theta'}{T_{\mathrm{vib}}}}. \tag{11–13}$$

In the study of vibrational excitation in oxygen molecules using light absorption in the region of $\lambda = 2245$. A good agreement of this expression with experimental results up to $T \sim 7000\,^{\circ}\mathrm{K}$ is obtained with $v = 4.6$; the quantity v has than the meaning of an effective level number.

Then it is necessary to solve simultaneously Eqs. (11–11) and (11–12) making use of Eqs. (11–10) and (11–13) in order to determine ϱ_2/ϱ_1 and T_{vib} for each time instant behind the shock front as a function of the experimentally obtained absorptivity distribution. This solution is performed numerically; specimens of distributions of ϱ_2/ϱ_1 and $\varphi(T_{\mathrm{vib}})$ behind a shock front for one of Generalov's experiments [41] are given in Fig. 66a. Having obtained the density ratio distribution it is possible to change from the laboratory to the natural time (Sect. 9). Knowing the magnitude of T_{vib}, we can now use Eq. (11–10) to find the distribution of $\varepsilon_{\mathrm{vib}}(T_{\mathrm{vib}})$ behind the shock front (Fig. 66b) and on the basis of the slope of the tangent to the curve $\varepsilon_{\mathrm{vib}}(t)$ at each point we can find the magnitude of $d\varepsilon_{\mathrm{vib}}/dt$. Finally, from the laws of conservation of mass and momentum fluxes, using Eq. (9–4) we obtain a relationship between the distribution of ϱ_2/ϱ_1 and the translational temperature T:

$$T = \frac{\varrho_1}{\varrho_2} T_1 + \frac{\mu}{R} V^2 \frac{\varrho_1}{\varrho_2}\left(1 - \frac{\varrho_1}{\varrho_2}\right). \tag{11–14}$$

This makes it possible, on the basis of the values of ϱ_2/ϱ_1 thus obtained, to find the temperature distribution in the relaxation zone behind the shock front (Fig. 66b) and to determine in each point the values of the equilibrium vibrational energy $\varepsilon_0(T)$. Thus all the terms of Eq. (9–30), with the exception of τ_{vib}, are determined. It remains to determine τ_{vib}; the distribution of

τ_{vib} in the same experiment is given in Fig. 66c. It is seen from the figure that the relaxation time for vibrational energy behind a shock front does not remain constant but increases, this is obvious, since at a constant pressure the relaxation time should actually increase with a drop in temperature.

Using the above approach the distribution of the absorptivity behind a shock front was used for obtaining the relaxation times for vibrational energy of O_2 in a wide temperature range (T=1200–10,000°K) and pressures (p=0.06–7 atm). It was discovered that the relaxation time is inversely proportional to the pressure of the gas behind the shock front (Fig. 67); thus

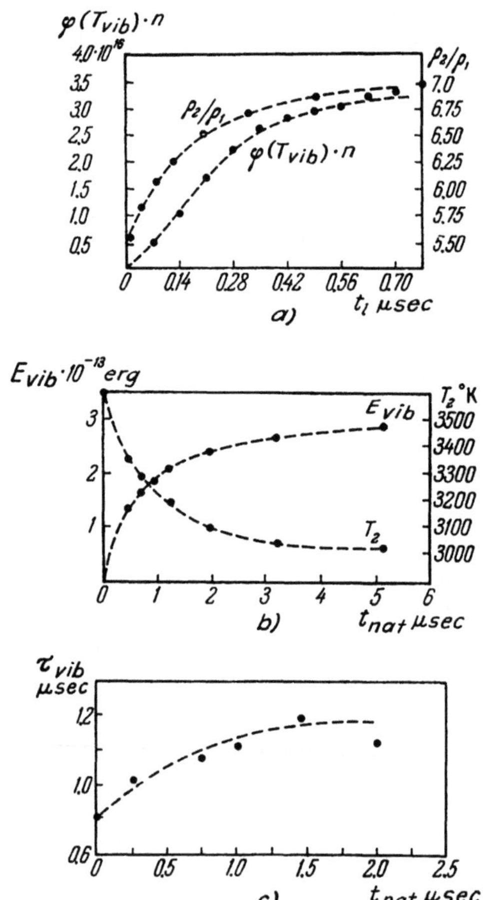

Fig. 66. Variations in the concentration of excited molecules $\varphi(T_{vib})\,n$ and the density (ratio) ϱ_2/ϱ_1 as a function of t_l (a), the vibrational energy E_{vib} and the translational temperature (b) and the vibrational relaxation time τ_{vib} (c) as a function of t_{nat} behind the front of a shock wave propagating in oxygen with the velocity V=2.5 km/sec with an initial pressure p_1=0.01 atm.

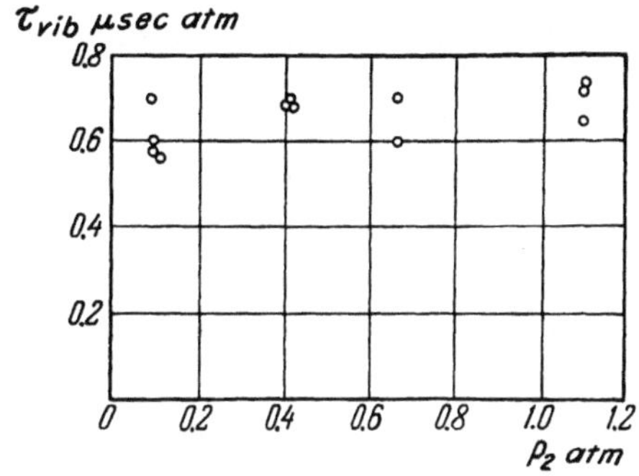

Fig. 67. Dependence of the product $\tau_{vib}p_2$ (the vibrational relaxation time and the pressure) on the pressure p_2 behind a shock front in oxygen at $T = 4300°K$.

the main contribution to excitation of molecular vibrations is made by paired collisions. Taking this into account it is possible to reduce all the results to atmospheric pressure (Fig. 68).

Let us turn to the consideration of the measured dissociation rates of O_2 and CO_2. As was already noted, the appearance of a section with reduced absorption on the oscillograms is due to dissociation of the O_2 and CO_2 molecules, *i.e.*, it is due to the reduction in the population of the vibrational levels resulting from the reduction of the number density of particles n_{O_2} (or n_{CO_2}) and of the gas temperature in conjunction with energy losses for dissociation. In accordance with the consideration presented in Sect. 9, it is simplest to determine the dissociation rate at the very beginning of the dissociation zone, where, by assumption, the greatest contribution is made by the direct reactions:

$$O_2 + O_2 \rightarrow 2O + O_2 \tag{11–15}$$

or

$$CO_2 + CO_2 \rightarrow O + CO + CO_2 . \tag{11–16}$$

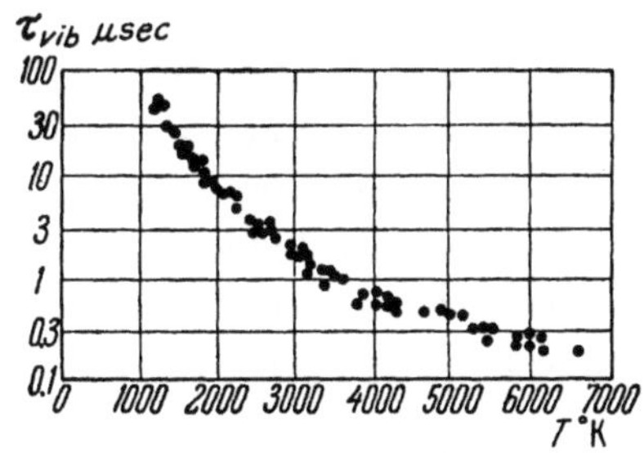

Fig. 68. The obtained relaxation times for molecular vibration in oxygen as a function of the temperature at atmospheric pressure.

We write the expression for the enthalpy h_2, setting

$$h_2 = \frac{\sum H_i(T)\,\xi_i}{\sum \mu_i \xi_i}, \tag{11-17}$$

where $H_i(T)$ is the molar enthalpy of the ith component of the gas at a temperature T. Let us consider further, for the sake of concreteness, the derivation of relationships which relate the observed absorption with the dissociation rate for the case of carbon dioxide, while for the case of oxygen only the final result will be given. Dalton's law for the mixture of CO_2, CO and O will be written as

$$\xi_{CO_2} + \xi_{CO} + \xi_O = 1\,. \tag{11-18}$$

The equation stating the fact that the ratio of the carbon and oxygen atoms is constant, has the form

$$\frac{\xi_{CO_2} + \xi_{CO}}{2\xi_{CO_2} + \xi_{CO} + \xi_O} = \frac{1}{2}. \tag{11-19}$$

Expressing, using Eqs. (11–18) and (11–19), ξ_{CO} and ξ_O in terms of ξ_{CO_2}, we find for the molecular weight of the mixture

$$\mu = \mu_{CO_2}\left(\frac{1 + \xi_{CO_2}}{2}\right) \tag{11-20}$$

and for the molar enthalpy of the mixture

$$H_2 = \frac{H_{CO} + H_O}{2} + \xi_{CO_2}\left(H_{CO_2} - \frac{H_{CO} + H_O}{2}\right). \tag{11-21}$$

Substituting these relationships and the expression for the law of conservation of energy and using the law of conservation of the mass flux, we get a relationship between ξ_{CO_2} and the gas temperature:

$$\xi_{CO_2} = \frac{C - H_{CO} - H_O}{2H_{CO_2} - H_{CO} - H_O - C}, \tag{11-22}$$

where $C = \frac{1}{2}\mu_{CO_2}V^2[1-(\varrho_1/\varrho_2)^2 + H_1 \approx \frac{1}{2}\mu_{CO_2}V^2 - H_1]$ is an almost constant quantity which depends on the shock-wave velocity and on the enthalpy ahead of the shock front. The sum in the square brackets, by virtue of the smallness of $(\varrho_1/\varrho_2)^2$ is close to unity. In fact, $(\varrho_1/\varrho_2)^2$ is of the order of 0.01; hence, it is possible to disregard not only the variation in its value but also the term $(\varrho_1/\varrho_2)^2$ itself. We represent the absorptivity in the form

$$A = 1 - e^{-\sigma(T) n_{CO_2} d} \equiv 1 - e^{-\sigma(T)\xi_{CO_2}\frac{p_2 d}{kT_2}}. \tag{11-23}$$

Differentiating Eqs. (11–22) and (11–23) in respect to time and referring the derivatives to the very start of the dissociation process ($\xi_{CO_2} \approx 1$, $T = T_a$, where T_a is the gas temperature calculated on the assumption that molecular excitation has ended and that no dissociation takes place), we obtain from Eq. (11–22) a relationship between $d\xi_{CO_2}/dt$ and dT/dt, and from Eq. (11–23) we get a relationship between dA/dt, $d\xi_{CO_2}/dt$ and dT/dt. The first of these relationships makes it possible to eliminate dT/dt and to directly relate $d\xi_{CO_2}/dt$ with the experimentally obtained value of dA/dt:

$$\frac{d\xi_{CO_2}}{dt} = -\frac{kT_a}{p_2 d}\,\frac{\dfrac{1}{1-A}\left(\dfrac{dA}{dt}\right)}{\sigma + \left[\left(\dfrac{d\sigma}{dT} - \dfrac{\sigma}{T}\right)\dfrac{C + H_{CO} + H_O - 2H_{CO_2}}{2\dfrac{dH_{CO_2}}{dt}}\right]}. \qquad (11\text{–}24a)$$

The time in this expression is the natural time (see Sect. 8).

Using the same calculations we obtain the following expression for the rate of dissociation of oxygen in the very beginning of the dissociation zone:

$$\frac{d\xi_{O_2}}{dt} = -\frac{kT_a}{p^2 d}\,\frac{\dfrac{1}{1-A}\left(\dfrac{dA}{dt}\right)}{\sigma + \left[\left(\dfrac{d\sigma}{dT} - \dfrac{\sigma}{T}\right)\dfrac{C + 2H_O - 2H_{O_2}}{2\dfrac{dH_{O_2}}{dT}}\right]}. \qquad (11\text{–}24b)$$

These relationships can be used to obtain the dissociation rate. Experiments show that the values of $d\xi_i/dt$ which are obtained at the same temperature are proportional to the pressure, *i.e.*, the dissociation of O_2 and CO_2 actually takes place according to the bimolecular mechanism (Fig. 69).

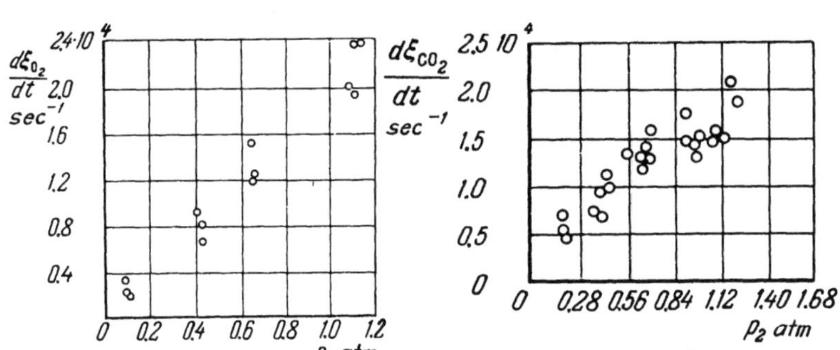

Fig. 69. Dependence of the dissociation rate of oxygen molecules a) and carbon dioxide molecules b) on the pressure at constant temperature (in oxygen $T = 4200°K$, in carbon dioxide $T = 4000°K$).

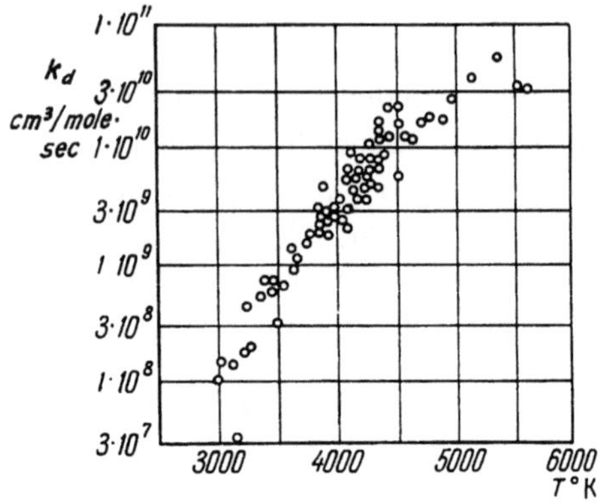

Fig. 70. Measured rate constants of bimolecular dissociation of carbon dioxide gas as a function of the temperature.

Knowing the values of $d\xi_i/dt$ it is possible to determine the dissociation rate constant k_d by using Eqs. (9–23), (9–22) and (9–29). The values of k_d for the direct dissociation of CO_2 thus obtained Eq. (11–16) in the range $3000 < T < 5500\,°K$ are shown in Fig. 70. These results can be approximated by a certain formula. Thus, assuming that k_d can be written in the form of the Arrhenius formula, we will get

$$k_d(CO_2, CO_2) \approx 1.5 \cdot 10^{14} e^{-\frac{87{,}000}{RT}} \ \text{cm}^3/\text{mole}\cdot\text{sec}, \qquad (11\text{–}25a)$$

where the activation energy is 87 ± 9 kcal/mole. On the other hand, as follows from Sect. 18, a better substantiated expression for k_d will be a relationship such as $k_d = P(D/RT)^n \times \sqrt{T}\exp(-D/RT)$, where D is the dissociation energy (in the given case this is the energy of breaking the bound CO–O, which is 125.7 kcal/mole). Then the results thus obtained can be approximated by the expression

$$k_d(CO_2, CO_2) \approx 3 \cdot 10^7 \left(\frac{D}{RT}\right)^6 \sqrt{T} e^{-\frac{D}{RT}} \ \text{cm}^3/\text{mole}\cdot\text{sec}. \qquad (11\text{–}25b)$$

Similar results were obtained for oxygen [102, 105, 43]. Thus, the approximating expression for dissociation rate constant for oxygen, according to [43] for the range $2600° \leqslant T \leqslant 7000\,°K$, has the form

$$k_d(O_2, O_2) = 4.6 \cdot 10^9 \left(\frac{D}{RT}\right)^4 \sqrt{T} e^{-\frac{D}{RT}} \ \text{cm}^3/\text{mole}\cdot\text{sec}, \qquad (11\text{–}26)$$

where $D = 118$ kcal/mole (see also Sect. 23).

The fact that we considered only the start of the nonequilibrium dis-

* The dissociation $CO \to C + O$ and other reactions with the participation of free carbon atoms at $T < 6000\,°K$ are disregarded here.

sociation zone has appreciably simplified the problem; actually [however], it is necessary to take into account secondary reactions and to seek values of k_d for the entire nonequilibrium region of the flow. To analyze processes in pure carbon dioxide gas at $T<6000\,°K$ we now consider the reactions Eqs. (9–24)—(9–26).* In this case it is more convenient to solve the reverse problem, *i.e.*, specifying different values of k_d, to seek the theoretical distribution of the absorptivity and to compare it with the experimentally obtained distribution. For this, Eqs. (11–18)—(11–21) should be replaced by analogous expressions which take into account the presence of four components, CO_2, CO, O_2 and O in the gas:

$$\left.\begin{aligned} &\xi_{CO_2} + \xi_{CO} + \xi_{O_2} + \xi_O = 1\,, \\ &\frac{\xi_{CO_2} + \xi_{CO}}{2\xi_{CO_2} + \xi_{CO} + 2\xi_{O_2} + \xi_O} = \frac{1}{2}\,, \\ &\mu = \mu_{CO_2}\frac{1 + \xi_{CO_2} + \xi_{O_2}}{2}\,, \\ &H_2 = \xi_{CO_2}\left(H_{CO_2} - \frac{H_{CO} + H_O}{2}\right) + \\ &+ \xi_{O_2}\left(H_{O_2} + \frac{H_{CO} - 3H_O}{2}\right) + \frac{H_{CO} + H_O}{2}\,. \end{aligned}\right\} \tag{11–27}$$

The enthalpies of the components in the region of 2000–6000 °K are easily approximated by the expression $H_i = b_i T + a_i$ (which gives an error which does not exceed several percent). This makes it possible to obtain an explicit dependence of the temperature on ξ_{CO_2} and ξ_{O_2} (V, the shock-wave velocity, is a constant parameter), by using Eqs. (9–1)—(9–3) and (11–27). The density is determined from Eq. (9–4); as before it is assumed that $p = \text{const}$.

Kinetic equations for the variations in the molecular fraction of CO_2 and O_2 are of the same form as Eqs. (9–27) and (9–28); the quantities W_i which are contained in these relationships can be represented as

$$W_{CO_2} = S_1 - S_2\,, \quad W_{O_2} = S_2 - S_3\,, \tag{11–28}$$

where

$$\left.\begin{aligned} S_1 &= -\left(\frac{p}{kT}\right) k_d(CO_2, CO_2)\left(\xi_{CO_2} - \frac{p}{K_1}\xi_{CO}\xi_O\right), \\ S_2 &= \left(\frac{p}{kT}\right) k_d(CO_2, O)\left(\xi_{CO_2}\xi_O - \frac{\xi_{CO}\xi_{O_2}}{K_2}\right), \\ S_3 &= \left(\frac{p}{kT}\right) k_d(O_2, M)\left(\xi_{O_2} - \frac{p}{K_3}\xi_O^2\right). \end{aligned}\right\} \tag{11–29}$$

The second terms in the parentheses characterize the contribution of the inverse reactions to the dissociation rate.* As the starting expression for $k_d(CO_2, CO_2)$ we use Eq. (11–25b), multiplying its right-hand side by the factor α_1 which is varied. The value of $k_d(CO_2, O)$ is unknown even in the first approximation, for which reason we assume for it a starting expression suggested by Kodrat'yev on the basis of analysis of the reaction rates of similar molecules, with the variable multiplier α_2:

$$k_d(CO_2, O) = \alpha_2 6\cdot 10^{12} e^{-\frac{35,000}{RT}} \text{ cm}^3/\text{mole}\cdot\text{sec}. \qquad (11\text{–}30)$$

The constant $k_d(O_2, M)$ is known with sufficient accuracy from numerous measurements in mixtures of oxygen with various gases (see Sect. 23).

Eqs. (9–27) and (9–28) were solved simultaneoulsy with an expression relating the temperature with ξ_{CO_2} and ξ_{O_2}, using an electronic computer. The absorptivity [Eq. (11–23)] was also simultaneously calculated. It was assumed that at the start of dissociation $(t=0)$ $\xi_{CO_2}=1, \xi_{O_2}=0$. The calculations were performed with parameters (initial pressure p_1, shock-wave velocity V), which correspond to conditions prevailing during shock-tube experiments.

As a result of these calculations there were obtained curves of the time dependence of all the thermodynamic variables, the gas composition and the

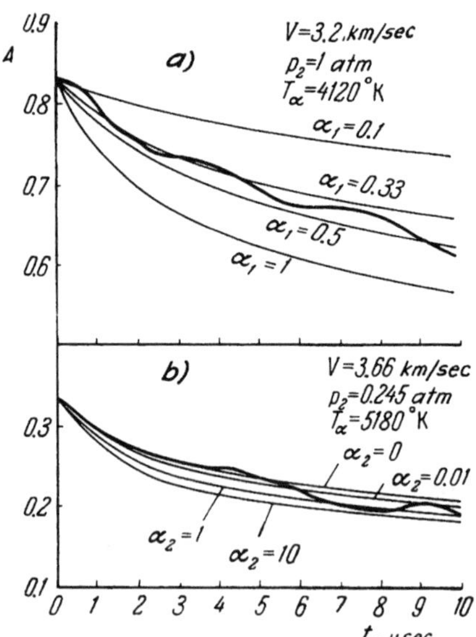

Fig. 71. The measured (heavy lines) and theoretical (thin lines) values of the absorptivity A ($\lambda=2380$ A) behind a shock front in CO_2 as a function of t_l. a) Varying the rate constant of the reaction $CO_2+CO_2 \to CO+O+O_2$; b) the same for the reaction $CO_2+O \to$ $\to CO+O_2$.

* The quantities K_i are in substance the equilibrium constants (see Sect. 18).

absorptivity after the shock wave has passed the observation holes of the shock tube. On the graphs of time dependence of the absorptivity (for different α_1 and α_2) were drawn results experimentally obtained under the same conditions and a comparison was made.

Specimens of such graphs are shown in Fig. 71. It has been calculated that the experimentally recorded absorptivity of CO_2 is sensitive to changes in $k_d(CO_2, CO_2)$ [the reaction $CO_2+CO_2 \rightarrow CO+O+CO_2$] and is only weakly dependent on $k_d(CO_2, O)$ [the $CO_2+O \rightarrow CO+O_2$ reaction]. This points to the fact that the first of these quantities can be determined (by comparison with experiment) with a higher accuracy than the second quantity. A comparison has shown that at $T_2 = 3000$–$4000\,°K$ and $p_2 \sim 0.1$–1 atm α_1 is close to 0.5–1.0, and α_2 can be substantially below unity and in any case does not exceed unity. For more accurate determination of $k_d(CO_2, O)$ it is necessary to simultaneously measure the distribution of the concentration of CO_2 and of some other component.

[4] The Visible Spectral Region

The arrangement for the study of excitation and decomposition of molecules by light absorption in the visible spectral region does not differ from that described above. Some peculiar features of the operation of such an arrangement can be demonstrated on the example of an installation used for the study of processes behind a shock front in molecular iodine [48]. To isolate the necessary spectral section in the visible region (in iodine of the

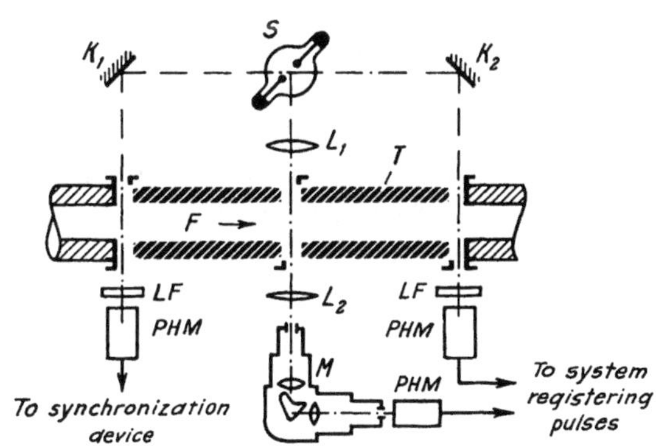

Fig. 72. Optical arrangement for the study of the character of absorption of visible radiation behind a shock front. *S* is an arc discharge light source; K_1 and K_2 are rotating mirrors; L_1 and L_2 are lenses; *LF* is a light filter; *M* is the monochromator; PHM I is the synchronizing signal photomultiplier; PHM II and PHM III are photomultipliers which give the signals necessary for measuring the shock wave velocity and the absorptivity; *T* is the shock tube (with a heating system for work with iodine vapor) and *F* is the shock front.

system $^1\Sigma_d^+ \rightarrow {}^3\Pi_{0+u}$) it is possible to use not only monochromators, but also light filters, since the spectrum is continuous. The arrangement which is described (Fig. 72) consists of three optical channels: the first is used for obtaining the synchronizing pulse, the second and third are used for the study of the distribution of the absorptivity of iodine behind a shock front and for studying the shock-wave velocity. The second channel is also used to measure the initial iodine concentration ahead of the shock front, since, as is known, molecular iodine has an appreciable absorptivity even at room temperature. In the first and third channels, use is made of light filters which isolate a spectral region several hundred angstroms wide near 5200 A (region of maximum iodine absorption), while a prism monochromator with a dispersion of 100 A/mm in this region is used in the second channel. Then photomultipliers are installed; the signal from the first photomultiplier which is produced by a sharp change in the absorptivity at the instant the shock wave appears in the observation section, is fed to the input of a synchronizing device, which starts up the scanning mechanisms of a double-trace oscillograph. Signals from the other two photomultipliers are fed to the input of this oscillograph. The same arc-discharge xenon lamp is used as the light source. In the visible spectral region the intensity of the light source fed by a direct current was sufficient, hence no additional current pulse was supplied to the lamp. Due to this the oscillograms obtained in these experiments have a form different from that obtained in experiments in the ultraviolet region (Fig. 73). Before the shock front arrives at the observation section no variable component of the signal appears; then when the shock front appears, the gas is compressed and heated, which produces a sharp change in the absorptivity of the iodine vapor and, consequently, results in the appearance of a signal on the oscillograph screen. The further shape of the signal depends on the shock-wave velocity and on the spectral section selected. When the wave velocity is sufficiently high, dissociation of iodine molecules takes place behind the front (the molecular concentration and the temperature are reduced, the total gas density is increased). In different spectral regions the absorption coefficient will either decrease with an increase in the temperature (transitions from the ground vibrational level of the starting electronic state $^1\Sigma_g^+$), or will increase (in the case of transitions from the excited level of the $^1\Sigma_g^+$ state). The specimens (Fig. 73a) show an increase in the absorptivity*, which is due to temperature reduction (spectral region near the absorption maximum) and density increase produced by dissociation. When the shock-wave velocity is high the time of absorption

* Let us recall that the oscillograph screen displays only the variable component of the signal; the polarity of signals shown in Fig. 73 is opposite of that for signals given in Figs. 63 and 64.

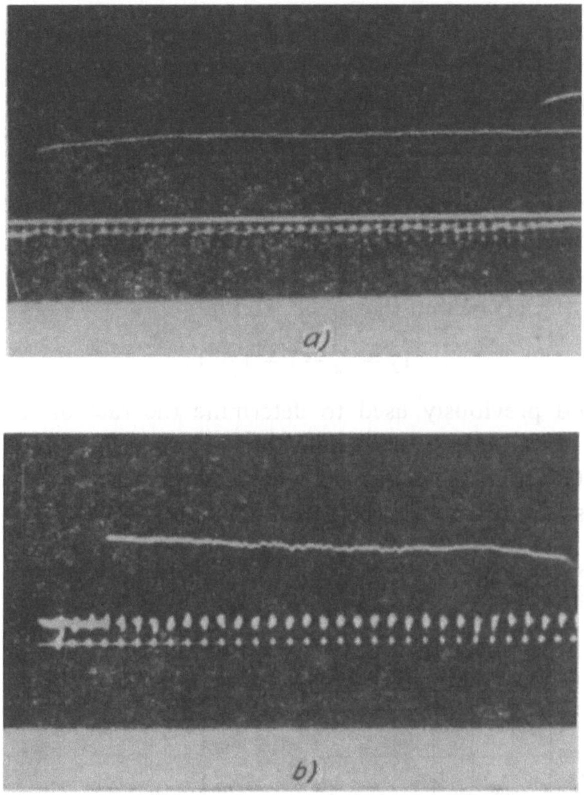

Fig. 73. Specimens of oscillograms of the change in intensity of probing visible radiation as a function of time following the passing of a shock front in iodine vapor. Experimental conditions: a) $\lambda = 5050$ A, $p_1 = 0.004$ atm, $V = 817$ m/sec, time marks were made each 5 μsec; b) $\lambda = 5050$ A, $p_1 = 0.0025$ atm, $V = 495$ m/sec, time marks every 25 μsec. The bottom shows zero lines.

increase is reduced. In strong shock waves the drop in the iodine molecules concentration can be so rapid and appreciable that, despite the drop in temperature and the density increase, the signal intensity is reduced. Conversely, for very low shock-wave velocities no change in the absorption may be observed during several tens of microseconds (Fig. 73b).

The procedure for calculating the absorptivity from oscillograms is similar to that described above for O_2 and CO_2 with the only difference that here it is necessary to take into account the light absorption which takes place before arrival of the shock wave. The illumination intensity is measured by preliminary calibration (a high-stability radiation source is needed for precise measurements). Measurements show that the absorption obeys Baire's law, and the magnitude of the absorption coefficient for $\lambda = 4660$,

5050 and 6600 A are in good agreement with experimental results obtained by Sulzer and Wieland by heating iodine vapor in a furnace and by calculation from a theoretical formula which they have derived [552] (Fig. 74). This agreement shows that up to high temperatures (here up to $T \sim 5600\,°K$), not enough time is available for appreciable dissociation of iodine molecules in the shock front and near it; a separation of the vibrational relaxation and dissociation zones is also observed in iodine.

This makes it possible to obtain the dissociation rate constant for molecular iodine $k_d(I_2, I_2)$ behind the shock front

$$I_2 + I_2 \rightarrow I + I + I_2 \tag{11–31}$$

by the method previously used to determine the rate of dissociation of oxygen and carbon dioxide. The results thus obtained embrace a temperature range from 1000 to 4000 °K (Fig. 75). In this temperature range the experimental values of $k_d(I_2, I_2)$ can be approximated by

$$k_d(I_2, I_2) = 2.3 \cdot 10^{11} \sqrt{T} \left(\frac{D}{RT}\right)^{\frac{3}{2}} e^{-\frac{D}{RT}} \text{ cm}^3/\text{mole} \cdot \text{sec}, \tag{11–32}$$

where D is the dissociation energy of iodine molecules which is equal to

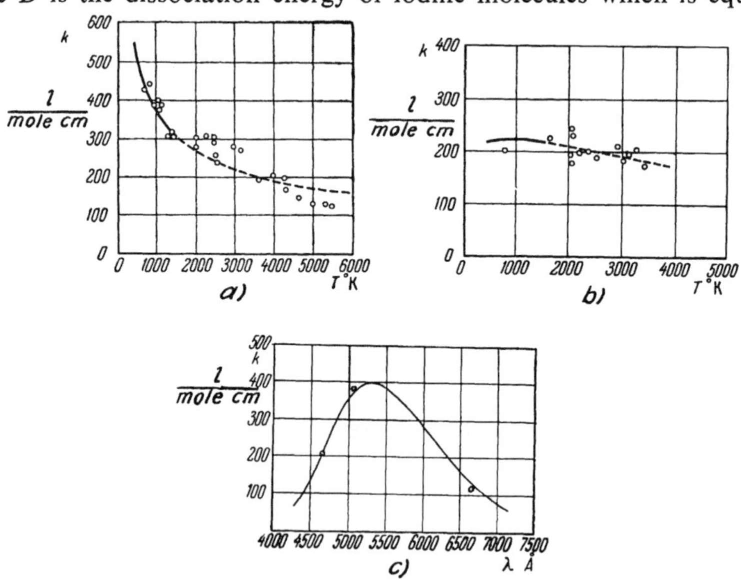

Fig. 74. Coefficient of absorption of iodine vapor k at high temperatures (preceding molecular dissociation) as a function of temperature for $\lambda = 5050$ A (a) and $\lambda = 4660$ A (b) and as a function of the wavelength at $T = 950\,°K$ (c). The solid line denotes experimental results due to Sulzer and Wieland [552] who have heated iodine vapor in a furnace; the dashed line denotes calculations using a formula suggested by the authors of [552], ○ denotes values obtained in a shock tube.

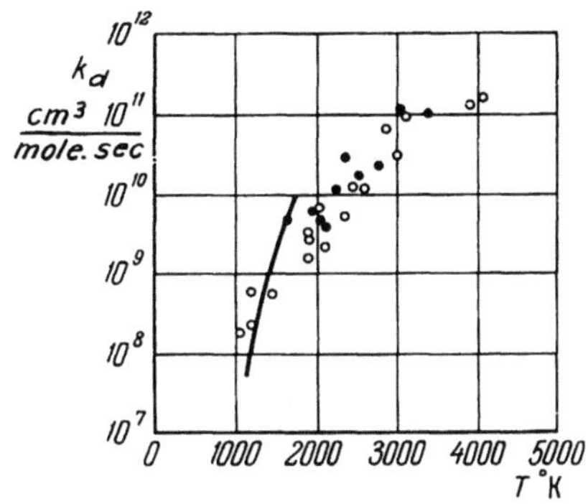

Fig. 75. The dissociation rate constant for iodine molecules as a function of the gas temperature, obtained in a pure gas using the arrangement shown in Fig. 72. ● is for $\lambda = 4660$ A, ○ is for $\lambda = 5050$ A. For comparison the results obtained by Britton et al [269], for a mixture of 1% I_2 + 99% Ar are shown by the solid line.

35.5 kcal/mole. If the experimental results are approximated by the Arrhenius formula $k_d = A \exp(-Q/RT)$, then the activation energy Q for the reaction (11.31) will be ~ 30 kcal/mole.

In all cases of investigation of dissociation of molecular oxygen, carbon dioxide and iodine, the measured dissociation rates for different wavelengths in each gas have yielded the same results. This serves as an additional proof of the validity of the method used.

In conclusion we make certain general remarks on the use of the spectral absorption analysis method for study of shock waves. When light is intensely absorbed it is possible to study processes under a very low partial pressure, *i.e.*, in mixtures with a low percentage content of the test gas in a medium which does not have a perceptible absorption in the spectral region under study. This makes it possible to best approach the obtaining of the thermal "bath," the concept of which is frequently used in the theory of excitation and dissociation of molecules at high temperatures (see Sects. 17 and 18). Such studies are performed in different spectral regions (see Table 5). From among these works we wish to point out the study of vibrational relaxation and dissociation of oxygen mixed with argon at temperatures up to 18,000 °K and nitric oxide and air mixed with argon (at T up to 8000 °K) in the vacuum ultraviolet spectral region, [276–278, 614–616]. Palmer and Horning [463] have studied the decomposition of bromine mixed with argon (visible spectral region), etc.

The use of an illuminating source with a line spectrum of components under study makes it possible in a number of cases to obtain a number of interesting results in the study of gas reaction kinetics [70, 86]. This approach was used successfully in the study of kinetics of decomposition of water vapor in a shock wave at temperatures of 2400–3200°K [244]. In this case, when the spectrum of the light source is not continuous, it should be expected that the dependence of the absorptivity on the concentration will not follow Baire's law. This was discovered, for example, in the study of the dissociation rate of hydrogen mixed with argon on the basis of the absorption of the far ultraviolet radiation in bands of the system of molecular hydrogen, $^1\Pi_u \rightarrow ^1\Sigma_g^+$ [468].

In all the above works the absorptivity was recorded by photoelectric methods, which made it possible to most simply and reliably perform the quantitative measurements. Gaydon, Kimbell and Palmer [329] have shown, however, that photographing of light absorption spectra at different time instants after passing of the shock front can also yield information about the kinetics of the processes. The method of study worked out by Gaydon and other investigators is similar to that which was used by Tumakayev and Lazovskaya [202] (see Sect. 10); *i.e.*, in each experiment only one photograph was obtained (in this case these were the absorption spectra) with the given

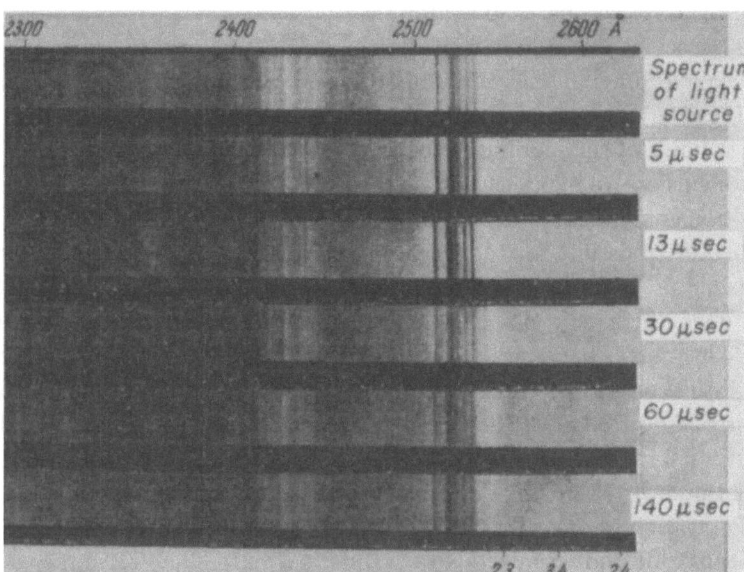

Fig. 76. Series of absorption spectra obtained by Gaydon et al [329], in the study of the kinetics of disintegration of SO_2 in a shock wave propagating in a mixture of 4% SO_2 + 96% Ar with a velocity of 2.07 km/sec, with $p_1 = 9$ mm Hg. The top photograph shows the spectrum of the light source. Then to the right is shown the time (in the laboratory measuring system) after the passing of the shock front. The measurements were performed on the basis of the SO bands (2, 3), (3, 4) and (2, 4) shown in the picture. a) Spectrum of the source; b) microseconds.

time lag after the wave-front passage. A set of such photographs* for the same shock-wave velocity makes it possible to reconstruct the picture of the development of processes in a volume behind a shock front (Fig. 76). The somewhat poorer resolving power in time in comparison with the photoelectric method is compensated by the possibility of following the changes simultaneously in the entire spectral region. Gaydon, *et al.*, have successfully measured the rate of disintegration of SO_2 at temperatures of about 3000–4000 °K.

Summing up the above discussion we can conclude that the use of absorption spectroscopy to analyze phenomena in shock waves is a very fruitful method for the study of kinetics of processes taking place in a gas at high temperatures; the possibilities of this methods have by far not been exhausted.

[12] OPTICAL STUDY OF GASES

One of the basic properties of a heated gas is the presence of optical emission. The study of the spectral distribution and its change with time can yield interesting information about the processes taking place in the gas. At present a large number of spectral studies of gas radiation in shock waves obtained in shock tubes is performed. The spectral composition and the time characteristics of radiation in air and its components, in inert and other gases was studied.

[1] Dependence of the Radiation on the Concentration of the Gas Components

In order to effectively use the information about the distribution of the radiation intensity in shock waves, it is necessary to relate the observed quantities with the concentration of gas components which contribute to the radiation. Let us consider here some basic relationships. We assume that the radiation of a gas heated in a shock wave is recorded by the optical system shown in Fig. 77. We place the origin of coordinates in the point of intersection of the optical axis of this arrangement with the surface of the radiation object and we direct the polar axis along the optical axis of the system. By definition, the radiant flux $d\Phi_\nu$ (in the frequency interval ν, $\nu + d\nu$), passing an area $d\sigma$ in the direction of the polar angles ϑ and φ within the limits of the solid angle $d\Omega$ per unit time, is

$$d\Phi_\nu = I_\nu(\vartheta, \varphi)\, d\nu \cos\vartheta \, d\sigma \, d\Omega\,, \qquad (12\text{–}1)$$

* The authors are grateful to Prof. A. G. Gaydon, who has supplied them with the photographs of the absorption spectrum which are shown in Fig. 76.

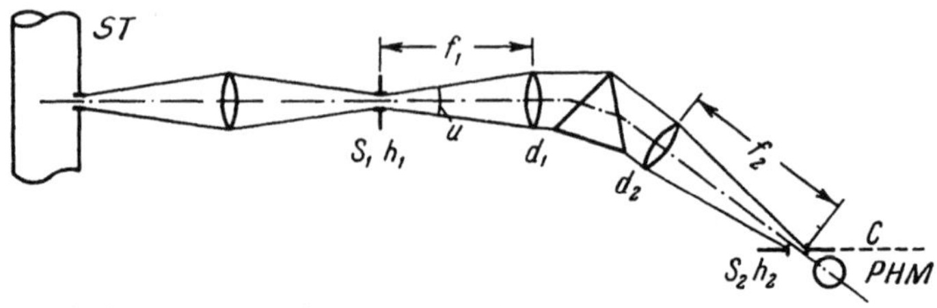

Fig. 77. Optical arrangement for the study of radiation of a gas in a shock wave. S_1 and S_2 are the width of the input and output slits of the instrument, h_1 and h_2 are the slit heights, u is an angular aperture, f_1 and f_2 are the focal distances of the collimator and the camera lenses, d_1 and d_2 are the effective diameters of these lenses, ST is the shock tube and PHM and the recording photomultiplier. The output slit is placed in the plane of the spectrum C.

where $I_\nu(\vartheta, \varphi)$ can be defined as the spectral intensity of the radiation. The total radiation flux is equal to the triple integral

$$\Phi = \int\limits_{\Sigma}\int\limits_{\Delta\nu}\int\limits_{\Omega} I_\nu \, d\nu \cos\vartheta \, d\sigma \, d\Omega \,, \tag{12–2}$$

where the integration is performed over the interval $\Delta\nu$ of the frequencies used, over the area of the radiating surface Σ and over the angular aperture Ω of the optical arrangement. It will be assumed that the image of the radiating surface is obtained in the plane of the input slit of the spectral instrument and is stopped down by the input slit. In order to be specific let us consider an arrangement with photoelectric recording of the radiation. Assuming that the radiation passing out of the test volume obeys Lambert's law, the radiant flux passing through the instrument and received in the receiver (when $f_1 = f_2$ and $d_1 = d_2$) can be written in the form

$$\Phi = s_1 h_1 \eta B \int\limits_{\Omega} \cos\vartheta \, d\Omega = \pi s_1 h_1 \eta B \sin^2 u \approx \frac{\pi}{4} s_1 h_1 \eta B \left(\frac{d_1}{f_1}\right)^2 \equiv \eta B \omega \,, \tag{12–3}$$

where s_1 and h_1 are the width and the height of the input slit,

$$B = \int\limits_{\Omega} \psi I_\nu d\nu \tag{12–4}$$

is the radiation intensity which is recorded, ψ is the apparatus function of the instrument which depends, among others, on the dispersion and the width of the input and output slits,

$$\omega = \frac{\pi}{4} s_1 h_1 \left(\frac{d_1}{f_1}\right)^2 \tag{12–5}$$

is a multiplier which takes into account the geometry of the spectral instrument, η is a coefficient of losses in the optical system as a result of aberration, reflection, scattering, etc. Let us now consider I_ν for cases of a discrete spectrum (radiation in the spectral line) and continuous spectrum (free-free and free-bound transitions). At present we disregard the disturbance of the radiation equilibrium as a result of the radiation which leaves the test volume of the gas.

In the line spectrum the spectral radiation intensity inside a luminescing gas in the interval ν, $\nu+d\nu$ for the $i\to k$ transitions is determined by the number of acts of spontaneous radiation $a_{ik}N_i$, absorption $b_{ki}N_kI_\nu$ and induced radiation $b_{ik}N_iI_\nu$, where N_i and N_k are the number densities of atoms which are in the excited state i and in the ground state k, respectively, a_{ik}, b_{ki} and b_{ik} are Einstein's coefficients which determine the probability of an optical transition per unit frequency interval and which are related by $(a_{ik}/b_{ik})=2h\nu^3/c^2$; $(b_{ik}/b_{ki})=g_k/g_i$, where g_i and g_k are the statistical weights of the corresponding levels. The change in the spectral radiation intensity in the direction of the optical axis y (across the shock-tube axis) will be [121, 215]:

$$\mathrm{d}I_\nu = a_{ik}N_ih\nu\,\mathrm{d}y - b_{ki}N_kh\nu I_\nu\,\mathrm{d}y + b_{ik}N_ih\nu I_\nu\,\mathrm{d}y\,. \tag{12–6}$$

Integrating over d, the tube diameter, we get

$$I_\nu = \frac{a_{ik}N_i}{b_{ki}N_k - b_{ik}N_i}\left[1 - e^{-h\nu(b_{ki}N_k - b_{ik}N_i)d}\right], \tag{12–7}$$

where $h\nu b_{ki}N_k$ defines the absorption coefficient k_ν considered in the preceding section. Usually, $g_kN_i \ll g_iN_k$, which makes it possible to neglect the induced radiation; when this cannot be done, the induced radiation is considered as a correction to k_ν.

Turning to the consideration of the total radiation of the entire spectral line, we assume that $a_{ik}=a_{ik}(\nu_0)\,\varphi_a(\nu)$, $b_{ki}=b_{ki}(\nu_0)\,\varphi_b(\nu)$ where ν_0 is the frequency of the center of the line, $\varphi_a(\nu)$ and $\varphi_b(\nu)$ are functions characterizing the contour of the radiation and absorption lines, respectively $(\varphi_a(\nu_0)=\varphi_b(\nu_0)=1)$. When $\Delta\nu \ll \nu$ and $\nu^3 \approx \nu_0^3$, as is pointed out by Yel'yashevich ([61], page 147), $\varphi_a(\nu)\equiv\varphi_b(\nu)\equiv\varphi(\nu)$, *i.e.*, the contours of the emission and absorption lines are identical.* Thus the recorded total radiation intensity B_m of an mth line of finite width corresponding to the transition under study, will be

$$B_m = \int\limits_{\Delta\nu} \psi I_\nu\,\mathrm{d}\nu = \frac{2h\nu_0^3}{c^2}\frac{g_k}{g_i}\frac{N_i}{N_k}\int\limits_{\Delta\nu} \psi\left(1 - e^{-k_{\nu m}d}\right)\mathrm{d}\nu\,. \tag{12–8}$$

For a rectangular apparatus function ($\psi=1$ for frequencies inside the line and $\psi = 0$ for frequencies outside of the line) in the case of single line

$$B_m = \frac{2h\nu_0^3}{c^2} \frac{g_k}{g_i} \frac{N_i}{N_k} W_{\nu m}, \tag{12–9}$$

where $W_{\nu m}$ is the total absorption of the line defined by Eq. (11–3). As was pointed out in the preceding section, for optically thin layers $W_m \sim N_k$. Thus the radiation intensity of the source B_m which is observed in this case is proportional to the number density [of atoms] in the upper electronic level N_i and is independent of the number density [of atoms] in the lower level N_k. In other cases (optically thick layers) the magnitude of B_m remains proportional to N_i, but its dependence on N_k may be complicated (see Sect. 11). In an inhomogeneous luminescing layer reabsorption may become appreciable as a result of temperature nonuniformity or due to a drop in the density of the excited atoms (or molecules) toward the boundary of the radiating volume [18]. It was assumed in deriving Eq. (12–9) that the particle distribution over the levels may not be equilibrium. If this distribution is a Boltzmann distribution, then Eq. (12–9) reduces to Kirchhoff's law and W_m becomes the absorptivity for the radiation.

The problem of the value of I_ν in the continuous spectrum is solved similarly. Since the Maxwellian velocity distribution for the electrons is usually established quite rapidly, then the spectral intensity of radiation of a layer with a thickness d in the direction perpendicular to the surface of this layer for free-free transitions in the field of an ion, is

$$I_\nu = \frac{2h\nu^3}{c^2} \frac{1 - e^{-k_\nu d}}{e^{\frac{h\nu}{kT_e}} - 1}, \tag{12–10}$$

where k_ν is the absorption coefficient for a continuous spectrum which is proportional to the square of the electron number density (in quasi-neutral plasma). In recombination transitions when no Boltzmann equilibrium exists with respect to the electronic level of ions and atoms, the expression for I_ν is not so simple; it is, however, possible to assume that in quasi-linear plasma under these conditions I_ν for optically thin layer is also proportional to the square of the electron number density ($I_\nu \sim N_e^2$). This relationship is used for measuring the electron number density (see Sect. 13).

Under certain conditions it is necessary to consider the possibility that the equilibrium in the test volume of a gas is disturbed by radiation leaving the given volume. Let us consider the case of a line spectrum [18]. The fact that

* For wide bands this is not valid, see [61].

radiation leaves the volume inevitably disturbs the Boltzmann distribution at the excited levels. The particle population of these levels is replenished by atomic or molecular collisions, which results in electronic excitation. Thus, if, β, the ratio of probability of such collisions to the probability of an optical transition is high, then the gas is insensitive to the leaving of the radiation. If β is small, then the deviation of equilibrium can be large. The value of β is determined by the actual mechanism of excitation of the electronic levels and is different for different levels. The determining quantity in β is the cross-section Q of the excitation of the upper electronic level on collision with other particles. Some data on the magnitude of Q for a number of levels of mercury, sodium, cadmium, etc., atoms colliding with electrons, atoms and molecules, were obtained in studies of radiation damping [123, 166, 40]. Collisions with atoms which are accompanied by the transformation of the translational energy into electron energy, are not too effective in exciting electronic levels; thus Q for ^2P level of the Na atom for the Ar-Na collision is less than one-tenth of the gas kinetic collision cross section. However, vibrationally excited molecules are quite effective in exciting electronic levels, since the transition from one form of internal energy to another is much easier. For the same example of exciting sodium lines the cross section for excitation (or, conversely, damping) by nitrogen molecules comprises $2-4\times 10^{-15}$ cm^2. This difference in the probability of excitation of electronic levels in collisions with nonexcited and excited molecules was used by Gaydon and Hurle to study the vibrational relaxation of nitrogen, carbon monoxide and carbon dioxide molecules behind a shock front when measuring the gas temperature by a pyrometer [387, 327 and 328].

Biberman has calculated δ, the ratio of the true radiation intensity B for the case of leaving radiation to the radiation intensity calcaluted by Kirchhoff's law, B_{vib} ($\delta = B/B_{\text{vib}}$). For lines he has obtained

$$\frac{\beta}{1+\beta}\left(1+\frac{1-q}{1+\beta}\right) < \delta < \frac{\beta}{1+\beta}\cdot\frac{1}{q}, \qquad (12\text{–}11)$$

where q is the average probability that a photon will leave the test volume. For transparent layers (without reabsorption) $\delta \to [\beta/(1+\beta)]$; for dispersion lines, when $k_0 d > 1$,

$$\delta = \frac{\beta\sqrt{k_0 d}}{0.75 + \beta\sqrt{k_0 d}}. \qquad (12\text{–}12)$$

Thus deviations from equilibrium due to the leaving radiation are reduced with an increase in the pressure and in the concentration of the radiating atoms.

[2] Recording Methods and Some Results

Let us now consider experimental methods for recording the radiation in shock waves. At the same time we shall present some results which characterize the optical radiation of a gas in the nonequilibrium region.*

The simplest method for the study of spectra in gases heated by a shock wave is the use of a spectrograph which records on a photographic film the

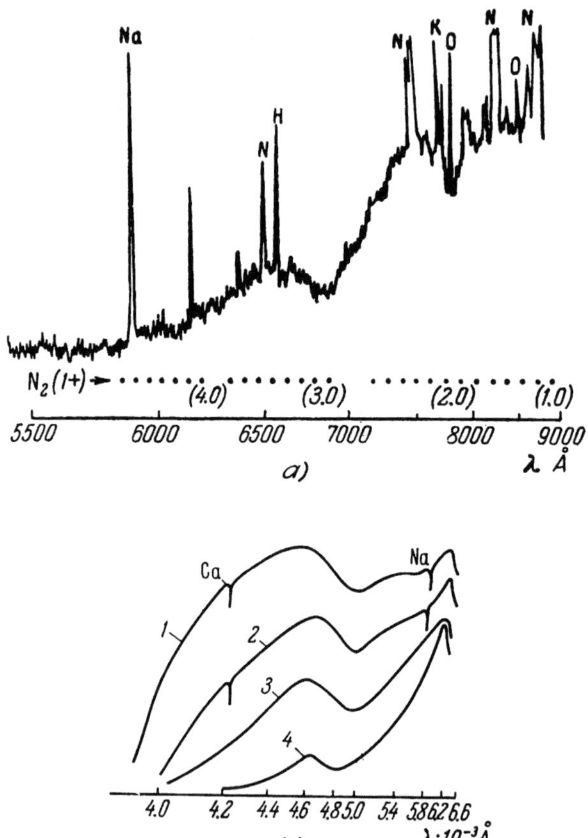

Fig. 78. Microphotograms of the radiation spectra of nitrogen (a) and xenon (b) behind a shock front. In the nitrogen spectrum ($T = 8000\,°K$, atmospheric density) lines of admixtures (sodium, potassium, etc.) are observed. Below is shown the position of the (1, 0)–(4, 0) bands of the first positive system of nitrogen bands. In xenon Curve 1 corresponds to $T = 18{,}500\,°K$, Curve 2 is for $T = 10{,}500\,°K$. The remaining curves correspond to: 3) solar comparison spectra at $T = 5700\,°K$ and 4) the spectrum of a tungsten ribbon filament lamp at $T = 2070\,°K$.

* Some more general problems are considered in Sect. 20.

total radiation of the heated volume of gas which is in the field of view of the instrument (in an incident or reflected wave). The earliest works have shown that the radiation spectra frequently contain a number of admixtures with a low excitation potential (sodium, calcium, etc.) [Fig. 78a]. Even after purification it is sometimes difficult to eliminate such lines as the sodium D lines in the spectrum.* At high temperatures (above 10,000 °K) continuous spectra are obtained (Fig. 78b) which are due to the ionization process. These measurements were basically performed in the visible and the close ultraviolet spectral region. Extending the study of the spectrum into the infrared and far ultraviolet regions requires the use of photoelectric radiation receivers (electron-optical converters, orthicons, etc.). Fig. 79 shows the relative energy distribution in the radiation spectrum behind a shock front in

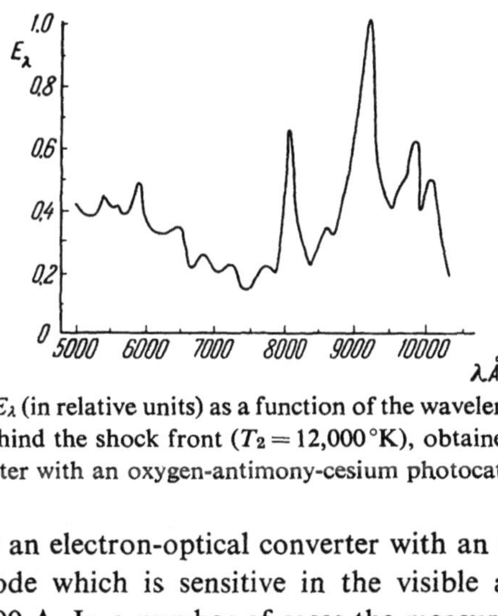

Fig. 79. The energy E_λ (in relative units) as a function of the wavelength λ in the radiation spectrum of xenon behind the shock front ($T_2 = 12{,}000$ °K), obtained by electron-optical converter with an oxygen-antimony-cesium photocathode.

xenon obtained by an electron-optical converter with an oxygen-antimony-cesium photocathode which is sensitive in the visible and near infrared regions up to 12,000 A. In a number of cases the measurement of the total radiation intensity of an entire spectral region of an entire spectrum is of practical interest; these studies aid in estimating the radiant heat transfer in comparison with the convective heat transfer behind a shock front in the supersonic motion of bodies under equilibrium conditions as well as in the presence of nonequilibrium flow regions. Thus, in measuring the absolute values of the radiation intensity of the nonequilibrium region behind the front of a shock wave propagating in a shock tube, a photoelement with a tungsten and molybdenum photocathode has made it possible to record the radiation of air in the region 300–1500 A, photographic film was used for the region of 2300–3500 A, a photomultiplier in the region of 3500–8000 A,

* Experiments [184, 405, 569] show that these admixtures enter the gas primarily from the shock-tube walls.

a bolometer with an aluminum element in the region of 1500–10,000 A. The radiation in the infrared region was recorded by PbSe semiconductor photo-resistances (in the region of 1–2.8 microns) and also by GeAu resistances (2–9 microns) with appropriate light filters [280].

Of greatest interest for the study of nonequilibrium phenomena is the study of the radiation intensity (of the given spectral composition) as a function of the distance from the shock front (along the flow). An experimental arrangement for they study of this distribution of radiation intensity in a shock wave with photoelectric recording is actually identical with that shown in Fig. 61, with the only difference that now no supplementary light source is necessary.* It was noted quite early [437, 576, 495] that the wave front sometimes glows much more intensely than the gas behind it. It was found that this, in a number of cases is due to the radiation of admixtures such as C_2 and CN, the excited molecules of which very rapidly (during a time which is sometimes less than one microsecond) form in the heated gas from carbon-containing impurities (Fig. 80). The study of the behavior of the radiation of the violet system of CN bands (in the region of $\lambda \sim 3900$ A) as an admixture to xenon, was undertaken by Roth [501], in order to determine the life of the vibrational levels of the $B^2\Sigma^+$ of a CN molecule at temperatures of 6300–9550°K. A study of radiation in bands of CN and C_2 molecules, which appear as a result of decomposition of BrCN (in a mixture with argon), has

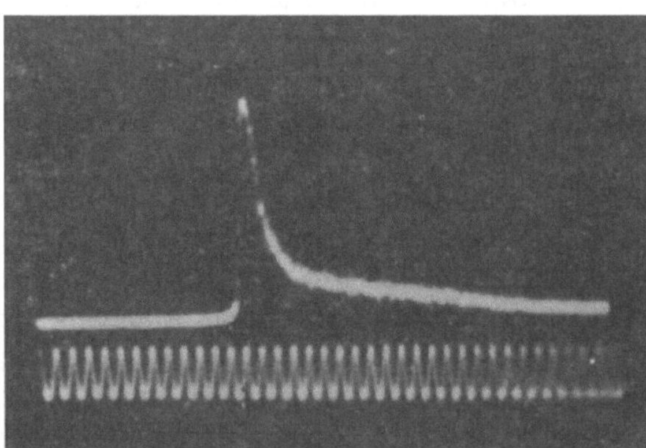

Fig. 80. An oscillogram of changes in the radiation intensity behind a shock front in a mixture of 10% N_2 + 90% Ar with an admixture of acetylene. The spectral instrument has isolated a spectral section corresponding to the head (1, 0) of the band of the red system of bands of the CN molecule ($\lambda = 8970$ A). The shock wave velocity is 2.84 km/sec, initial pressure 6 mm Hg, frequency of oscillation of the calibrating sinusoid was 500 kc/s.

* This light source can be used for adjusting and setting up the optical and electrical systems. The oscillogram shown in Fig. 45 was obtained in precisely such an installation.

yielded information on the kinetics of processes in BrCN at temperatures of 2500–7000 °K [469]. From among other works in which the study of changes in the radiation intensity behind a shock front in mixtures has yielded a number of valuable results concerning the kinetics of transformations in gases, we wish to point out the study of dissociation of carbon dioxide and nitrogen peroxide which were added to argon [264, 417, 418].

In addition to the study of spectral characteristics of impurities, successful investigations were also performed of the time characteristics of the radiation of the main gas in a volume behind a shock front. Thus, for example, the radiation intensity behind shock fronts in air, oxygen and nitrogen was studied in [74, 224, 278, 280, 355, 401, 401a, 405, 618, 619]; the measurements were performed in various shock tubes. The experimental arrangement used in these works is shown in Fig. 81. The photomultipliers PHM

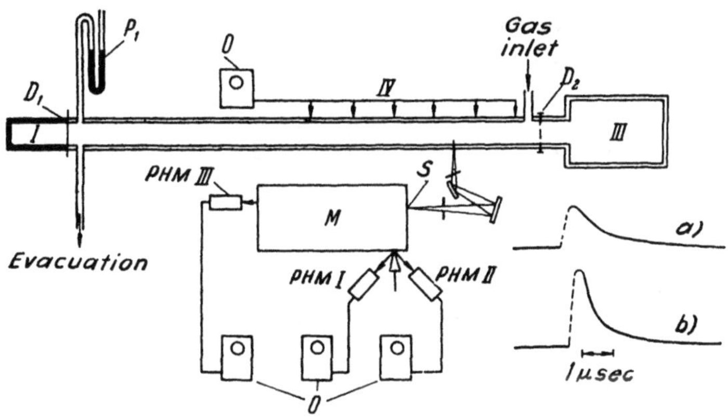

Fig. 81. Experimental arrangement for the study of radiation of shock waves. I is the high-pressure chamber, II is the low-pressure chamber, III is the receiving chamber, IV is a system for measuring the shock-wave velocity, D_1 and D_2 are diaphragms, P_1 is a manometer for measuring the initial pressure, M is a monochromator with an input slit S, PHM are photomultipliers, O are pulsed oscillographs. In the right bottom corner are samples of oscillographic records of radiation of the first positive system of nitrogen bands in the range of 5500–10,000 A when a shock wave propagates in pure nitrogen with an initial pressure $p_1 = 2$ mm Hg with a velocity of 5.18 km/sec (a) and 5.54 km/sec (b). 1) Gas inlet; 2) evacuation; 3) 1 μsec.

I and II have recorded the radiation in two narrow spectral regions, PHM III has recorded some fraction of the total radiation which has passed through the inlet slit of the monochromator without spectral resolution. The resolving power in time of the optical system was limited by the width of the inlet slit of the instrument, which has acted as an diaphragm at the observation section of the shock tube; for a shock-wave velocity of 5 km/sec the resolving time was 0.03 microsec. To isolate a spectral region corresponding to the radiation of the first positive band system of the nitrogen molecule use was made of a filter which isolated a region of $\lambda > 5500$ A; the signal was taken off PHM III. Specimens of the oscillograph curves which were obtained in the study of shock waves in nitrogen are also shown in Fig. 81. It can be seen that immediately following the shock front the radiation intensity exceeds appreciably the equilibrium level. This intensity increase was observed also in experiments for measuring the luminescence of the second positive system of the N_2 molecule and the β and γ systems of the NO molecule. The increase in the radiation intensity behind the shock front is, apparently, related to the excitation of molecular vibrations and with the rapid transfer of vibrational energy to the electronic degrees of freedom. The subsequent radiation intensity drop is due primarily to molecular dissociation.* Unlike the above, in oxygen (the Schumann-Runge band system) no radiation maximum is observed, but rather the radiation level increases gradually to that of equilibrium [401].

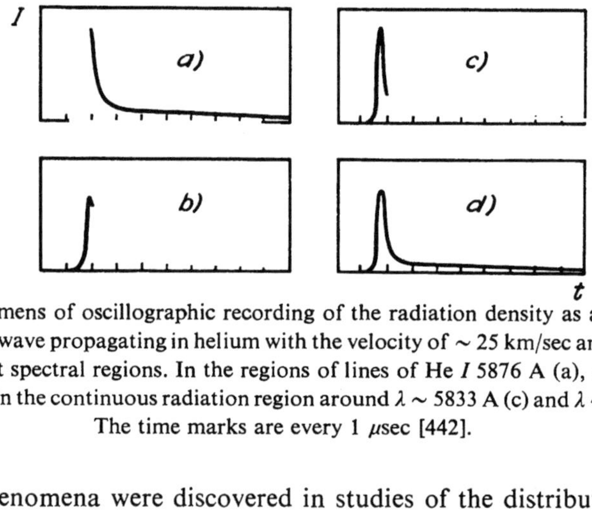

Fig. 82. Specimens of oscillographic recording of the radiation density as a function of time in a shock wave propagating in helium with the velocity of ~ 25 km/sec and $p_1 = 1$ mm Hg for different spectral regions. In the regions of lines of He *I* 5876 A (a), and of He *II* 3203 A (b) and in the continuous radiation region around $\lambda \sim 5833$ A (c) and $\lambda \sim 3264$ A (d). The time marks are every 1 μsec [442].

Similar phenomena were discovered in studies of the distribution of the radiation power in inert gases. A pulse-type magnetohydrodynamic shock tube was used to measure the character of changes in the radiation intensity of a number of individual lines of neutral and ionized helium and of the continuous spectrum near these lines using shock-wave velocities up to 25 km/sec. The specimens of oscillograms thus obtained are shown in Fig. 82

* This behavior of the radiation intensity behind a shock front with time after the passing of the front is similar to the distribution of the absorptivity of O_2 and CO_2 described in the preceding section, with the only difference being that here we are considering the excitation of electronic, rather than vibrational, molecular levels.

[442, 606]. Roth and Gloersen [502, 510, 340] have studied the changes in the radiation intensity in xenon using moderate wave velocities (up to 2 km/sec) in order to study the mechanism and rate of excitation of electronic levels of the xenon atom.

One of interesting results of a number of works [340, 535, 587] is the discovery that the gas ahead of a strong shock wave radiates. In certain cases this radiation may be due to diffusion of radiation from the discharge (in pulse-type shock tubes) [587]. However, this phenomenon was also discovered in ordinary shock tubes when studying strong shock waves in inert gases. In this case the radiation from a volume of gas behind the shock front is transported into a gas region ahead of the shock. Absorption of this radiation results in exciting the gas ahead of the shock and to its scintillation. Simultaneously, photoelectric emission from the shock-tube walls is observed. Gloersen [340], measuring the delay time in establishing the equilibrium radiation level behind a shock front in xenon, has found that it is independent of the pressure, but changes when the shock-tube diameter is changed. This points to the appreciable contribution of diffusion of the radiation and of photoelectric effects (at the tube walls) and shows that the contribution of atom-atom collisions to the rate of excitation of electronic levels is small.

A large amount of information concerning the kinetics of excitation of vibrational degrees of freedom of polar molecules can be obtained by study of the intensity of infrared radiation behind the shock front. However, up to recently, the development of these experiments was retarded by the inavailability of sensitive radiation receivers with a good resolving power in time. Thus, in the work by Windsor, Davidson and Taylor [607], which is devoted to measuring the time of vibrational relaxation of the CO molecule using radiation in the region of 2–2.8 microns, the time resolution of the receiver was about 30 microseconds.* Lately, it has become possible to use semiconductor receivers from germanium, alloyed with gold and from antimony indium. These receivers have a resolving power of at least $10^{-7}-10^{-6}$ secs. Using alloyed germanium receivers, Hooker and Millikan [380] succeeded in investigating the vibrational relaxation of CO molecules in a wide temperature range. The resolution time of the entire recording arrangement used by them was less than 2 μsec and was not determined by the receiver, but rather by the width of the light beam used in the test section of the shock tube. This system was used to study changes in the radiation intensity with time in the flow behind a shock front in carbon monoxide and other gases in a spectral region of up to five microns.

* It appears that certain errors were made [607] in estimating the temperature of the test gas (in the reflected wave). Results due to Hooker and Millikan [380] are more reliable.

[3] Temperature Measurement

One of the important applications of spectral methods in the study of the state of the gas behind a shock front is their use in temperature measurement. The existing spectral methods of temperature measurement are divided into the absolute (pyrometric) and the relative, which are based on the study of the relative intensity of bands and lines in the vibrational-rotational electronic spectra.

The temperature of a gas behind a shock front by the pyrometric method is, as a rule, measured by studying the radiation and absorption of light in lines of admixtures especially added to the main gas (the admixtures consisting of indium, barium and sodium) [183, 184, 206–208, 106, 295–297]. This method is based on assumptions to the effect that a thermal equilibrium exists between the excited atoms of the admixture and the primary gas, that the Kirchhoff's law is applicable and that the disturbance of equilibrium as a result of radiation leaving the heated gas is small. Simultaneous measurement of the radiation and absorption powers of the gas behind the shock front makes it possible, when the above assumptions are justified, to measure the temperature distribution along the flow behind the shock front.

For this purpose two optical channels through which the radiation propagates are isolated in the test section perpendicular to the shock-tube axis. Light from an external source placed behind the shock tube, which is used for measuring the absorptivity, is passed along the first channel through the tube; simultaneously, radiation of the gas in the tube also flows into this channel. Only the radiation of the test gas is passed through the second channel; it serves for measuring the radiation power. To separate the beams the input slit of the spectral instrument used is divided into halves along its height;* the image of the outside source is obtained in one-half of the slit (first channel), and the image of the center of the observation section is obtained in the entire slit. Two photomultipliers were placed behind the output slit of the instrument; the signals received are recorded by a double-trace oscillograph. Using Eqs. (12–3)—(12–5) we can write the following for I, the signal amplitude on the oscillograph screen

$$I' = k'\Phi'_{x+l} = k'(\eta'_x\omega'_x B_x + \eta'_l\omega'_l B'_l) = k'\eta\omega(B'_x + \vartheta B'_l), \qquad (12\text{–}13)$$

(first channel)

$$I'' = k''\Phi''_x = k''\eta''_x\omega''_x B''_x = k''\eta\omega B''_x, \qquad (12\text{–}14)$$

(second channel)

* Mak [429], Gaydon and Hurle [327, 328] have used interference light filters to isolate the needed spectral region.

where B is the total intensity of the recorded radiation, given by Eq. (12–4), and k is a coefficient which takes into account the sensitivity of the photocathode of the photomultiplier, the amplification of the photomultiplier and the amplification of the oscillograph. The subscripts x and l refer to the test gas and to the external light source, respectively, and the primes pertain to the quantities recorded off the first and second channels. The right-hand sides of Eqs. (12–13) and (12–14) are written for the particular case of identical apertures $\omega_x = \omega_l$ and on the assumption that the losses after the radiation leaves the test gas are the same, *i.e.*, that $\eta_x = \eta_l$; here the factor ϑ was introduced to take into account the losses of light in the system from the outside source to the inner surface of the shock tube. If the spectrum of the external source (in the isolated frequency interval) is continuous, as a result of light absorption in the test gas, the consideration of the attenuation of the source intensity makes it possible to represent B_l' in the form $B_l' = B_l(1 - A)$, where B_l is the intensity of the source radiation and A is the absorptivity of the gas (11.7). The hypothesis which follows is based on the assumption that $B_x' = B_x''$, which may not be valid in the case of differences in the admixture concentration along the flow cross section. The relationship between B_x and the sought temperature is given by Kirchhoff's law, which follows from equations such as Eqs. (12–9) and (12–10) and which relates the spectral intensity of radiation I_ν to Planck's function I_ν^0 for a black-body radiator:

$$I_\nu = \frac{2h\nu^3}{c^2} \cdot \frac{A}{e^{\frac{h\nu}{kT_x}} - 1} \equiv I_\nu^0(\nu, T_x) \cdot A\,, \tag{12–15}$$

where the absorptivity A represents here the radiation absorptivity. For $\lambda T \leqslant 0.3$ cm °K (practically $T < 6000$ °K) Planck's function with an accuracy better than 1% can be represented in the form of Wien's radiation law and to write, using wavelengths for convenience,

$$I_\lambda^0 = 2c_1 \lambda^{-5} e^{-\frac{c_2}{\lambda T}}, \tag{12–16}$$

where $c_1 = hc^2 = 5.88 \times 10^{-6}$ ergs-cm²/sec and $c_2 = (hc)k = 1.438$ cm-degree are the first and second radiation constants. The total intensity of black-body radiation $B^0(\lambda, T)$ recorded by a given spectral instrument is then given by Eq. (12–4). Thus for the case at hand, $B_x = B^0(\lambda, T_x) \cdot A$.

Absolute values of intensity can be obtained by comparing with a certain standard temperature source (for example, tungsten ribbon lamps, the brightness temperature T_{st} of which for a wavelength λ_{st} is obtained by calibration and depends on the current intensity through the lamp). Using various versions of calibration of recorded signals by comparing to a standard temperature source, it is possible to differentiate between several variations of the above temperature-measurement method.

a) Calibration of the gas radiation directly by replacing the test section of the tube by a standard ribbon lamp. In this case, on the assumptions made, we can find from Eqs. (12–13), (12–14) and (12–16) that

$$T_x = T_{st}\left[1 - \frac{\lambda T_{st}}{c_2}\ln\frac{m\alpha_{\lambda\lambda st}}{1 - \dfrac{k''I' - k'I''}{k''I_l'}}\right]^{-1}, \qquad (12\text{–}17)$$

where I_l' is the amplitude of the signal from the source only, m is the ratio of the measurement and calibration signals and $\alpha_{\lambda\lambda_{st}}$ is a coefficient which takes into account changes in the brightness temperature of the standard when changing from λ_{st} to λ.

b) Calibration of the external source on the basis of the brightness temperature (T_l) by replacing it by a standard lamp (generalized method of spectral line reversal [182, 183]). Then under the same conditions, using a DC source

$$T_x = T_l\left[1 + \frac{\lambda T_l}{c_2}\ln\frac{1 - \dfrac{k''I'}{k'I''}}{\vartheta}\right]^{-1}. \qquad (12\text{–}18)$$

The brightness temperature of the source is given by

$$T_l = T_{st}\left[1 - \frac{\lambda T_l}{c_2}\ln n\alpha_{\lambda\lambda st}\right]^{-1}, \qquad (12\text{–}19)$$

where n is the ratio of the source to the standard signals. In substance both these methods are identical.

To raise the upper limit of the temperatures which can be measured by the above method, use is sometimes made of a pulsating external light source [126 and 106]. Other modifications of the pyrometric method are also possible (use of secondary passing of the natural radiation in the continuous spectrum through the radiating volume using a mirror instead of a secondary source, the use of sources with discrete spectra when the test gas spectrum is continuous, etc.). A particular case of simplification is the possibility of performing temperature measurements as $A \to 1$ (*i.e.*, of black-body radiation), either in the continuous spectrum (very strong waves), or in reabsorbed parts of individual strong lines.* This makes it possible to work with one channel [instead of the previous two], which appreciably simplifies the experimental arrangement.

Clouston, Gaydon and Glass [295] have measured the gas temperature in

* Sometimes this requires a too high density of the admixture atoms [182]. In this case the admixtures may exert an appreciable effect on the state of the primary gas.

a shock tube by the nongeneralized method of line reversal, which requires achieving "reversal" of lines, *i.e.*, equalizing of the signals from the external source and the natural luminescence of the gas in the tube. The advantage of this method is the need of only one recording channel, while the disadvantage consists in the fact that it is possible to measure the temperature only at those points where the "reversal" is observed.

As before, in the consideration of photoelectric recording methods, the main temperature measurement errors are due to photomultiplier noise and the finite thickness of the oscillograph beam. Some contribution to the error is also made by improper calibration, fluctuation of the external source intensity and other factors. A detailed analysis of errors in measuring temperature by the pyrometric method using admixtures was performed by Fayzullov [205], who has shown that under certain conditions it is possible to measure temperature in the region of $T \sim 4000\,^\circ$K with an accuracy of up to $\pm 100\,^\circ$K.

The applicability of the admixture method to the analysis of nonequilibrium phenomena is substantially limited by the appreciable time which is needed for the evaporation, dissociation and excitation of the admixture atoms, which usually exist in the gas in the form of small crystals of salts. Thus, for example, when an admixture of sodium is present in nitrogen and in air this interval, when $T \sim 4000\,^\circ$K, comprises 10–20 microseconds; in certain cases it may be as high as 50 microseconds [207, 295, 297]. Thus, temperature measurement near the shock front by this method is practically impossible. To eliminate this shortcoming, it is necessary to measure the temperature on the basis of light radiation and absorption in the primary gas, rather than in the admixtures; which was done by Model', *et al.* [126], and by others. Of special interest is the method used by McLean, *et al.* [442], for measuring the temperature on the basis of the absolute value of the total radiation of helium lines with an accuracy of about 1%; a high time resolution was then obtained. Temperature measurements on the basis of the radiation of the primary gas are possible only at very high temperatures (above 10,000 °K).

The error in temperature measurement due to the disturbance of equilibrium produced by leaving of the radiation, as was already noted is estimated by the expression $\delta = (B/B_{\text{vib}})$ and gives temperature values on the low side; Biberman has shown that its relative magnitude is

$$\frac{\Delta T}{T} = \frac{\dfrac{\lambda T}{c_2} \ln \delta}{1 + \dfrac{\lambda T}{c_2} \ln \delta} \approx \frac{\lambda T}{c_2} \ln \delta\,. \tag{12–20}$$

In nitrogen, oxygen and other polyatomic gases ΔT is insignificant. This has enabled Sobolev, Fayzullov, Kudryatsev, and others [205, 91], to successfully measure the gas temperature in the equilibrium state. Gaydon and Hurle [326–328], using sodium and chromium atoms as the admixtures, have studied the change in temperature behind a shock front in oxygen, carbon dioxide and a mixture of hydrogen with argon in a state different from complete equilibrium. They have also used a two-beam system which does not differ in principle from that described above (the monochromator was replaced by two interference light filters). Specimens of temperature values obtained in these experiments are shown in Fig. 83. On the basis of these results Gaydon and Hurle have estimated the characteristic time for the dissociation of O_2, CO_2 and the H_2+Ar mixture at temperatures of 2000–3000°K. Consequently, the temperature-measurement method described above can be successfully used in the study of nonequilibrium phenomena of long duration, which exceeds the time needed for establishing an equilibrium between the gas and the admixtures.

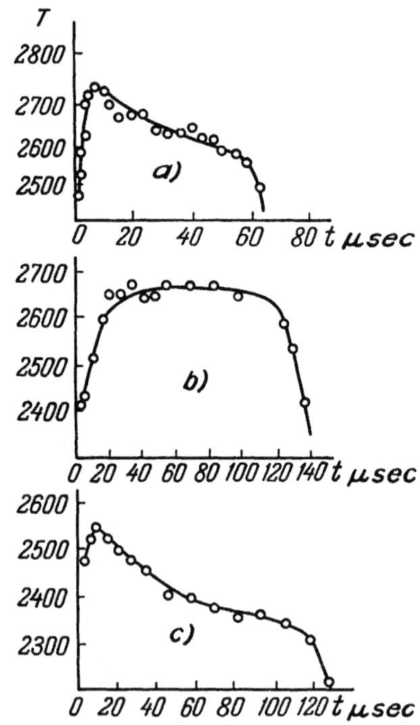

Fig. 83. Specimens of the time dependence of the gas temperature behind a shock front obtained by Gaydon and Hurle by the pyrometric method on the basis of the radiation and absorption of light by *D* lines of sodium. The test gases: (a) carbon dioxide, b) carbon monoxide, c) a mixture of 50% H_2 + 50% Ar.

Temperature measurements performed behind a shock front in argon on the basis of the radiation and absorption of light in the resonance line of ionized barium have shown that the temperature is appreciably lower than the expected theoretical value [205]; thus, at $T_2 \sim 4500\,°K$ and $p_2 \sim 0.5$ atm this divergence was

$$\Delta T \sim 1000\,°K\,.$$

As the pressure is increased, as should have been expected, the divergence drops. These measurements have made it possible to estimate the effective cross section for excitation of resonance radiation of the barium atom upon collision with argon atoms, which was found to be 4×10^{-17} cm^2. Studies of this sort open up possibilities for measuring the probabilities of excitation of individual electronic levels under different conditions by using shock waves.

Temperature-measurement methods based on the relative distribution of the intensity of lines and bands are based on the assumption that a Boltzmann distribution exists with respect to the levels, *i.e.*, that at least a partial equilibrium exists in the gas volume behind the wave front. Thus, when the electronic levels of the admixture atoms, which are in thermal equilibrium with the primary gas, are in equilibrium [among themselves], the ratio of the intensity of two lines corresponding to transition to the same level, in the absence of reabsorption [184], is

$$\frac{B_1}{B_2} = \frac{P_1}{P_2}\frac{g_1}{g_2}\frac{\lambda_2}{\lambda_1} e^{-\frac{E_1 - E_2}{kT_x}}, \qquad (12\text{–}21)$$

where P_i is the probability of a transition from the level under consideration, g_i is the statistical weight and E_i is the energy of the corresponding starting level.

For the case of very strong reabsorption in both lines

$$\frac{B_1}{B_2} = \left(\frac{\lambda_2}{\lambda_1}\right)^2 \sqrt{\frac{P_1 g_1 \Delta\lambda_{B^1}}{P_2 g_2 \Delta\lambda_{B^2}}}\, e^{-\frac{E_1 - E_2}{kT_x}}, \qquad (12\text{–}22)$$

where $\Delta\lambda_{\text{inter.}\,i}$ is a half-width, which is determined by the interaction of atoms in the gas volume (for the corresponding lines). The use of this method is limited by the need to know exactly the transition probability and the degree of reabsorption of the lines used.

Parkinson and Nicholls [467] have measured the rotational temperature by the intensity distribution in electron rotation bands of CN, AlH and CuH behind a reflected wave in argon. Only measurements in the (0, 1) band of the ultraviolet system of CN have yielded satisfactory (within 5%) agreement with calculations in the region of 6000–9000 °K. The use of relative methods of temperature measurement could give interesting information on the equilibrium of gas behind the shock front, however, no extensive studies in this direction were performed.

The translational temperature can be measured by studying the Doppler widening of spectral lines. As is known [125, 204], the Doppler half-width is equal to

$$\Delta\nu'_D = \frac{2\sqrt{\ln 2}}{c}\sqrt{\frac{2RI}{\mu}}\nu' = 0.71\cdot 10^{-7}\sqrt{\frac{T}{\mu}}\nu'\ \mathrm{cm}^{-1}. \qquad (12\text{–}23)$$

However, the Doppler widening can be isolated only under low pressures, at a sufficiently high temperature and in the absence of appreciable ionization, when the collisions and the Stark effect do not appreciably affect the line width. These conditions cannot always be satisfied in a shock tube. This method has not as yet been used for the study of nonequilibrium phenomena in shock waves.

[13] MEASURING THE ELECTRON CONCENTRATION

The existing methods for measuring the electron concentration in a shock wave are substantially a development of known methods for study of gas discharge plasma; among them are probing, spectral, induction, radio frequency and other methods [5, 173]. Measurement of the electron concentration N_e in a shock wave makes it possible to solve problems of kinetics of thermal ionization of gases at high temperatures.*

[1] The Methods of Probes

One of the earliest methods of plasma studies is the known method of Langmuire's probes, which is described in detail in a number of instructions and textbooks on gas electronics (for example [76, 53, 223]). It is based on the determination of the volt-ampere characteristics of a current through an insulated conductor which is introduced into the plasma; this characteristic can be used to determine the electron concentration in the plasma. One of the prerequisite conditions for the use of the method of Langmuire's probes is the assumption to the effect that the mean-free path of electrons in the plasma is considerable greater than the geometric dimensions of the probe, so that all the electrons which arrive in the region of influence of the probe arrive at the probe. However, in the gas behind a shock front propagating in a shock tube the gas pressure is so high that the dimensional ratio is usually

* Here and subsequently the electron concentration N_e is meant to denote the electron number density.

opposite of that desired. Under these conditions the ordinary theory of Langmuire's probes becomes invalid; all kinds of corrections for the increased gas density and for taking into account diffusion near the body of the probe [263, 527] are insufficient for measuring the absolute values of N_e. An additional complication is introduced by the boundary layer which forms at the probe when the gas behind the shock front flows past it. This makes the use of these probes for shock-tube measurements possible only for relative measurements or for qualitative estimates. In the latter case the probes are used as ionization indicating sensors of the instant of shock-front passage. An attempt made by Lin and others [425] to use probes for measuring the electrical conductivity of argon was unsuccessful; the electrical conductivity obtained was by a factor of many thousands lower than expected. This result is hardly attributable to the absence of ionization equilibrium. A detailed analysis of the possibilities for using probes in shock-tube measurements is given by Petschek and Byron [474]. They assume that the potential of a probe in a shock tube is determined by the sum of three components—φ_x—the potential of the test gas, the potential φ_f which arises as a result of gradients of flow variables in the boundary layer attendant to the flow past the probe, and also the potential φ_e, which appears as a result of differences in the mobilities of electrons and ions, which results in charging the probe. An evaluation of these components points to the fact that the total potential of the probe in the case of low electron concentrations is independent of N_e; in the case of high electron concentrations the change in φ, the potential of the probe, with distance is given by

$$\frac{e}{kT_e}\frac{\mathrm{d}\varphi}{\mathrm{d}x} \approx \frac{\mathrm{d}\ln Ne}{\mathrm{d}x}. \tag{13–1}$$

It is here assumed that the electron temperature gradient is highly insignificant for large N_e.* In experiments described in [474] the probe was an insulated conductor with a sharp end placed into the flow. The signal from the probe was fed to an oscillator. Upon the passing of a shock-wave front the potential of the probe undergoes a step increase. Then the signal remains constant for some time, since for low electron concentrations φ is independent of N_e. Then as N_e increases, the probe becomes sensitive to changes in the electron concentration and the signal increases until the equilibrium is reached. As was shown in [474], the velocity which was obtained is in accordance with the hypothesis that in the region under study the overwhelming contribution to the processes of establishing equilibrium ioniza-

* This assumption is verified by calculations [25].

tion is made by electron-atom collisions. Using the probe method together with other [methods] Petschek and Byron have measured the time needed to establish equilibrium ionization behind a shock front in argon.

Boronin and Ignat'yeva [31], in order to measure the electron concentration in the reflected wave in argon have first read off the volt-ampere characteristic of the probe, by supplying to it a repeating sawtooth-shaped wave. In analyzing the results these authors, unfortunately, were forced to make a large number of substantial assumptions (an assumption to the effect that the electron and ion temperatures are equal, that the plasma is homogeneous). This work lacks a quantitative estimate of the contribution of thermal diffusion processes near the probe. The resolving power of the method of dynamic characteristics is not too high (~ 10 microseconds), which makes its use in the study of ionization kinetics in a shock wave difficult.

Weyman and Gloersen [595, 340] have used an electrostatic probe in the form of a wire loop around the glass section of the shock tube to detect electrons in the region ahead of the shock wave. The ground-to-probe potential was used in the experiments. Such probes made it possible to detect an appreciable electron concentration in the low-pressure chamber several hundred microseconds before the shock wave arrived (in argon [595], in xenon [340]). Most probably the observed charges are a result of the photoelectric effect from the walls and in the gas volume produced by diffusion of the radiation from the heated region behind the shock front [94].

[2] Microradiowave Techniques

The ionization kinetics in shock waves moving with relatively moderate velocities are most conveniently studied by using ultrahigh frequency waves (microradiowaves). One of such possibilities is the study of absorption and reflection of radio waves in the gas behind a shock front.

The absorption of a plane electromagnetic wave in an unbounded and homogeneous quasi-linear plasma can be characterized by the change in the energy flux $\boldsymbol{S}=(c/4\pi)\,[\boldsymbol{E}\times\boldsymbol{H}]$ by a factor of e over a path $1/\beta$, *i.e.*, by

$$S = S_0 e^{-\beta y}, \tag{13–2}$$

where β is the absorption coefficient. According to [50],

$$\beta = \frac{2\omega}{c}\sqrt{-\frac{\varepsilon}{2}+\sqrt{\left(\frac{\varepsilon}{2}\right)^2+\left(\frac{2\pi\sigma}{\omega}\right)^2}}, \tag{13–3}$$

where ω is the angular frequency of the radio emission, ε is the dielectric permittivity, σ is the plasma conductivity in a high-frequency field* and c is the speed of light in a vacuum. In the absence of an external magnetic field σ is related to N_e by

$$\sigma = K_\sigma\left(\frac{\omega}{\nu_{\text{eff}}}\right)\frac{e^2 N_e \nu_{\text{eff}}}{m(\omega^2 + \nu_{\text{eff}}^2)}, \tag{13–4}$$

where ν_{eff} is the effective collision frequency between an electron and heavy particles, $K_\sigma(\omega/\nu_{\text{eff}})$ is a slowly varying quantity which is practically equal to unity (see [501], page 70). For comparatively low N_e which are precisely those dealt with in the described experiments $(4\pi\sigma/\varepsilon\omega) \ll 1$ and $\sqrt{\varepsilon} \approx 1$; hence, from Eqs. (13–3) and (13–4) we have

$$\beta = \frac{4\pi e^2}{mc} N_e \frac{\nu_{\text{eff}}}{\omega^2 + \nu_{\text{eff}}^2}. \tag{13–5}$$

The quantity ν_{eff}, to within an arbitrary constant, which is close to unity, is equal to

$$\nu_{\text{eff}} = \bar{v}_e \sum_j n_j Q_j$$

where $\bar{v}_e$ is the average thermal velocity of electrons in the gas which corresponds to the electron temperature T_e, n_j is the number density of heavy particles of the jth kind and Q_j is the cross section of electron scattering at heavy particles, averaged over the Maxwellian electron distribution. Unfortunately, the values of Q_j for the majority of gases are not known exactly at present, which makes the successful use of the method of ultrahigh frequencies difficult. Possible values of Q_j for some gases (O_2, N_2, NO, N, O, Ar) are given by Lamb, Lin and Schulz [413, 527]. Comparison with experimental results shows that the value of Q_j for collisions of electrons with argon atoms given in these works is, apparently on the low side ($Q_j \sim 2 \cdot 5 \cdot 10–17$ cm^2 at $T_e = 4000\,°$K).

One of the possibilities of direct determination of Q_j is the simultaneous measurement of β for two wavelengths with the same value of N_e [100, 12], or simultaneous study of the absorption and reflection of one wavelength, [391]. A second possibility for determining Q_j is simultaneous measuring of N_e by the ultrahigh-frequency method and by some other method (for example, measuring the conductivity by the induction method [413]). In this manner the value of $1.5–2.0 \times 10^{-16}$ cm^2 (at $T_e = 4000\,°$K) was obtained [423]. Sometimes Q can be determined by simultaneous measuring of

* We wish to note that σ in a high-frequency field is related to the static conductivity σ_0 by

$$\sigma_0 = \sigma\left(1 + \frac{\omega^2}{\nu^2{}_{\text{eff}}}\right).$$

β and the magnitude of the phase shift in the ultrahigh-frequency wave when it passes through the ionized gas [100].

In practice one deals not with an infinite plasma but with layers of finite thickness. In this case it is necessary to take into account the reflection of the ultrahigh-frequency wave from the boundaries. The reflection coefficient R of ultrahigh-frequency waves is of substance only for high electron concentrations, which makes it possible to increase the upper limit of values of N_e which could be measured by a factor of two to five in comparison with measurements of N_e by absorption [424, 391]. A particularly simple relation exists between R and N_e in the one-dimensional case of reflection from an unbound medium, which surrounds the output of a rectangular waveguide:

$$|R|^2 = \left|\frac{1-\sqrt{\bar{\varepsilon}}}{1+\sqrt{\bar{\varepsilon}}}\right|^2 . \tag{13–6}$$

Arrangements for measuring the absorption coefficient of ultrahigh-frequency waves are similar to those used in optical measurements. In one of the versions the waveguide axis is directed across the shock-tube axis [430, 391, 424, 10, 526, 360a]. To one side of the test section of the tube is placed a source radiating the ultrahigh-frequency waves (usually this is a klystron generator), with the entire equipment for channeling the ultrahigh-frequency waves which is needed for synchronization with the waveguide and the load; to the other side of it is placed a crystal detector, the signal from which is fed to an oscillograph.

An object of particular care is the design of the antenna horn, which should ensure the propagation in the test gas of a sufficiently plane ultrahigh-frequency wave without perceptible scattering; to ensure good spatial resolution the cross section of the horn along the tube axis should be minimal. Jahn [391] assumes that when working with a wavelength of 1.25 cm the front of an ultrahigh-frequency wave emerging from the antenna horn will not differ from a plane front with an accuracy of $\frac{1}{16}$ of a wavelength, if the antenna will have the shape of a four-sided pyramid with a rectangular base, which serves as the entering aperture into the shock-tube test section and which has dimensions of 25×50 mm. The cross section of the waveguide used by Manheimer-Timnat and Low, had dimensions of 12.7×25.4 mm (in some experiments even 2.54×25.4 mm) [430]; however, the authors do not present an analysis of the configuration of the radio waves ($\lambda = 3$ cm) in the test gas. The large size of the input aperture in the test section of the tube limits the resolving power of the installation with time. An attempt to improve the resolving power was made by Lobastov who has introduced into the construction of the antenna conical dielectric rods and polystyrene lenses [100]. The working diameter of the beam thus obtained was about 20 mm, which has determined the resolving power of the installation (about 6 microseconds for M ~ 10). The diameter of the beam used should be in any case appreciably smaller than the length of the plug. If these quantities differ little from one another, then it is difficult to understand what produces the observed change in the intensity of the ultrahigh-frequency waves; the time needed for establishing the maximum absorption of radio waves is here more likely determined by the time of passages of the heated gas plug, rather than by the time needed for establishing equilibrium ionization. These doubts can be expressed, for example, when considering the results given in [430], since in the experiments described there (tube diameter about one centi-

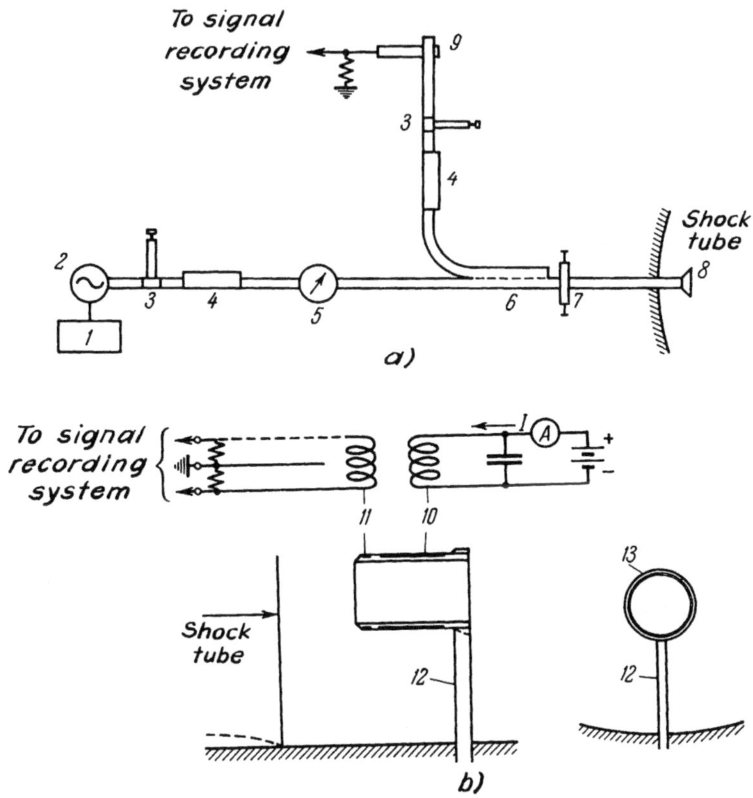

Fig. 84. Schematic drawing of the experimental arrangement of Lin, Neal and Fyfe, [424], used for measuring the electron concentration behind a shock front on the basis of microwave reflection (a) and changes in the magnetic field intensity (b). 1) power supply; 2) klystron; 3) unit for tunning and matching the generator and receiver with the waveguide; 4) attenuator; 5) wave meter; 6) directional coupler; 7) two-way switch; 8) quartz window; 9) detector; 10) solenoid for producing a constant magnetic field; 11) measuring coil; 12) supporting bracket; 13) holder. A) To the signal recording system; B) Shock tube; C) Shock wave.

meter, initial pressure 1–3 mm Hg for experiments with air) the plug dimensions may be very small.* To check the ratio of the plug length to the antenna dimensions it is possible to simultaneously obtain films of the flow pattern using a shlieren system [11]. Among other errors in determining N_e by the given method is the distortion of the field in the test section on the appearance of the ionized gas due to a certain inhomogeneity in the radiowave propagation pattern, effect of the boundary layer behind the shock front, appearance of charges at the tube walls as a result of friction between the gas and the wall, etc. The effect of the field inhomogeneity can be eliminated by calibrating the absorption and reflection of the signal. For this a stationary charge is passed through a filled (with hydrogen [430], or other gas) tube and the electron concentration is measured by some other method; the results of this calibration are compared with theoretical values.

A second version, which is free of the majority of the above shortcomings, was developed by Johnston and Kornegay [395], who have studied the kinetics of the appearance of electrons in xenon behind the shock front, reflected from the end face of the tube. Here the shock tube served as a waveguide and the ultrahigh-frequency waves have propagated along the shock-tube axis. A source of ultrahigh-frequency waves was placed in the low-pressure chamber behind the end wall, which consisted of a Plexiglass partition in the waveguide and had a sufficient mechanical strength and a low absorption of ultrahigh-frequency waves. The detector was placed in the high-pressure chamber. The observation time varied from 0.5 to 100 microseconds. In these experiments the thickness of the absorbing layer was changed in such a manner that the observed absorption index was determined by the integral $\int_0^x \beta \, dx$. The installation has a satisfactory time resolution; experiments using it made it possible to determine the ionization rate of xenon at 4200–7600 °K. In a similar arrangement a glass shock tube was slid into a rectangular waveguide [460]; here the character of the reflection of ultrahigh-frequency waves from the shock wave and from the interface and the rate of electron diffusion ahead of the shock front were studied.

An example of successful use of the microwave method in the study of kinetics of thermal ionization in a shock wave propagating in air is the work by Lin, Neal and Fyfe [424]. This success was due to a fortunate selection of experimental conditions. Thus, the shock-tube diameter (610 mm) exceeded manyfold the length of the ultrahigh-frequency waves which were used, which made it possible, in measuring the coefficient R, to assume that the ionized gas occupies the entire half-space. The use of a tube with a large diameter has made it possible to work with low pressures ($p_1 = 0.02$–0.2 mm

* Estimates performed for this case using results noted in Sect. 5 show that the maximum length of the plug in experiments of [430] could not exceed 2–6 cm.

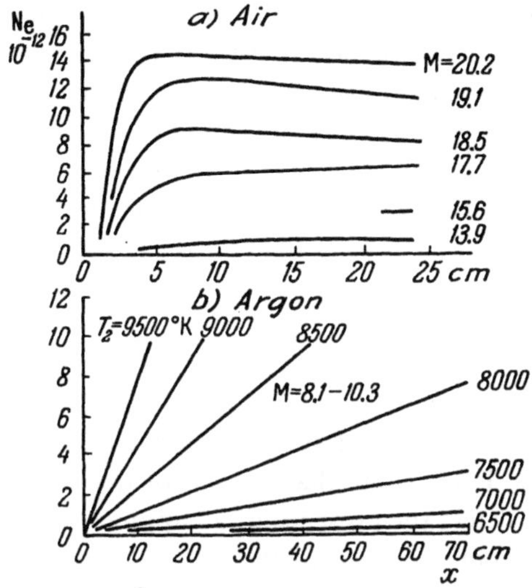

Fig. 85. The electron concentration N_e (in cm^3) behind the shock front as a function of the distance from the shock front in air (a) and argon (b) obtained by the microradio-wave method [424, 360a]. The initial pressures: in air $p_1 = 0.02$ mm Hg, and argon $p_1 = 5$ mm Hg.

Hg), when the nonequilibrium ionization zones are larger. The increase in the ratio of the size of the nonequilibrium region to the waveguide size has appreciably increased the resolving power of the installation and has made it possible to perform measurements with the waveguide axis perpendicular to the shock-tube axis. Another important advantage of working with low pressures is the low collision frequency $\nu_{\rm eff}$ in comparison with the radio-wave frequency ω, so that $(\nu_{\rm eff}/\omega) \lesssim 1.5 \cdot 10^{-2}$ for $p_1 = 0.02$ mm Hg. This makes it possible in the determination of R to avoid the aforementioned difficulties, which are due to the need to know exactly the collision cross sections of electrons with heavy particles. In this case all the relationships used in calculations simplify, since $\varepsilon = 1 - (4\pi e^2 N_e / m\omega^2)$. Measurements of N_e on the basis of R are here fully reliable for values of $(2 \cdot 10^{12}/\lambda^2 \leqslant N_e \leqslant (1 \cdot 10^{13}/\lambda^2))$, ($\lambda$ is in cm). This has enabled Lin, Neal and Fyfe, using $\lambda = 2.94$, 1.84 and 0.855 cm, to measure the distribution of N_e along the flow behind the shock front in air in a range of Mach numbers from 14 to 20 (Fig. 85a). The experimental setup used by them is shown in Fig. 84a. To reduce the effect of the boundary layer and of the shock-front curvature the waveguide was moved into the tube channel by 10 cm. The resolving power of this arrangement can be characterized by the fact that the smallest measured size of the ionization development zone for M = 20.2 and $p_1 = 0.02$ mm Hg was found to be equal to ~ 1.2 cm. Among other works in which the ionization kinetics in air were studied by microradiowaves we wish to point out experiments of Manheimer and Low [430] and by Bazhenovaya and Lobastov [12 and 13]. These experiments were performed using lower shock-wave velocities and higher initial pressures. As was pointed out, considering the experiments of [430] the danger arises that the dimensions of the heated plug are too small; in experiments of [12 and 13] the time for reaching the maximum absorption of radio waves is by a factor of two to three shorter than the duration of the flow in the plug, but only just about exceeds the resolving power of the installation (6 microseconds).

Harwell and Jahn [360a] have performed a successful study of the ionization kinetics of argon, krypton and xenon for $\mathsf{M} = 7$–10 and $p_1 = 3$–17 mm Hg in a tube 50×50 mm in cross section using a waveguide directed across the shock-tube axis. Let us note that the use of the microwave method has made it possible to detect the presence of an appreciable electron concentration (above 2×10^{13} cm^{-3}) ahead of the front of a very strong shock wave [410].

Another method for measuring N_e by ultrahigh-frequency methods is the so-called resonance method [52, 173], the substance of which consists in recording changes in the quality factor of the resonator the central part of which receives the ionized gas. Using this method it is possible to measure N_e up to 10^{13}–10^{14} cm^{-3} (see, for example [232]). Brandt and Kurtmulayev [32] have attempted to use this method to measure the electron concentration behind a shock front in a shock tube. At present no successful experiments have been performed for using the resonance method for the study of thermal ionization kinetics in shock waves. This is due to difficulties in improving the resolving power, since this method is advantageous only with small-diameter shock-tube channel. Here it is possible to work only with sufficiently high gas pressures, which is responsible for the very short size of the non-equilibrium ionization zone.

[3] The Magnetic Induction Method

The magnetic induction method worked out by Lin, Resler and Kantrovitz [425] has been used for obtaining a number of estimates of the time needed for establishing ionization equilibrium. In this method a magnetic field is introduced into the shock tube; the appearance of ionized gas in this section of the tube results in displacing the magnetic lines of force and in the appearance of an EMF in a special measuring coil. The constant magnetic field is produced by a solenoid. If the test section of the tube is made from quartz, glass, Plexiglass or some other insulating material, then both coils can be placed at the outer surface of the tube along its perimeter.

In another version of this method both coils were placed inside a large-diameter tube [424]. The displacement of the lines of force and, by virtue of the same fact, the magnitude of the EMF which is induced in the measuring coil and recorded by an oscillograph, depends on the conductivity of the gas behind the shock front, *i.e.*, on N_e. It was found that the most convenient way for establishing a unique relationship between N_e and the observed EMF is calibration of the oscillograph readings using a metal rod with a known conductivity, which was shot through the measuring section of the tube with a known velocity. The results of measuring the conductivity σ_0 show that the use of this method is most successful when $N_e \sim 2 \times 10^{14}$ cm^{-3}. For analyzing the process of appearance of free electrons in the gas behind the shock front it is important to establish a relationship between the conductivity distribution curve in the gas and the shape of the observed pulse. A detailed consideration of this problem is given in [425] for a large variety of cases. Thus the above method makes it possible to determine σ_0 without using (unlike the microradiowave method) the given electron scattering cross sections. However, in order to calculate N_e from the measured σ_0 one must know the scattering cross sections, since in weak, slowly varying waves approximately

$$\sigma_0 \approx \frac{e^2}{m\nu_{\text{eff}}} N_e. \tag{13-7}$$

To determine the time of establishing ionization equilibrium it is sufficient to know only the behavior of changes in N_e, *i.e.*, to perform relative measurements. Hence in this case one can restrict himself to measuring the dependence of σ_0 on the distance from the shock front, assuming that ν_{eff} does not vary appreciably in the nonequilibrium region. As in the case of the ultrahigh-frequency wave method, the greatest successes in the use of the magnetic induction method for the study the kinetics of appearance of charges behind a shock front in air were achieved in [424]. The schematic drawing of their installation is given in Fig. 84b, the solenoid and the measuring coil were placed in a thin-walled nylon tube placed on a metal supporting bracket inside the shock tube. The nose part of the nylon tube is made so that the aerodynamic disturbance which was produced distorted primarily the external part of the gas flow. The distortions which resulted from magnetohydrodynamic interactions between the plasma and the magnetic field of the solenoid, according to estimates under experimental conditions, were found to be negligible even for stronger waves. In the case under consideration the calibration was performed by two coaxial aluminum cylinders which have embraced both sides of the induction tube.

The distance δ_σ at which a constant conductivity level behind a shock front is established is determined from oscillograms by comparing the shape

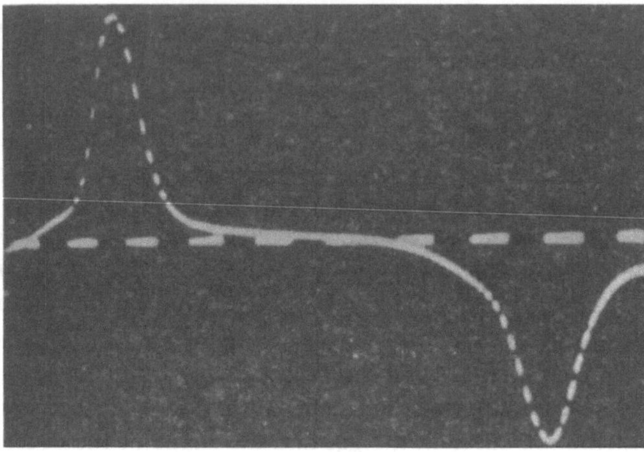

Fig. 86. Specimen of a calibrating curve for the time dependence of the EMF, obtained upon shooting through a metal rod past the measuring coil (in the induction method for measuring the electron concentration N_e.

of the calibration and measurement curves assuming that in the shock wave itself σ increases linearly. For this the widening of the measured signal is measured and compared with the half-width of the calibrating signal (Fig. 86). The corresponding values of σ_σ can be obtained from relationships given in [425] (Fig. 87). It is remarkable that the values thus obtained are close to those obtained in the study of the ionization process by the microradiowave method.

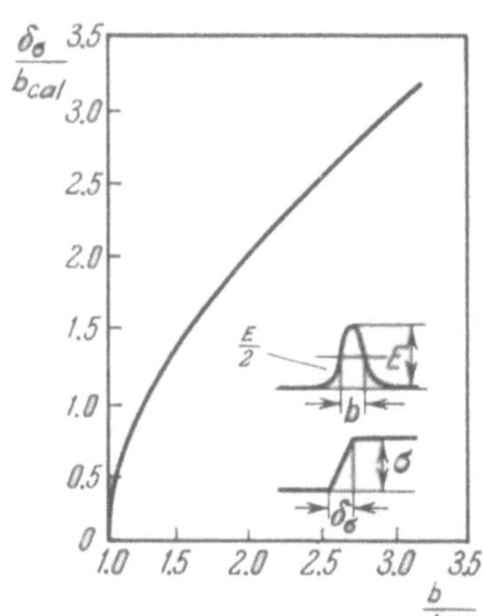

Fig. 87. Calculated dependence of δ_σ (distance at which a constant conductivity leve, is established) on the width of the observed signal b, referred to the width of the calibration signal b_{cal}.

The resolving power of the above method is characterized by the quantity b_{cal}, *i.e.*, by the half-width of the potential pulse recorded on calibration, relating this quantity with the effective length l_{cal} of those regions of the magnetic field which participate in the formation of the calibrating pulse. In experiments described in [413, 424] l_{cal} was approximately 2.4 cm. The actual resolution, as the ability of measuring the small dimensions of the zone in which constant conductivity is established is, apparently, better than the above value, which enabled the above authors to measure δ_σ up to magnitudes of 2 cm for $\mathsf{M}=20.2$ with sufficient reliability.

The above method was used for measuring the conductivity and estimating the time needed for establishing equilibrium in even stronger shock waves which propagated in deuterium with a velocity of 100 km/sec [212].

[4] Using the Stark Effect

One of the methods used for measuring large electron concentrations in plasma is the study of the width and displacement of atom lines produced by the Stark effect, *i.e.*, as a result of interaction of the radiating atoms with the microfields of the electrons and ions which surround them. In the last few years this method has also come into use in studies performed in shock tubes [475, 1], particularly when using very high shock-wave velocities [305, 442].

As is known, a linear Stark effect is observed for hydrogen atoms and for hydrogenlike ions, while a quadratic Stark effect is peculiar of other atoms. Hence it is necessary to differentiate between methods for measuring N_e in hydrogen plasma and in the plasma of other gases. In a particular case, hydrogen can exist as some admixture to the primary gas, let us say, argon and, nevertheless, N_e can still be measured on the basis of the Stark effect in hydrogen, *i.e.*, under linear effect conditions.

The basic theory of the Stark effect has been worked out by Lenz, Weiskopf, Lindholm and Holtzmark, and the results they have obtained have become an integral part of a number of textbooks and monographs. However, it was shown in the last few years that in the case of the linear as well as of the quadratic Stark effect certain requirements are needed. A systematic presentation of the present theory of the Stark effect can be found, for example, in the book by Sobel'man [180].

For the case of widening of hydrogen lines a correction was made for the thermal motion of ions (V. I. Kogan [80]), the effect of electrons (Griem, Kolb and Shen [349, 350]) and the combined effect of electrons and ions (see [410]). The expression for the line contours thus obtained are complicated and final results can be obtained only numerically—separately for each line. These calculations were performed in [349, 351]; they can be regarded as valid for electron concentrations of 10^{14}–10^{17} cm^{-3}, correct within 10% and in good agreement with experimental results. Here N_e is then determined by comparing the contour of the observed line with a number of calculated contours of the same line, obtained for different values of N_e and by selecting a contour for calculations which is closest to the experimentally obtained contour [1, 78]. Alyamovskiy and Kitayeva [1], using a shock tube, have obtained an accuracy of at least 15–20% in measuring N_e. McLean and others [442] have measured N_e by the widening of lines of ionized helium with an error of up to 50%. Griem, however, assumes that in the study of contours of H_β lines with $N_e \sim 2 \times 10^{17}$ cm^{-3} and $T \approx 18{,}000\,°K$ it is possible to obtain an accuracy in measuring N_e of not less than $\pm 5\%$. Comparison with experimental data shows that it is possible to appreciably reduce the error by measuring the widening of a number of lines at the same time, determining N_e and then averaging the result thus obtained.

In the case of the quadratic Stark effect the line with γ and the shift of the maximum Δ, according to the Weiskopf-Lindholm theory, are determined by

$$\gamma = 11.4 C_4^{\frac{2}{3}} v^{\frac{1}{3}} N, \quad \Delta = 9.8 C_4^{\frac{2}{3}} v^{\frac{1}{3}} N, \quad \gamma/\Delta = 1.16, \tag{13–8}$$

where C_4 is the constant of the quadratic Stark effect, v is the relative velocity of the atom and the exciting particle, and N is the number density of the exciting particles. These formulas, however, are valid only for small v. Mandel'shtam and Mazing [120] have found that the experimental results obtained in argon plasma do not conform to Eqs. (13–8). The subsequent development of the theory of the quadratic Stark effect has made it possible to overcome these disagreements by taking into account nonelastic collisions of electrons with atoms and the reduction of the effect of rapid traverses by electrons by the amount of widening (within the framework of the non-steady-state theory due to Sobel'man-Vaynshteyn [181]). In this case the widening and shift of the lines are determined by the parameter

$$\beta = \frac{Z^2 (2\pi c \Delta E_q)^{\frac{3}{2}} C^{\frac{1}{2}}}{v^2}, \tag{13–9}$$

where Z is the ionic charge, ΔE_q is the distance from the ion under consideration to the nearest exciting ion (cm^{-1}), so that Eqs. (13–8) now contain instead of γ and Δ the correction factors $I'(\beta)$ and $I''(\beta)$, respectively. Specific values of β and C_4 for a number of lines of inert gases can be found in [120, 116, 117], and the values of functions $I'(\beta)$ and $I''(\beta)$ are given in [181]. Satisfactory agreement with experimental data is here noted.

The setting up of experiments for study of the Stark effect in plasma behind a shock front is complicated by the need of simultaneously obtaining (after passing of the shock wave) the dependence of the radiation on the wavelength (over the spectrum) as well as on time with a high resolving power. The photographic method is most frequently used in experiments [475, 1, 442, 416]. In this case either the spectral region under study is projected on a moving film, or the spectrum is scanned in time by moving the image of the spectrum over a stationary film, using a disk with inclined slits which rotates in front of the input slit. The time resolution which is thus obtained is not too high. Thus, in experiments by Petschek, *et al.*, [475], it comprised about 12–24 microseconds (in terms of the natural time); the authors have found that the measured N_e ($N_e \sim 10^{16} - 10^{17}$ cm^{-3}) corresponds to equilibrium values. This is understood, since the resolving power in time of the arrangement under experimental conditions is comparable with the time needed to establish ionization equilibrium (obtained from later measurements by other methods [474]).

The use of photoelectric recording methods is more promising. For this it is necessary to divide the image of each spectral line into a number of sections and to direct each of them to the photomultiplier. A number of methods is available for successfully overcoming this problem, such as the use of multicomponent slits and mirrors, hair optics for breaking up the ray, inclined slits, interferometer techniques, etc. [410]. Measuring the image of one line into seven segments has made it possible to study the behavior of the contour of a number of lines in shock waves in helium and deuterium [305]. Malyshev and others [118] suggest the use of the Fabry-Perot standard [interferometer] with a monochromator and an electron-optical amplifier for obtaining contours of spectral lines with a high time resolution.

[5] Recording the Optical Radiation

As was noted in the preceding section, the intensity of the continuous radiation produced by free-free and free-bound electronic transitions in quasi-neutral plasma is proportional to the square of the electron number density. It is known that the radiation intensity in this case depends also on the electron temperature (see, for example [22]). However, in the region where N_e is sufficient for observing the radiation the electron temperature does not vary appreciably (see [25]). Hence for the change in I, the radiation intensity, with time behind a shock front we have

$$\frac{1}{I}\frac{\mathrm{d}I}{\mathrm{d}t} = \frac{2}{N_e}\frac{\mathrm{d}N_e}{\mathrm{d}t}. \tag{13–10}$$

The arrangement for measuring the distribution of the radiation intensity behind a shock front is similar to that described in the preceding section. Particular attention must here be given to reducing the level of light scattered from the tube walls, in order to increase the sensitivity of the installation. This is important in the study of weak radiation at the initial ionization stage. As before, it is necessary to provide sufficiently high time resolution.

Using such an optical arrangement, which includes collimating devices, a spectral device and photomultipliers, Petschek and Byron [474] have studied the behavior of changes in N_e behind a shock front in argon and have measured the rate of appearance of electrons in the process of establishing equilibrium ionization. For checking purposes the measurements were performed in different spectral sections which were free from lines and they have yielded the same results.

It should be noted that the method of study of changes in N_e by the continuous radiation can be used only in the case when it is produced by free-bound or free-free transitions. Before it is used for the study of ionization kinetics in molecular gases it is necessary to determine the origin of the spectrum, since in this case the radiation at moderate temperatures can be produced by bound-bound transitions in molecules or ions.

One of the possibilities for measuring the concentration of charged particles is the study of ionic line intensities, since the intensity of these lines is related to the ion concentration by an equation such as (12–7) and for optically thin layers it is proportional to the ion concentration. These measurements were performed by McLean and others [442] for certain helium lines.

Niblett and Blackman [462], and Haught [361] have studied ionization kinetics by filming the radiation of the entire spectrum; particular care was paid to the study of the distribution patterns of the radiation of cesium vapor, small amounts of which were added to the argon, behind the shock front. For temperatures of 4500–9000 °K behind the shock front Haught has obtained films of the total radiation in the visible and in the near infrared regions of the spectrum. Control photographs of the spectral radiation distribution (without development in time) have shown that the recorded radiation originates primarily in the linear spectrum of cesium, and the admixture lines (potassium rubidum) cannot appreciably affect the results. In the beginning a very weak radiation is observed behind the shock front, then its level rises appreciably, reaching a maximum and the drops to some constant level (Fig. 88a). Haught relates the first period τ_1 to the development of atom-atom processes of excitation and ionization of cesium atoms; as these processes develop, the upper excited levels of the cesium atoms are filled, which produces intensive radiation. During the second period τ_2 of decisive importance to the ionization kinetics are collisions of atoms with the appearing electrons; ionization of the cesium atoms results in reducing the population

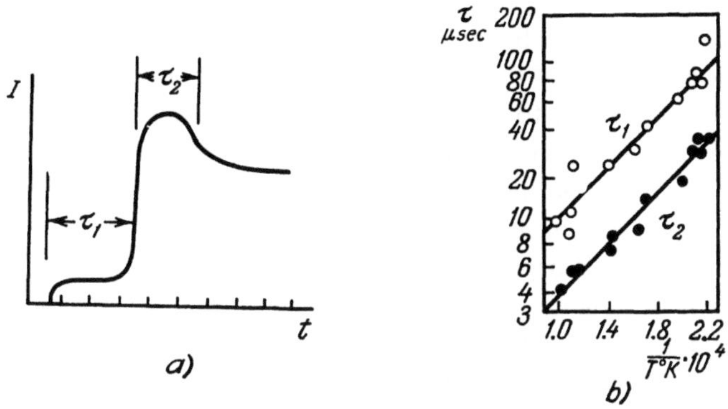

Fig. 88. Radiation intensity of a moving gas as a function of t_{nat} behind a shock front in a mixture of cesium vapor with argon (a) and the excitation times τ_1 (open circles) and τ_2 (blackened circles) as a function of the reciprocal of the temperature (in terms of the natural time) (b), [361].

of the excited atomic states and reduction in the radiation level to the equilibrium value. The picture thus described is in accord with the presently held view of ionization kinetics in shock waves (see Sect. 19); the values of τ_1 and τ_2 obtained by Haught are presented in Fig. 88b.

[6] The Interferometer Method

Finally, high values of N_e ($N_e \sim 10^{16}$ cm^{-3}) are measured by the Mach-Zehner interferometer [227–229, 406]. The refraction index for the electron gas n_e in the visible spectral region in the absence of appreciable light absorption is $n_e = \sqrt{\varepsilon}$; determining ε by known relationships, we obtain with a satisfactory approximation

$$n_e = 1 - \frac{2\pi e^2}{m\omega^2} N_e = 1 - 4.475 \cdot 10^{-30} \lambda^2 N_e, \qquad (13\text{–}11)$$

where λ is given in angstroms. Thus for an electron gas the quantity $n_e - 1$ is proportional to the square of the wavelength. The main difficulty is measuring the particle density in a mixture of gases by using an interferometer which was noted in Sect. 10, which are the need to know the specific refractions and the gas component composition, can be overcome in this case, since the refraction of the electron gas differs appreciably from the refraction of the other components. Thus, according to Alpher and White [228], $n-1$ for argon and an electron gas for $\lambda = 5413$ A is $+1.06 \times 10^{-23} N_{Ar}$ and $-13.33 \times 10^{-23} N_e$, respectively, *i.e.*, $n-1$ for electrons not only appreciably exceeds this value for atoms but also is of opposite sign. The presence of ions in the mixture does not change the picture too much; the specific refraction of argon ions is 0.67 of the specific refraction for atoms. This is the fact which makes it possible to assume that the light dispersion in an ionized gas is determined basically by electrons, and the contribution made by atoms and ions into the dispersion has the character of a correction. Since the refraction index in a medium made up of atoms and ions with a wavelength far removed from the absorption lines does not vary appreciably, and that for electrons depends highly on λ, the effect of atoms and ions on the final result can be appreciably reduced by measuring the difference in the values of n for two different wavelengths λ_1 and λ_2. Then $n(\lambda_1)-n(\lambda_2)\approx n_e(\lambda_1)-n_e(\lambda_2)=4.475\cdot 10^{-30}(\lambda_2^2-\lambda_1^2)\,N_e$. This is the manner in which it is possible to isolate the effect of electrons and to measure changes in N_e by an interferometer.

The layout of such an installation is described in Sect. 10. Performing experiments in argon with $10<\mathrm{M}<20$, Alpher and White [228] have obtained values of N_e in the equilibrium region ($N_e\sim 10^{16}-10^{17}$ cm^{-3}) which are very close to the calculated values; the probable error in determining N_e was about 10%. They have discovered nonequilibrium regions behind the shock front. The error in calculating N_e in the nonequilibrium region can become more appreciable due to inaccuracies in determining the position of the front on each interferogram (λ_1 and λ_2), which results in displacing the curves of the refraction index distributions for each wavelength and in changes in their difference. Unfortunately, detailed studies of ionization kinetics by the interferometer method were not as yet performed; at the same time this method is highly promising for measurements in sufficiently large nonequilibrium zones with large values of N_e.

[14] OTHER METHODS OF MEASUREMENT

A large number of methods for the study of the structure of shock waves and nonequilibrium phenomena in shock waves exists in addition to those described in the preceding sections. Some of them are used for measuring the same gas characteristics which were discussed above (Sects. 11 and 12), but without using methods of spectral analysis. Other works consider the effect of physicochemical processes on the gasdynamic picture of the flow behind a shock front. In the present section we shall only briefly point out the characteristic features of these methods.*

* A detailed presentation of the measurement methods discussed in this section is beyond the scope of this book, since the results obtained by these methods are practically not used in the subsequent discussion. They can be found in the sources quoted in the text.

[1] Thermal Measurements

Sherman and Talbot [532, 534, 566], have used an thermoanemometer in their successful study of the structure of a shock front in air and helium. A thermoanemometer [hot-wire anemometer, as commonly known; editor] is a thin wire heated by an electric current and placed in the gas flow perpendicular to the direction of flow. The quantity of heat given off by such a wire depends on its geometry, physical characteristics of the medium and the flow conditions. If the quantity of supplied heat [electrical current] is held constant, then the interaction between the wire and the flow changes the temperature of the wire [constant-current hot-wire anemometer; editor]; in another arrangement the heating is adjusted so that the temperature of the wire should remain constant [constant-temperature hot-wire anemometer; editor]. The wire thus serves as a thermoanemometer probe. Among the variables which can affect the readings of such a probe are the flow velocity, density and the temperature of the gas [211].

Sherman and Talbot have studied a standing plane shock wave which is formed in the supersonic flow of a sufficiently rarefied gas in a wind tunnel. The diameter of the filament wire used in the experiment was small in comparison with the molecular mean-free path. To obtain the distribution of variables across the shock front the probe was moved along the flow. A specimen of a temperature distribution across the shock front in air with $M = 1.89$ obtained by this method is shown in Fig. 89. Hot-wire anemometric probes are also used in the study of shock-tube flows [376]. Using film sensors for measuring the thermal flows, Millar and Price have studied the structure of shock waves in argon up to $M = 14$ [445]. Thermal film sensors are used extensively in the study of heat transfer, of the behavior of the boundary layer in shock tubes, for measuring shock wave velocities, etc.,

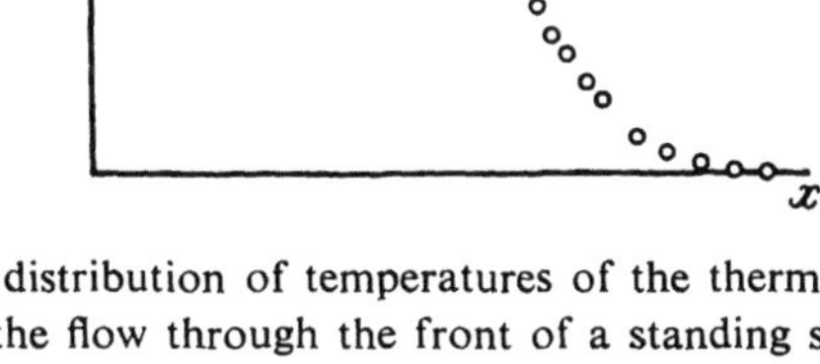

Fig. 89. Shape of the distribution of temperatures of the thermoanemometer filament upon moving it along the flow through the front of a standing shock wave in rarefied air with $M = 1.89$.

Bromberg [273] points out that by measuring the wall temperature of a shock tube by using thermal film sensors it is possible to analyze processes in the main body of the flow behind the front, however, this method has not as yet been used for study of relaxation processes.

[2] Chemical Analysis

Since the rate of certain reactions is highly dependent on the temperature, then by sharply cooling a previously heated gas it is possible to retard the development of reactions, which results in "freezing" certain reaction products. Following this, analysis of the final state of the gas will make it possible to determine the reaction rate preceding the cooling. This method is known for a long time in chemical physics [70]. This process is particularly simply achieved in a shock tube, heating the gas by a shock wave and cooling it in a rarefaction wave. A special shock-tube design was created for this purpose (Fig. 90) [336, 81]. The gas is heated behind the front of a wave reflected from the end wall of the shock tube. To obtain a sufficiently intensive rarefaction wave which would follow the shock wave, a large volume III is placed behind the high-pressure chamber and is separated from it by a second diaphragm D_2.* Before performing the experiment, this volume is evacuated to a minimal pressure. The tube design contains special devices, *i.e.*, "plungers," which can be used to first pierce the diaphragm D_1 and then D_2. The plungers are actuated by detonation waves which arrive from an auxiliary detonation tube filled with a combustible mixture. The required time difference in piercing the diaphragms is determined by the difference in the length of the supply tubing. After D_1 bursts, a shock wave propagates in

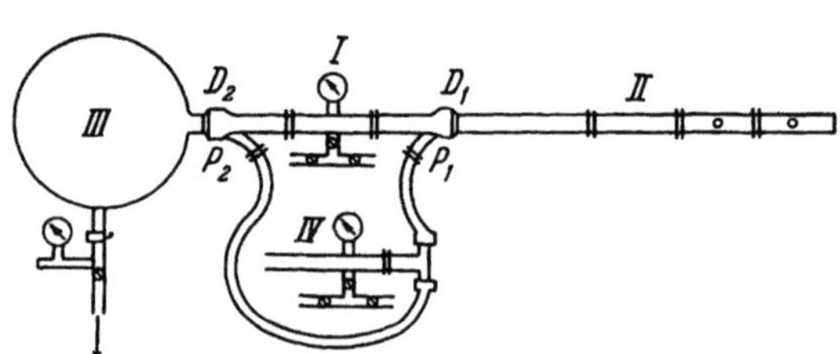

Fig. 90. Schematic drawing of a shock tube used for heating and rapid cooling of the test gas and for subsequent chemical analysis of the frozen reaction products. I is the high-pressure chamber; II is the low-pressure chamber containing the test gas, III is a receiver which serves to form a rarefaction wave, IV is an auxiliary detonation tube which is needed for actuating devices P_1 and P_2 which pierce the diaphragms D_1 and D_2 [81].

* A similar shock tube, but without volume III was developed by Jacobs, *et al.* [390].

the low-pressure chamber, and after D_2 is burst it is followed by a rarefaction wave. To resolve chemical kinetics problems using these tubes it is necessary to know to what temperature was the gas heated in the reflected wave and during what time did this heating take place. The temperature in the reflected wave is usually determined from calculations on the basis of the velocity of the incident wave [486]; to reduce the effect of deviations from one-dimensional flow, the test gas is usually introduced as an admixture to an inert monatomic diluting gas. The gas heating duration is measured by a piezoelectric sensor which is placed near the end wall of the shock tube. A specimen of an oscillogram of pressure obtained by Kozlov and Knorre is shown in Fig. 91. After the pressure is increased first in the incident and then in the reflected wave, a certain practically constant pressure level is established; then, after the cooling rarefaction wave arrives the pressure begins to drop. The temperature also drops simultaneously.

After the rarefaction wave has passed, a gas sample is rapidly taken for analysis into a prevacuumed vacuum pipet. Then the gas composition can be analyzed by using the most varied methods such as the mass spectrograph, volumetric and other methods. In the study of process kinetics in a gas heated by a shock wave it is most advantageous to use chromatographic analysis, which is distinguished by its sufficient sensitivity to small doses of the test gas and by satisfactory accuracy. A specimen of a chromatogram obtained by Kozlov and Knorre by analyzing the decomposition of hydrocarbons is shown in Fig. 92. This method was used to study the kinetics of formation of nitric oxide in air at temperatures of 2000–3000°K [336, 337], decomposition of carbon dioxide [264], hydrocarbons, ethane, methane, ethylene, acetylene, etc. [79, 82, 335, 543]. Results obtained in the study of kinetics of nitric oxide formation behind a shock front in air, obtained by the "freezing" method are presented in the fifth chapter.

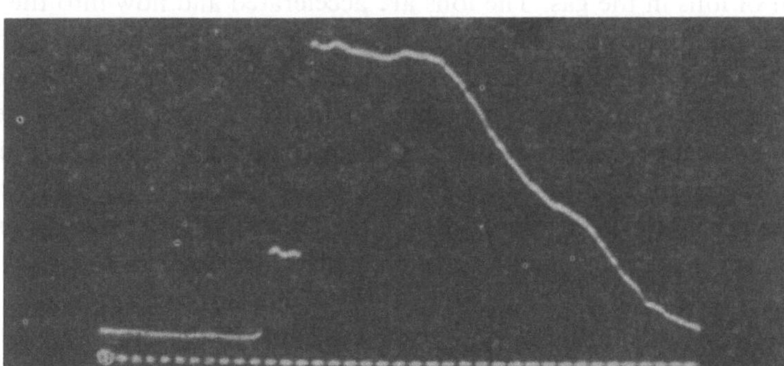

Fig. 91. Specimen of an oscillogram of gas (argon) pressure near the end of a shock tube depicted in Fig. 90. The time marks are every 100 microseconds.

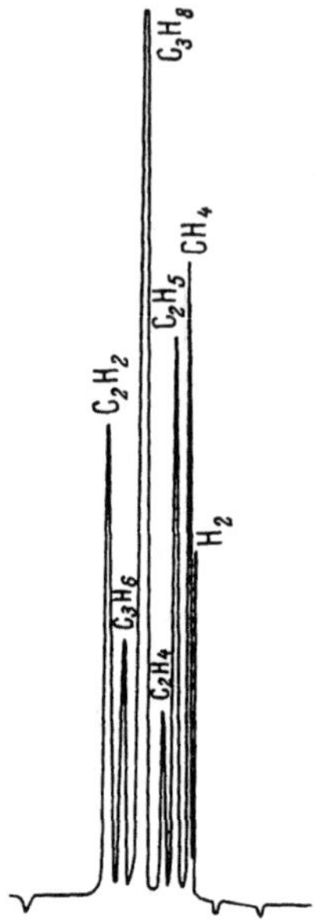

Fig. 92. Specimen of a chromatogram on which the result of analysis of gas composition attendant to the study of thermal decomposition of methane in a shock tube shown in Fig. 90 is recorded. The analysis time is measured along the abscissa axis and the concentration is laid off on the ordinate axis.

Breadley and Kistiakowsky have studied the gas composition behind a reflected shock wave using a time-of-flight mass spectrometer [266–268]. The experimental arrangement is as follows: the gas heated by the shock wave flows from a narrow aperture in the end wall of the tube in the evacuated chamber of the mass spectrograph. At a distance of about 1 mm from the aperture the gas flow intersects a beam of electrons, which results in the formation of ions in the gas. The ions are accelerated and flow into the transit chamber and then move with velocities which are inversely proportional to the square root of the mass of the corresponding ion. Thus, ions which have left the aperture at the same time arrive at different times, which depend on their masses, at the detector situated at the end of the chamber. Scanning the signal thus obtained in time, it is possible to obtain a mass spectrum of particles removed from the volume behind the reflected shock wave. In order to avoid superposition of signals, the ionizing electron beam is chopped off at a frequency sufficient for completing the scanning of the spectrum. In the Breadley and Kistiakowsky experiments this frequency was 20 kc/s, which has made it possible to obtain a complete mass spectrum every 50 microseconds; the time used by Diesen and Felmlee [313, 314] is smaller by a half. Naturally, this value determines the resolving power in time of the installation. Thus, if the characteristic time of the reaction under study is less than 100 microseconds, then the use of the above method will be ineffective. On the other hand, in the study of very slow reactions the change in the mass spectrum with the experiment time is very small, which establishes another limit for the method applicability. Using the above method it was possible to achieve successes in the study of the decomposition of nitrous oxide at temperatures of 1780–2000 °K [267]. The decomposition of nitromethane [266], oxidation and pyrolysis of acetylene [268], as well as the effect of the hydrogen-deuterium isotope exchange in the reaction $CF_3 + CH_2D_2$ [396], was also studied. Diesen and Felmlee have studied the dissociation of chlorine at temperatures of 1700–3200 °K and the decomposition of hydrazine at $T = 1200$–2500 °K [313, 314].

[3] Gasdynamic Experiments

A number of works presents the study of the effect of physicochemical processes on the flow variables in the shock tube: they contain measurements of the flow velocity, of rate of propagation of disturbances in the gas, of the velocity of the reflected shock wave, etc. Unfortunately, the available methods for measuring the flow velocity by using photoscanning of the shlieren pattern of the process [7] or the gas radiation [361] are not sufficiently effective. One of the first attempts to study the rate of propagation of disturbances in the flow behind the shock front was attempted by Marlow and others [432]. They have measured the velocity and temperature behind the shock front by intersecting the gas flow by a beam of ultrasonic radiation, which was generated by a piezoelectric source. The ultrasonic wavelength and the inclination of the beam due to the fact that it was carried away by the flow were recorded by instantaneous photography through the shlieren system; it was not possible to obtain distributions of flow variables by this method. In other works disturbances were produced by inserting a wedge into the flow behind the shock front and by photographing the flow around it (using a shlieren system) [38, 179, 14]. This method was used by Semenov to determine the dissociation energy of nitrogen by the angle of inclination of an attached oblique shock [179]. Some judgements on the rate of the excitation and decomposition of carbon dioxide molecules can be made on the basis of results obtained by Bazhenova and Naboko, who have measured M_2* in the flow behind the front of a passing shock wave on the basis of the Mach angle α obtained at a wedge one of whose planes coincided with the

* The Mach number M_2 which is given here pertains to conditions behind the front of a passing shock wave and is equal to the ratio of the flow velocity to the sound speed in it.

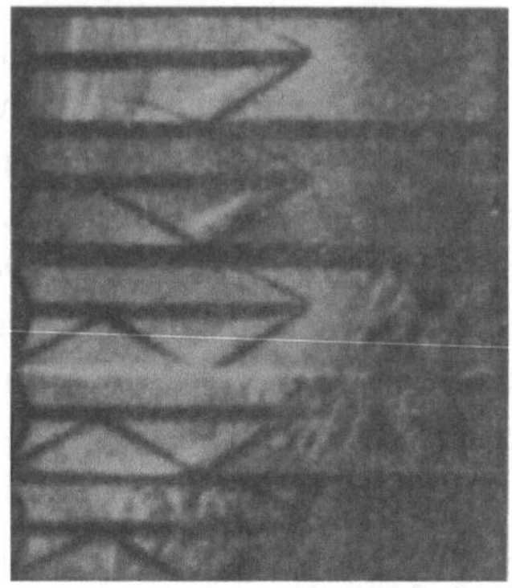

Fig. 93. Shlieren photograph of the flow past a wedge behind a shock front propagating in carbon dioxide at a speed of 2.2 km/sec. On the photograph the shock moves from the right to the left [14].

direction of flow of the gas (Fig. 93) [14]. In this case, as is known, $\alpha = \arcsin (1/M_2)$. The values thus obtained were compared with theoretical values calculated on different assumptions on the state of the fluid and on the behavior of sound propagation in the flow. The advantage of the above method consists in the fact that the study is performed in the main body of the flow, which is free of the boundary layer; the disadvantages consists in the not too high spatial resolving power, since the results pertain to a large volume of gas behind the shock front.

The time needed for establishing equilibrium in the gas can be obtained by measuring the velocity of the shock wave reflected from the end wall of the shock tube. Such attempts were made by Strehlow and Cohen [550], who have estimated the relaxation time for the vibration of nitrogen and oxygen molecules. Bazhenova and Zaytsev have studied the dissociation of carbon dioxide [9]. For this they have obtained continuous shlieren photographs of the gas flow near the end of the shock tube. These photographs make it possible to measure the velocity of the incident and reflected shock wave (see Sect. 7). Naturally, the velocity of the reflected wave depends on the state of the gas in a volume behind the reflected wave. By comparing the measured and calculated velocities of the reflected wave an estimate was obtained of the time of decomposition of carbon dioxide molecules at temperatures up to 3000 °K. Among the factors which complicate the interpretation of results obtained by these methods we must point out the effect of the boundary layer at the shock-tube walls (see Sect. 6). Soloukhin [190] has placed at the end of the shock tube a large-diameter chamber and has obtained an appreciable reduction in the boundary-layer effect (Fig. 94a). A specimen of the continuous photograph of the flow in the center of the tube ahead of a

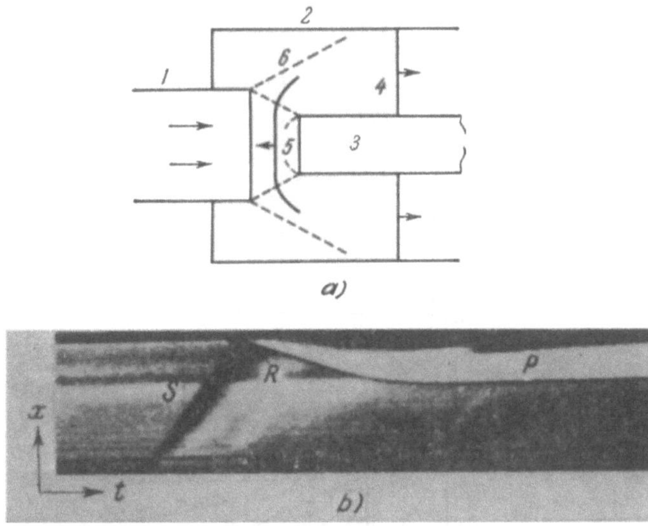

Fig. 94. Arrangement for measuring the velocity of the shock wave reflected from the end wall of a cylinder placed in a wide chamber (to eliminate the interference of the boundary layer) (a) and a continuous photograph of the shlieren pattern of the reflection (b). 1) Shock tube; 2) wide chamber; 3) retarding cylinder; 4) front of the passing wave; 5) front of the reflected wave; 6) Mach lines; *S* is the trace of the incident wave, *R* is the trace of the reflected wave and *P* is the trace of the standing wave in the flow around the cylinder [190].

cylinder placed in the additional chamber is presented in Fig. 94b. All these results [550, 9, 190] are at present only estimates, since no consideration is available of the complex mechanism of continuous effect of the changing (in the process of relaxation) state of the gas in the volume behind the reflected shock wave on the wave velocity. Particularly, difficulties arise in analyzing the transformation kinetics in systems where several reactions may take place simultaneously.

One of the variables which is sensitive to changes in the rate of physico-chemical processes is the distance through which the shock front moves away from the forward part of a blunt body in supersonic flow. Schwartz and Eckerman [529] have measured the distance through which the shock wave has moved away attendant to the supersonic motion of spheres in gaseous chlorine using different initial pressures (Fig. 95a). For $\mathsf{M} > 3$ the measured departure distance corresponds to the equilibrium excitation of vibrations in chlorine molecules; for $2 < \mathsf{M} < 3$ and an initial pressure of 4 mm Hg, this quantity corresponds to the state in which the molecular vibrations are "frozen." It may be assumed for rough estimates that a transition from one state to another takes place at a velocity $V \approx (\varepsilon/\tau)$, where ε is the departure

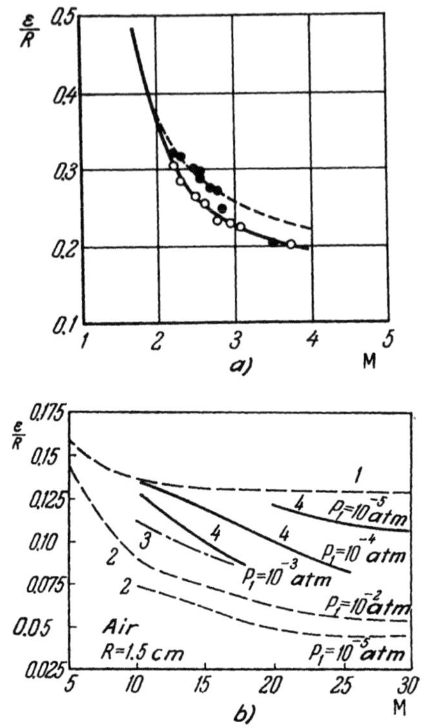

Fig. 95. Distance between the front of a departed shock wave and the surface of a sphere, referred to the radius R of the sphere, as function of the Mach number of the incoming flow in chlorine [529] (a) and in air [201] (b). Designations: 1) Only molecular rotation is excited; 2) equilibrium state; 3) vibrational degrees of molecular freedom in equilibrium with the translational degrees; 4) nonequilibrium excitation of vibrations, decomposition and ionization. ○ denotes experiments by Schwartz and Eckerman with $p_1 > 20$ mm Hg, while ● denotes their experiments with $p_1 = 4$ mm Hg.

distance and τ is the relaxation time of the process under study for certain average conditions behind a shock front [156].

One of the examples of a more systematic accounting of the effect of kinetics of physicochemical transformations on the flow variables in supersonic flow past a sphere are calculations performed by Telenin and Stulov [201]. Specifying a shape of a shock front, they have determined the flow field behind the front and the shape of a body which should correspond to this flow; if this shape was different than a sphere, then the problem was solved in the next approximation, until an agreement was obtained between the shape of the shock front and the sought shape of the body, *i.e.*, a sphere. They have considered flow in air with velocities such that not only excitation of vibrations, but of dissociation and ionization are also possible. They have used a complex mechanism of the excitation, dissociation and ionization reactions (see Chapter 5); the values of the reaction rate constants were taken from [426]. The vibrational relaxation times for O_2 and N_2 were determined on the basis of Eqs. (22–6) –(22–8), disregarding the vibrational energy transfer between the nitrogen and oxygen (see Sect. 22). The values obtained for ε with air flowing past a sphere with a radius of 1.5 cm (at an initial temperature of 250 °K) are given in Fig. 95b. The results remain practically the same for the product $p_1 R$, which enables one to find ε for other values of the sphere radius R. Some experimental results concerning the departure distance in nonequilibrium flow of air and carbon dioxide past a sphere were obtained by Maslennikov, *et al.* [121a].

4

Relaxation Processes in Shock Waves

[15] ESTABLISHING A MAXWELLIAN DISTRIBUTION

In the preceding chapters we have considered the experimental study of nonequilibrium phenomena in shock waves. This and the next chapter are devoted to the theoretical analysis of relaxation processes in shock waves and of results obtained in their experimental study. The various processes will be studied in the order they appear with an increase in the temperature. First, we shall consider the establishing of the Maxwellian distribution, then the rotational and vibrational relaxations, thermal dissociation and ionization kinetics, and, finally, nonequilibrium radiation. In the fifth chapter we shall consider the nonequilibrium properties of air.

The process of establishing an equilibrium with respect to the translational degrees of molecular freedom which results in a Maxwellian distribution, is one of the most rapid relaxation processes which take place in gaseous systems. The relaxation time, during which, in the first approximation, a local Maxwellian distribution is established in each volume element of the system, is of the order of time needed for traversing a mean-free path (see Sect. 2). When applied to strong shock waves this means that the shock-wave thickness which is produced by processes of establishing equilibrium with respect to the translational degrees of freedom, will be of the order of a mean-free molecular path. To describe relaxation processes which take place during a time (or over a length) of a mean-free molecular path, resort must be had to methods of gaskinetics. However, in weak shock waves, the relative change in the macroscopic parameters over a mean-free path is small. In this case the shock-wave structures can be described in the gasdynamic approximation.

The structure of the relaxation zone with a non-Maxwellian distribution is best considered using a monatomic gas. In this case no side effects occur which are due to the excitation of rotational and vibrational degrees of freedom. The ionization and electronic excitation may also be neglected in the consideration, since these processes proceed at substantially lower rates. In polyatomic gases the process of establishing a Maxwellian distribution will be similar to that for a monatomic gas, if it is possible to disregard all the relaxation phenomena which are due to the participation of internal degrees of freedom. In this sense the results presented in this section are general in character.

From the viewpoint of gasdynamics, the determination of a shock-wave structure reduces to solving the equations of conservation of the mass, momentum and energy fluxes in the relaxation zone. Since we are interested in the thickness of the shock wave, then these equations should not be written in the form of equality of the corresponding variables to both sides of the shock front, but rather should be a statement of their constancy across the entire thickness of the relaxation zone. Here to obtain finite values of the shock-front thickness, it is necessary to consider the additional momentum and energy fluxes produced by internal friction and heat conductions. The physical reason for this is obvious. The passing of a gas through a shock front is accompanied by a rise in the entropy. To describe the structure of a shock layer in which the entropy increases, it is necessary to take into account irreversible processes, which is achieved by considering the internal friction and heat conduction. From the viewpoint of molecular kinetics, the introduction of the additional momentum and energy fluxes is equivalent to considering small perturbances of the equilibrium distribution function.

The application of the conservation laws to the mass, momentum and energy fluxes results in the equations of gasdynamics. For a plane shock wave in a coordinate system moving with the shock, these equations in the steady-state case are

$$\left.\begin{aligned} \varrho v &= \text{const}\,, \\ \varrho v^2 + p + p_{xx} &= \text{const}\,, \\ \varrho v(\varepsilon + \tfrac{1}{2}v^2) + v(p + p_{xx}) + q_x &= \text{const}\,, \end{aligned}\right\} \tag{15–1}$$

where $\varepsilon = \frac{3}{2}(RT/\mu)$, $p = (\varrho RT/\mu)$ [the equation of state for a perfect monatomic gas], p_{xx} is a component of the tensor of viscous stresses, q_x is the heat flux, ϱ is the density, v is the velocity, T is the temperature, μ is the molecular weight and R is the universal gas constant.

Solving Eq. (15–1) we find ϱ, T and v as functions of the position x. Knowing these expressions it is possible to find the shock-wave thickness. The shock thickness L is usually determined by the velocity or density change in the following manner:

$$L_v = \frac{v_1 - v_2}{\left|\dfrac{dv}{dx}\right|_{\max}} \quad \text{or} \quad L_\varrho = \frac{\varrho_2 - \varrho_1}{\left|\dfrac{d\varrho}{dx}\right|_{\max}}\,, \tag{15–2}$$

where v_1, ϱ_1 and v_2, ϱ_2, respectively, are the velocity and density of the super- and subsonic flows far from the shock wave. In specific calculations of L_v or L_ϱ it should be remembered that the maximum velocity and density gradients are obtained, generally speaking, in different points within the shock waves. The definitions given by Eq. (15–2) are not the only ones possible. In particular, Grad [342] has suggested another expression, based on the integral properties of the velocity distribution, for determinging L. Subsequently, however, we shall use the definition of Eq. (15–2). The certain indeterminacy, which is due to the fact that L is not uniquely defined, turns out to be practically inconsequential.

To solve Eq. (15–1), it is necessary to specify the form of q_x and p_{xx}. Within the Navier-Stokes gasdynamic approximation

$$p_{xx} = -\frac{4}{3}\eta\frac{dv}{dx}, \quad q_x = -\lambda\frac{dT}{dx}, \tag{15–3}$$

where η and λ are the coefficients of viscosity and thermal conductivity. In this form the problem was stated by Rayleigh [485] and Taylor [568], and in different particular cases, which correspond to certain selections of η and λ, by Rayleigh [485], Taylor [568], Becker [245], Thomas [573] and others. A detailed presentation of these works can be found in [186].

The most general solution of Eqs. (15–1)—(15–3) when the viscosity coefficient is a power function of the temperature, and the theoretical value of the Prandtl number

$$\mathsf{Pr} = \frac{c_p\eta}{\lambda} = \frac{4\lambda}{9\gamma - 5} = \frac{2}{3} \quad \left(\gamma = \frac{c_p}{c_c}\right)$$

was used was obtained by Gilbarg and Paolucci [330]. The reciprocals of the shock thickness l_1/L_v calculated in the above work are presented in Table 6; they were obtained by using different values of s in the viscosity relationship $\eta \sim T^s$; l_1 is the mean-free path ahead of the shock wave

$$l_1 = \frac{15}{6}\frac{\eta_1}{\varrho_1\sqrt{\dfrac{2\pi R T_1}{\mu}}}.$$

Table **6.** The Ratio l_1/L_v for Different Values of $s(\eta \sim T^s)$

M \ s	1/2	0.647 (He)	0.816 (Ar)	1
1.2	0.070	0.068	0.067	0.067
1.4	0.136	0.132	0.128	0.122
1.7	0.222	0.210	0.199	0.187
2.0	0.292	0.270	0.248	0.224
2.5	0.381	0.344	0.303	0.264
3.0	0.437	0.377	0.314	0.261
4.0	—	0.410	—	—

The results presented in Table 6 were obtained in the Navier-Stokes gas-dynamic approximation, *i.e.*, under the conditions given by Eq. (15–3). However, as is known, Eq. (15–3) is only valid to within quantities of the order of l_1/L [220], *i.e.*, for small relative changes in the macroscopic quantities over a mean-free path. Expressions for p_{xx} and q_x which take into account terms of the order of $(l_1/L)^2$ were obtained by Barnett. The shock-wave structure in the Barnett approximation, *i.e.*, when terms of the order of $(l_1/L)^2$ are taken into account in determining p_{xx} and q_x, is calculated by Wang Chang [588], Zoller [625], Talbot and Sherman [534]. In [588], the shock thickness is determined also in the following, so called super-Barnett approximation, in which terms of the order of $(l_1/L)^3$ are taken into account in the expression for p_{xx} and q_x. Results of these calculations for Maxwellian molecules, which are point centers of repulsion forces which vary in proportion to r^{-5}, are presented in Table 7 [624]. (We recall that for Maxwellian molecules $\eta \sim T$.)

It can be seen from Table 7, that the taking into account of subsequent approximations in the gaskinetic consideration of the shock-wave structure does not appreciably change the results. Representation of p_{xx} and q_x as a series expansion in terms of powers of l_1/L is equivalent to solving the Boltzmann equation by the Chapman-Enskog method. The series obtained in such an expansion converge so slowly that their applicability for $\mathsf{M}>1.2$ is in doubt [454].

The thirteen-moment approximation due to Grad [342], also practically reduces to a gasdynamic description. In this approximation the relationship between p_{xx} and q_x is described by two equations [534]:

$$\left.\begin{aligned}
&\frac{d}{dx}(vp_{xx}) + \frac{8}{15}\frac{dq_x}{dx} + \frac{4}{3}(p + p_{xx})\frac{dv}{dx} + \frac{p}{\mu}p_{xx} = 0, \\
&\frac{d}{dx}(2vq_x) + \frac{22}{5}q_x\frac{dv}{dx} + 2RT\frac{pd_{xx}}{dx} + 7p_{xx}\frac{d(RT)}{dx} - \\
&\quad - \frac{2p_{xx}}{\varrho}\frac{d}{dx}(p + p_{xx}) + 5p\frac{d(RT)}{dx} + \frac{4}{3}\frac{p}{\mu}q_x = 0.
\end{aligned}\right\} \tag{15–4}$$

Table 7. The Ratio $4\, l_1/L_v$ for Maxwellian Molecules

M	The Navier-Stokes approximation	The Barnett approximation	The super-Barnett approximation
1.01	0.0141	0.0141	0.0141
1.02	0.0281	0.0281	0.0281
1.05	0.0697	0.0695	0.0695
1.10	0.1372	0.1361	0.1359
1.15	0.2023	0.1983	0.1979
1.20	0.2645	0.2551	0.2540

These equations are valid for Maxwellian molecules, however, they can with a satisfactory approximation be also used for other molecular models [534]. Solvings Eqs. (15–1) and (15–4) simultaneously, one can find the shock thickness in terms of Grad's thirteen-moment approximation. The results of calculations together with data obtained by other authors are shown in Fig. 96. The available experimentally obtained data are also shown in this figure.

It can seen from Fig. 96 that the thickness of weak shock waves (Mach numbers from 1 to 1.2) is practically identical in all the approximations and is in satisfactory agreement with the few available experimental data. In the range of $1.2 < \mathsf{M} < 2$, the experimentally obtained values of the shock thickness are most satisfactorily described by the Navier-Stokes and Barnett approximations. However, the accuracy of experimental data is such that, at present, no preference can be given to either of these approximations.

As the shock-wave strength is increased, the gasdynamic description of the shock-wave structure encounters certain difficulties. Formally the boundary-value problem for each of the above theories can be solved only in a certain range of Mach numbers, which is determined by the behavior of singular points of the system of equations. For the Navier-Stokes approximation this region is $1 \leqslant \mathsf{M} \leqslant \infty$, for the Barnett approximation it is $1 \leqslant \mathsf{M} < 2.1$, for the thirteen-moment approximation it is $1 \leqslant \mathsf{M} < 1.65$. Formally the Navier-Stokes approximation is suitable for describing shock-wave structures in the entire range of Mach numbers. Actually, however, its applica-from transport equations in which the term describing the transport balance due to collision does not vanish. Mott-Smith has considered transport equa-

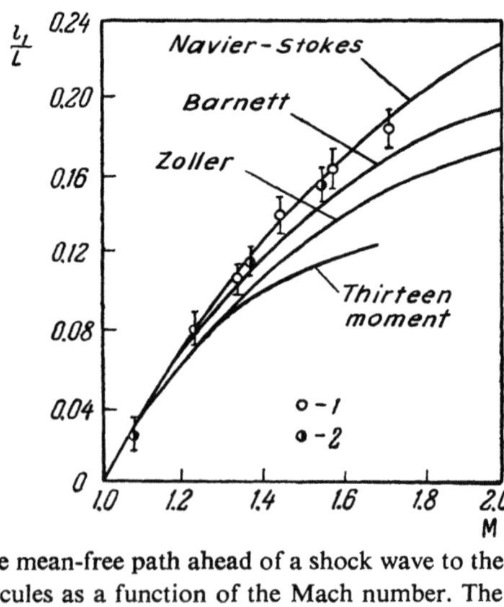

Fig. 96. Ratio of the mean-free path ahead of a shock wave to the shock-front thickness for Maxwellian molecules as a function of the Mach number. The points denote experimental values obtained by Sherman and Talbot (1) and Horning (2) for argon.

bility is also limited to low Mach numbers, since in strong shock waves the gradients of macroscopic quantities become so large, that there is no point in speaking about a gasdynamic approximation [453] (see Sect. 1). The distribution function in each point of the shock front differs so much from the equilibrium distribution that this deviation can no longer be expressed in terms of the heat flow and the momentum flux produced by internal friction.

To describe the structure of strong shock waves, resort must be had to gas-kinetic methods based on the solution of Boltzmann's equation; here even such well worked-out methods of solutions, such as Grad's method of moments or the Chapman-Enskog method is found to be inapplicable in the given case, since they cannot describe a state in which the local distribution function differs greatly from the equilibrium distribution. To describe the structure of a strong shock wave, it is necessary to have a method which would take into account as early as in the first approximation, the finiteness of disturbances. This method for solving the Boltzmann equation was proposed by Mott-Smith [454]. His considerations reduce to the following. Since the thickness of a strong shock wave is commensurable with the mean-free path, it may be expected that a large number of molecules from super- and subsonic flows, which obey their own Maxwellian distribution, penetrate the shock wave. Thus one of the characteristics of the velocity distributions in a strong shock wave should be the presence of two maxima which are determined by the temperatures of the super- and subsonic flows. On the basis of these considerations, Mott-Smith has proposed in the first approximation to seek the solution of Boltzmann's equation in the form of a sum of two Maxwellian functions, which, in the one-dimensional case, has the form

$$f^{(0)} = \nu_\alpha(x) f_\alpha + \nu_\beta(x) f_\beta , \tag{15–5}$$

where f_α and f_β are Maxwellian functions corresponding to the super- and subsonic flows, $\nu_\alpha(x)$ and $\nu_\beta(x)$ are the number densities for particles of the super- and subsonic components, referred to the number density of the undisturbed gas in the super- and subsonic regions, respectively. Eq. (15–5), as can easily be seen, is not a solution of the Boltzmann equation; hence, to determine the unknown parameters of Eq. (15–5), Mott-Smith has used the transport equation.

The transport equations for the summational invariants for collisions 1, c_x (the molecular velocity component along the axis) and c^2 (square of molecular velocity) in the case of steady flow, reduce to ordinary Rankine-Hugoniot relationships, which express the density, velocity and temperature of the subsonic flow in terms of values of these parameters in the supersonic flow. The functions $\nu_\alpha(x)$ and $\nu_\beta(x)$ which remain unknown are determined tions for the functions u^2 and u^3 and has found that

$$\nu_\alpha(x) = \nu(-x), \quad \nu_\beta(x) = \nu(x), \quad \nu(x) = \frac{1}{2}\left(1 + \text{th}\frac{2x}{L}\right). \tag{15–6}$$

The coordinate origin was placed in the center of the shock. The quantity L is a function of the Mach number of the incoming flow and determines the shock-front thickness. In the case of the bimodal distribution under consideration, both definitions of the shock thickness (on the basis of the density and velocity) give the same results, hence, $L = L_v = L_\varrho$. The form of the function $L(\mathsf{M})$ depends on the specific selection of the function in the transport equation and on the molecular model. However, the form of Eq. (15–6) is found to be insensitive to the selection of these parameters. The corresponding shock thicknesses, calculated with the function u^2 for various intermolecular interaction potentials are given in Fig. 97 [457]. The known experimental data were also drawn on this figure. It can be seen from this figure that the Mott-Smith method describes satisfactorily the available experimental results. The selection of the Mott-Smith solution in the form of Eq. (15–5) is arbitrary in the sense that no methods exists for evaluating this approximation. In addition, as can be seen from Table 8, the thickness of a shock wave depends on the form of the function in the transport equation. To remove this arbitrariness, Rosen [499, 500] has formulated a variational principle for determining L using the trial function (15–5). The variational integral was

$$I = \int_{-\infty}^{\infty} \int_{c} \left[\bar{f} u \frac{df}{dx} + \left(\frac{\partial \bar{f}}{\partial t}\right)_{\text{col}} f \right] dx\, dc, \quad \text{as} \quad \bar{f} \to f, \tag{15–7}$$

where $(\partial f/\partial t)_{\text{col}}$ is the collision term in the Boltzmann equation. The unclear

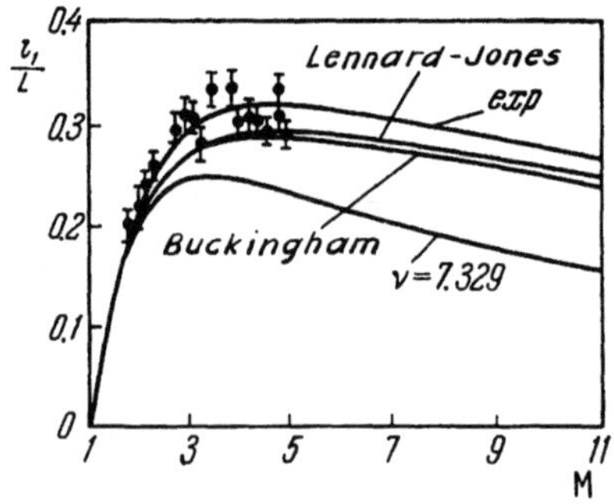

Fig. 97. Reciprocals of shock thicknesses in argon in mean-free paths ahead of the shock front calculated by the Mott-Smith method for different potentials of intermolecular interaction (exponential repulsion potential, Lennard-Jones and Buckingham potentials, the potential of repulsion forces $F \sim r^{-\nu}$) as a function of the Mach number. The dots denote experimentally obtained values.

point in Rosen's formulation remained the selection of the variational integral I. Subsequently, Gustafson [353a] showed that the variational principle is equivalent to the Mott-Smith method if $\partial f/\partial(n_\alpha v_\alpha)$, where n_α is the particle number density in the undisturbed super- or subsonic flow, is taken as the transfer function in the latter. The shock thickness, calculated by Rosen for a model of solid elastic spheres is also presented in Table 8. The problem of selecting the solution in the form of Eq. (15–5) which remained unclear is considered by Sakurai [514–516]. In particular, Sakurai has shown that it is possible to select an L such that Eq. (15–5) will be the asymptotic solution of Boltzmann's equation for large Mach numbers in a substantial region of the velocity space and has pointed out a method for finding the following terms. The corresponding values of the shock thickness are also presented in Table 8.

Table **8.** Ratio l_1/L for a Molecular Model of Solid Elastic Spheres. $(l_1/L)_r$ denotes data due to Rosen [499, 500], $(l_1/L)_{u^2}$ and $(l_1/L)_{u^3}$ denote data due to Mott-Smith [454], $(l_1/L)_s$ denote data due to Sakurai [514, 515]

M_s	(l_1/L)	$(l_1/L)_r$	$(l_1/L)_{u^2}$	$(l_1/L)_{u^3}$
∞	0.703	—	0.468	0.628
10	0.685	—	0.455	0.600
5	0.630	—	0.419	0.527
4	0.596	0.495	0.397	0.478
3	0.520	0.414	0.346	0.397
2.5	0.474	0.355	0.304	0.332

The work by Mott-Smith has stimulated an entire series of studies in which the idea of bimodal solution is developed.

In particular, in [592] the solution to the Boltzmann equation is no longer written as a superposition of two Maxwellian functions, but in the form of an integral of the Maxwellian functions taken with a certain weight. The two-fluid model is developed in [333, 624, 334]. In this model not one Boltzmann equation is considered but two, correspondingly for the super- and subsonic component, taking into account the transfer between them. The corresponding shock thicknesses observed are close to those obtained by Mott-Smith.

In theoretical calculations of the shock thickness one usually calculates the ratio l_1/L, where l_1 is the mean-free path (for a model of solid elastic spheres) in the gas which is at rest ahead of the shock wave. This unit of measurement is not mandatory. In particular, it has been suggested in [455, 456] that the mean-free path be calculated on the basis of experimentally obtained values of the viscosity coefficient, while in [623, 457] it is proposed that the unit for measuring the shock thickness by the Mott-Smith method be the mean-free path in the shock-wave center. In this case the reduced shock thickness will depend very little on the selection of the molecular forces model.

Up to now we have considered pure gases. The establishing of a Maxwellian distribution in a gas mixture is slightly more complicated. The thickness of weak shocks in gas mixtures can be calculated in the gasdynamic approximation. The method of calculations is similar to that presented at the beginning of this section and differs from the other only by the fact that the diffusion fluxes are considered. This calculation was performed in [304, 533, 60], in which it was shown that the diffusion results in widening of the shocks. The latter effect can be observed experimentally [235]. Table 9 presents results of measuring the shock thickness in mixtures of He and Ar of different concentrations. The initial pressure and temperature were 1 atm and 296 °K, L_{exp} is the experimentally obtained shock thickness, L_b [basic] is the shock thickness calculated in [330] without taking diffusion into account, and L_d is the shock thickness when diffusion is taken into account.

Table 9. Shock Thickness in an Ar–He Mixture

Mixture composition by volume	M	L_{exp}	L_b [basic]	L_d	T °K	g/cm³
Ar – 95%; He – 5%	1.23	8.5	8.5	8.5	362	$1.57 \cdot 10^{-3}$
Ar – 95%; He – 5%	1.55	4.0	4.0	3.2	458	$1.57 \cdot 10^{-3}$
Ar – 60%; He – 40%	1.23	16.8	10.7	19.8	362	$1.052 \cdot 10^{-3}$
Ar – 60%; He – 40%	1.55	8.9	5.1	7.7	458	$1.052 \cdot 10^{-3}$

It can be seen from Table 9 that the shock thickness increases with an increase in the admixture concentration. Theoretical results obtained within the framework of the gasdynamic approximation are valid only for weak shocks, and on the condition that the components of the mixture should have close masses. In the opposite case the state of the gas will not be close to equilibrium, and the initial prerequisite conditions for the validity of gasdynamic approximations will not be fulfilled. For analysis of the thickness of strong shocks in a gas mixture, resort must be had to methods of gaskinetics. No work was performed in this respect with neutral components, but the following may be presented as general considerations. In a binary mixture, consisting of a heavy gas with a small admixture of a light gas, the process of establishing the Maxwellian distribution, which determines the shock thickness will take place as follows. First, during a time of the order of an average time between collisions, the Maxwellian distribution is established in the heavy gas. Then the process of establishing the Maxwellian distribution starts in the light gas; here the relaxation time of the latter process is greater by a factor of M/m than the relaxation time of the first process (M and m are the mass of atoms of the heavy and light components, respectively). If the concentration of the light gas is not low, then each of the gases first establishes its own Maxwellian distribution. Then the temperatures of both distributions begin to equalize, which results in establishing a single Maxwellian distribution. In the process of temperature equalization, due to differences in the mean-thermal velocities of the molecules of the light and heavy gases, the gas components will be divided up in a manner such that in the region adjoining the shock front the concentration of the light component will be higher. In the cases when the temperatures of the light and the heavy gas are close to one another, this effect can be described within the framework of the ordinary gasdynamic approximation as is done in [304, 71].

[16] ROTATIONAL RELAXATION

In the study of relaxation processes which take place behind a shock front in di- and polyatomic gases, in addition to the process of establishing the Maxwellian equilibrium, it is also necessary to consider relaxation of the internal energy and, in particular, of the energy of the rotational degrees of freedom.* The characteristic time of rotational relaxation at ordinary temperatures, as was shown in Sect. 2, is appreciably smaller than the vibrational relaxation time and the time of establishing the dissociation and ionization equilibrium. In conjunction with this, in the study of rotational relaxation it is possible to regard all other internal degrees of freedom as frozen. The situation is different with respect to the division of equilibrium-establishing processes between the translational and rotational degrees of freedom. The characteristic time for establishing the Maxwellian equilibrium is smaller than, but of the order of the rotational relaxation time (see Sect. 2). The only exception are light gases, *i.e.*, hydrogen, deuterium, etc., for which the rotational relaxation time is appreciably greater than the time for establishing the Maxwellian equilibrium. Subsequently, for simplicity of calculations, we shall assume that the vibrational relaxation takes place under conditions when a Maxwellian distribution exists for the translational degrees of freedom. This assumption, of course, introduces a certain arbitrariness into the results and, in particular, makes it possible to determine only the rotational

* Rotational relaxation is encountered also in other fields of physics. In particular, in polyatomic gases it determines the second viscosity and the correction to the thermal conductivity coefficient (such as Aiken's correction) [51]. Anomalous rotational excitation is observed, for example, on electric discharges in gases, and also in photochemical reactions [87]. In addition to shock waves, rotational relaxation is studied by the method of dispersion and absorption of ultrasonic waves [368].

relaxation time. However, this does not change the physical picture of the process. At the same time, it should be noted that at present, no calculations are available of the simultaneous process of establishing the equilibrium with respect to rotational and translational degrees of freedom.

The width of the relaxation zone behind a shock front in diatomic gases is sometimes determined in the gasdynamic approximation by analogy with a monatomic gas. The specific character of a diatomic gas in this kind of consideration manifests itself only in the selection of the ratio of specific heats γ [186]. This approach cannot be considered satisfactory, since it does not take into account the departure from equilibrium with respect to the internal degrees of freedom. Generally, within the framework of gasdynamics it is possible to describe the structure of an equilibrium region in diatomic gases by introducing the bulk viscosity coefficient (or a corresponding thermal conductivity coefficient). Rigorous calculations of this kind are not available at present, although a number of works [157, 330] emphasize the role of bulk viscosity. However, a gasdynamic description of the relaxation zone which is related to the excitation of internal degrees of freedom is valid only for small Mach numbers, since the bulk viscosity coefficient takes into account departure from equilibrium with respect to the internal degrees of freedom only in the limiting case of $(v\tau/L) \ll 1$, where v is the macroscopic velocity, L is a characteristic dimension and τ is the relaxation time [197, 200] (see also Sect. 25). In a strong shock wave under conditions considered here this parameter is of the order of unity.

This requires that the description of the structure of the relaxation zone in which the equilibrium with respect to the rotational degrees of freedom is established be given by using methods of gaskinetics. At present, calculations of this sort are not available. Instead of this the relaxation zone structure is usually described within the framework of ordinary gasdynamics by introducing a new variable which characterizes the deviation of the energy of the rotational degrees of freedom from its equilibrium value [190]. The temperature of the rotational degrees of freedom is usually taken as this variable. The structure of the relaxation zone in this case will be described by ordinary gasdynamic equations, supplemented by the relaxation equation. For simplicity, the gasdynamic equations do not take into account the viscosity and thermal conductivity of the gas. In the relaxation zone behind the shock front, the gasdynamic equations have the form

$$\begin{gathered} \varrho v = a\,, \\ p + \varrho v^2 = b\,, \\ \frac{p}{\varrho} + \frac{v^2}{2} + c_{v_{\mathrm{trans.}}} T + c_{v_{\mathrm{rot}}} T_{\mathrm{rot}} = c\,, \end{gathered} \tag{16–1}$$

where a, b and c are constants which are determined from initial conditions, and $c_{v_{trans}}$, T and $c_{v_{rot}}$, T_{rot}, respectively, are the specific heats and temperatures of the translational and rotational degrees of freedom. The term rotational degree of freedom in the relaxation zone is used to denote the so-called kinetic temperature, which is determined from the condition that the average rotational energy of a unit mass is equal to $c_{v_{rot}}T_{rot}$.

Eqs. (16–1) are completed by an additional relaxation equation, which is usually written in the form [90]

$$\frac{dT_{rot}}{dt} = \frac{1}{\tau_{rot}}(T - T_{rot}), \tag{16–2}$$

where $\tau_{rot} = \tau_{rot}(p, T)$ is the relaxation time.

Simultaneous solving of Eqs. (16–1) and (16–2) for a known value of τ_{rot} makes it possible to determine the distribution of parameters in the zone where equilibrium is established with respect to the rotational degrees of freedom and also the width of this zone.

From the physical viewpoint, the writing of the relaxation equation in the form of Eq. (16–2) is not obvious. In addition, the explicit form of the functional behavior of the relaxation time $\tau_{rot} = \tau_{rot}(p, T)$ is unknown. In conjunction with this, this section considers problems related to the derivation of Eq. (16–2). To derive a relaxation equation it is desirable to select some molecular model which takes into account the rotational degrees of freedom. Historically, the first models of this kind were of solid ovaloids, sphero-cylinders, rough and loaded spheres [51]. Within the framework of the first two models, molecular collisions are described as collisions of solid bodies of an arbitrary oval shape (the model of ovaloids) or solid, cylindrically shaped bodies with half-spheres at the end (sphero-cylinders model). In the rough spheres model the molecules on colliding engage one another without slipping in a manner such that the relative velocity at the point of contact changes sign after the collision. The term loaded spheres is meant to denote smooth solid spheres, the center of mass of which does not coincide with the geometric center.

A consideration of rotational relaxation for these models was performed in [520], where Eq. (16–2) was derived and the relaxation time was determined. It was assumed in deriving Eq. (16–2) that the distribution function of the translational and rotational molecular energies has the form of a product of the Maxwellian distribution function with a temperature T by a Boltzmann distribution function with a temperature T_{rot}, and that then $T \neq T_{rot}$. The equation was obtained by calculating the amount of energy which is transferred to the rotational degrees of freedom per unit time. It was found that the form of the relaxation equation under these assumptions is independent of the molecular model, the specifics of which affect only the magnitude of τ_{rot}. In particular, for the loaded spheres model in which molecules with mass m are considered to be spheres with a diameter d with a center of gravity displaced by a distance δ from the geometric center, the following expression was obtained for the relaxation time (on the assumption that $\delta \ll d$):

$$\tau_{rot} = \frac{9}{20}\left(\frac{I}{m\delta^2}\right)\tau_0, \tag{16–3}$$

where

$$\tau_0 = \frac{1}{4nd^2}\sqrt{\frac{m}{\pi kT}}$$

is the mean-free path time, I is the moment of inertia with respect to an axis passing through the center of masses perpendicular to the axis of symmetry. This result coincides with the known result due to Jeans [392]. Results obtained in [520] for the rough spheres model agree with Widom's calculations [602]. In particular, for rough spheres with a mass m_α, which constitute a small admixture to a monatomic gas with a mass m_β, the number of collisions Z, which is needed for establishing rotational equilibrium ($Z = \tau_{rot}/\tau_0$), is

$$Z = \frac{3}{8}\left(\frac{m_\alpha + m_\beta}{m_\alpha}\right)\frac{(1+b)^2}{b}, \tag{16–4}$$

where

$$b = \frac{4I}{\mu d^2}, \quad \mu = \frac{m_\alpha \cdot m_\beta}{m_\alpha + m_\beta}, \quad d = d_\alpha + d_\beta$$

is the collision diameter. The values of Z are given in Table 10.

Table **10.** Number of Collisions Z which is Needed for Establishing Equilibrium with Respect to the Rotational Degrees of Freedom of α Molecules, which Constitute a Small Admixture in a Monatomic Gas β

α \ β	He	Ne	Ar	Kr	Xe
GeH_4	5	23	51	119	197
SiH_4	6	24	52	116	204
CH_4	8	34	77	169	302
CD_4	4	18	39	29	159
$C(CH_3)_4$	2	3	5	11	17
CF_4	2	2	4	8	15
SF_6	2	2	3	6	10
CCl_4	2	2	3	5	7
$SiBr_4$	4	2	2	2	3

Reference [602] also quotes an expression obtained in [589] for the number of collisions needed for establishing rotational equilibrium in a pure diatomic gas consisting of rough spheres

$$Z = \frac{3}{8}\frac{(1+2b_1)_2}{b_1}, \tag{16–5}$$

where $b_1 = 4I/m_1d_1$, m_1 is the reduced mass of the colliding molecules ($m_1 = m/2$) and d_1 is twice the molecular diameter. Values of Z calculated from Eq. (16–5) are given in Table 11.

Table **11.** Number of Collisions Z Needed for Establishing Rotational Equilibrium in a Pure Gas

	$Z_{theor.}$	$Z_{exp.}$		$Z_{theor.}$	$Z_{exp.}$		$Z_{theor.}$	$Z_{exp.}$
GeH_4	63	—	CD_4	12	—	SF_6	7	—
SiH_4	28	—	$C(CH_3)_4$	7	—	CCl_4	6	—
CH_4	18	14–17*	CF_4	6	—	$SiBr_4$	5	—

* Taken from [402].

The derivation of Eq. (16–2) which was performed in [520] is based primarily on the assumption that the function of molecular distribution with respect to rotational levels has an equilibrium form, with a time-dependent rotational temperature. It is precisely this circumstance which has made the obtaining of Eq. (16–2) possible. However, in a rigorous consideration of rotational relaxation, it is necessary to start with an integro-differential equation for the molecular distribution function with respect to the rotational levels or, if quantization of the rotational energy is taken into account, with a system of ordinary differential equations [366]. Thus, in the general case, instead of Eq. (16–2), there appears one integro-differential equation or a system of ordinary differential equations. This situation is a general one and is independent of the model selection. The problem of transforming these systems into Eq. (16–2) has been studied only in certain particular cases for equations which describe the establishing of equilibrium with respect to translational and vibrational degrees of freedom [539]. In addition, in the above works the interaction between the translational and rotational degrees of freedom of colliding molecules is introduced artificially by roughnesses or by an asymmetric mass distribution, while in fact the energy transfer between translational and rotational degrees of freedom most likely takes place due to the fact that the interaction potential is not spherical [220, 51]. A more realistic molecular model is analyzed by Parker [465]. He has considered the problem of energy transfer between the translational and rotational degrees of freedom on collision of two rigid rotators. In his calculations, which were performed within the framework of classical mechanics, he has assumed that the rotators lie in one plane; he has also assumed that each rotator has two centers of repulsion which in general do not coincide with the center of the nuclei, and that one of the two centers is situated in the center of gravity of the molecule. The interaction between the centers of attraction and repulsion of different molecules were described by a potential expressed in the exponential form.

The total interaction potential has the form

$$V_{12} = Ce^{-\alpha R}(1 + \varepsilon \cos 2\theta_1)(1 + \varepsilon \cos 2\theta_2) - Be^{-\frac{\alpha R}{2}}, \qquad (16\text{–}6)$$

where R is the distance between the centers of gravity of the molecules, θ_1 and θ_2 are angles which determine the orientation of molecules with respect to an axis passing through the centers of gravity of the molecules, ε is an asymmetry parameter, which is a function of the position of the repulsion centers, the distance between which is d^*, $\varepsilon = 2I_2(\alpha d^*/2)/I_0(\alpha d^*/2)$, where I is a Bessel function.* As is easy to see, the potential of Eq. (16–6) is asymmetric. This is precisely the fact which makes the transfer of energy between the translational and rotational degrees of freedom possible. Solving the dynamic problem of the collision of two such rotators, Parker determines the magnitude of the rotational energy which is obtained by the nonrotating rotator in one collision. The energy thus found is averaged over all the possible velocities of arriving molecules on the assumption that the latter obey the Maxwellian distribution, and over all the possible orientations. This is the manner in which the total increment of rotational energy per unit time, $\mathrm{d}E_{\mathrm{rot}}/\mathrm{d}t$ is determined.

* The first term in Eq. (16.6), which approximates the potential of repulsion forces V_{rep}, is obtained as follows. The potential V_{rep} between the molecules AB and CD has the form

$$V_{\mathrm{rep}} = A\,[\exp(-\alpha r_{AC}) + \exp(-\alpha r_{AD}) + \exp(-\alpha r_{BC}) + \exp(-\alpha r_{BD})],$$

where r_{ij} is the distance from the ith repulsion center of one molecule to the jth repulsion center of another. If one assumes that

$$r_{AC} = R + (d^*/2)(\cos\theta_1 + \cos\theta_2), \quad r_{BC} = R - (d^*/2)(\cos\theta_1 - \cos\theta_2),$$
$$r_{AD} = R + (d^*/2)(\cos\theta_1 - \cos\theta_2), \quad r_{BD} = R - (d^*/2)(\cos\theta_1 + \cos\theta_2),$$

then

$$V_{\mathrm{rep}} = 4Ae^{-\alpha R}\,\mathrm{ch}\left(\frac{\alpha d^*}{2}\cos\theta_1\right)\mathrm{ch}\left(\frac{\alpha d^*}{2}\cos\theta_2\right).$$

The expression for V_{rep} thus found is identical to the first term in Eq. (16–6), if the hyperbolic cosines are expanded in a series of which only the first two terms are retained:

$$\mathrm{ch}\left(\frac{\alpha d^*}{2}\cos\theta\right) = I_0\left(\frac{\alpha d^*}{2}\right) + 2I_2\left(\frac{\alpha d^*}{2}\right)\cos 2\theta + \cdots \approx I_0\left(\frac{\alpha d^*}{2}\right)(1 + \varepsilon\cos 2\theta).$$

The further consideration reduces to the following. It is postulated that the process of establishing equilibrium is described by a relaxation equation of the form (see Sect. 2)

$$\frac{\mathrm{d}E_{\mathrm{rot}}}{\mathrm{d}t} = \frac{E_0(T) - E_{\mathrm{rot}}}{\tau_{\mathrm{rot}}}, \qquad (16\text{–}7)$$

where $E_0(T)$ is the equilibrium value of the rotational energy of a unit volume at a temperature T.

It follows from Eq. (16–7) that

$$\tau_{\mathrm{rot}} = \frac{E_0(T)}{\left(\frac{\mathrm{d}E_{\mathrm{rot}}}{\mathrm{d}t}\right)_{E_{\mathrm{rot}}=0}} . \qquad (16\text{–}8)$$

Since $E_0(T)$ is known, and $\mathrm{d}E_{\mathrm{rot}}/\mathrm{d}t$ is determined by solving the dynamic problem, then τ_{rot} is easily obtained. It is more convenient to use the number of collisions Z which are needed for establishing rotational equilibrium ($Z = \tau_{\mathrm{rot}}/\tau_0$):

$$Z = \frac{Z_R^{\infty}}{1 + \frac{\pi^{\frac{3}{2}}}{2}\left(\frac{T^*}{T}\right)^{\frac{1}{2}} + \left(\frac{\pi^2}{4} + \pi\right)\left(\frac{T^*}{T}\right)}, \qquad (16\text{–}9)$$

where $Z_R^{\infty} = \frac{1}{16}(\alpha d/\varepsilon)$, d in the internuclear distance within the molecule, and T^* characterizes the depth of the intermolecular potential hole [dissociation energy; editor]. The number Z for Cl_2, N_2 and O_2 was calculated in [465]. Calculation of Z using Eq. (16–9) requires that the values of α, d^*, d and T be known. The value of d is known with a high degree of accuracy, T^* is determined from the intermolecular interaction potential ($T^* = E^*/k$, where E^* is the depth of the intermolecular potential hole). The values of α and d^* are calculated from experimental data on vibrational relaxation.

For Cl_2 $\alpha = 3.51$ A^{-1}, $d = 1.988$ A, $d^* = 0.617$ A, $T^* = 300\,^{\circ}$K. From these data we have $Z_R^{\infty} = 47.1$ and Z (300 °K) = 4.90. The experimental value of Z found by Andersen and Horning by studying the thickness of a shock front in a shock tube has an upper limit of 5.5 [234]. For N_2 $\alpha = 4.09$ A^{-1}, $d = 1.094$ A, $d^* = 0.557$ A, $T^* = 80\,^{\circ}$K. From these data we have: $Z_R^{\infty} = 15.7$ and Z(300 °K) = 4.01. The experimental value due to Andersen and Horning is 5.5. Ultrasonic measurements of Z yield the following values [301] (see Table 12).

For O_2 $\alpha = 4.10$ A, $d = 1.207$ A, $d^* = 0.605$ A, $T^* = 90\,^{\circ}$K. From these data we have that $Z_R^{\infty} = 14.4$ and Z(300 °K) = 3.45. According to Andersen and Horning $Z \sim 5$. Results of ultrasonic measurements are given in Table 13 [301].

The deliberations presented above show that satisfactory agreement exists

Table **12.** Rotational Relaxation in N_2 Based on Data of Ultrasonic Measurements

T °K	$\tau_{rot}\cdot 10^{10}$ sec	Z	References
302	8,2	6	[626]
293	4,7	3	[466]
273–333	6 ± 2	4–6	[570]
333	20	13	[352]
373	11	7	[345]
Room	—	5.3	[348]
303	—	4.7	[377]

Table **13.** Rotational Relaxation in O_2 Based on Data of Ultrasonic Measurements

T °K	$\tau_{rot}\cdot 10^{10}$ sec	Z	References
302	$20\div 48$	$12\div 30$	[571]
293	3.7	2	[466]
298	22	14	[472]
273–333	5 ± 2	$2\div 4$	[464]
313	21.8	12	[300]
373	13	7	[345]
Room	—	4.1	[348]
303	—	4.1	[377]

between the calculated and measured values of Z. However, the latter cannot serve as a criterion for the correctness of the model or calculation methods selected, since all more or less reasonable calculations, given the present level of development of the theory, will yield quantities of the same order of magnitude.

The main shortcoming of Parker's work consists in the fact that it actually does not consider the statistical aspect of the problem. As is known (see Sect. 2), two problems arise in the theoretical consideration of relaxation processes: one being the problem of the result of the elementary act and the second being that of formulating the appropriate equations of gaskinetics. The first problem is considered by Parker in sufficient detail, while the second is not considered at all. Instead of it he postulates the form of the relaxation equation.

The most systematic scheme for calculating rotational relaxation is formulated by Widom [602]. However, the model he has selected, as was pointed out above, meets with serious objections. A rigorous consideration of rotational relaxation of heavy rotators which constitute a small admixture to a light inert gas, has also been performed in [175a].

Up to now the consideration of rotational relaxation as well as of the process of establishing the Maxwellian distribution, was performed within the framework of classical concepts. This is justified as long as the characteristic effect S for the given problem is appreciably greater than the Planck constant $\hbar$

$$S \gg \hbar . \tag{16–10}$$

For establishing a Maxwellian equilibrium $S \sim m\bar{c}d$, where d is the molecular diameter, m is the molecular mass and $\bar{c}$ is the average value of the absolute molecular velocity. In the temperature range under study

$$m\bar{c}d \gg \hbar , \tag{16–11}$$

hence the concepts of classical physics can always be used to describe the establishing of equilibrium with respect to the translational degrees of freedom in the given problem.

In processes which are involved in establishing rotational equilibrium, in addition to Eq. (16–11), it is required also to satisfy the condition

$$\frac{kT}{\Omega} \gg \hbar \quad \text{or} \quad \frac{\Delta E_{\text{rot}}}{k} \ll T , \tag{16-12}$$

where Ω is the angular speed of the rotator and $\Delta E_{\text{rot}} = \hbar\Omega$. The condition Eq. (16–12) at room temperatures is satisfactorily fulfilled for heavy molecules and poorly satisfied for light molecules, for example, for H_2, D_2, etc. Hence rotational relaxation in light gases must be calculated by quantum-mechanical methods. This is particularly true with respect to calculating the probabilities of energy transfer between the translational and rotational degrees of freedom. Such a calculation for H_2 molecules was performed in [511, 554, 556, 559, 560, 564, 565, 246, 274]. In [560], the probabilities of excitation of rotational levels for D_2 and HD were also calculated. A detailed consideration of the above works can be found in [563a].

The theoretically calculated cross sections for excitation of various rotational levels $Q(T)$ are presented in Table 14 [558]

$$Q(T) = \pi R_c^2 F(T) = \pi R_c^2 \beta^2 \left[\frac{F(T)}{\beta^2} \right],$$

where R_c is the closest intermolecular distance in head-on collisions, which is by order of magnitude equal to the collision diameter, β is a parameter which takes into account the angular asymmetry of intermolecular interaction potential ($\beta = 0.075$). It was shown in [558] that $R_c^2 \beta^2 \approx 0.1$. The values of F/β^2 for the rotational transitions $j \rightarrow j+2$, where j is the rotational quantum number, are given in Table 14.

Table **14.** Values of $F(T)/\beta^2$ for H_2 and D_2

Rotational transitions	100°K		200°K		300°K	
	H_2	D_2	H_2	D_2	H_2	D_2
0–2	$2.7 \cdot 10^{-3}$	$9.0 \cdot 10^{-2}$	$5.9 \cdot 10^{-2}$	$5.4 \cdot 10^{-3}$	$9.0 \cdot 10^{-1}$	1.13
1–3	$1.7 \cdot 10^{-5}$	$4.0 \cdot 10^{-3}$	$2.6 \cdot 10^{-3}$	$6.8 \cdot 10^{-2}$	$1.7 \cdot 10^{-2}$	$2.1 \cdot 10^{-1}$
2–4	$1.7 \cdot 10^{-7}$	$2.5 \cdot 10^{-4}$	$1.6 \cdot 10^{-4}$	$1.16 \cdot 10^{-2}$	$2.0 \cdot 10^{-3}$	$5.4 \cdot 10^{-2}$

Experimentally obtained values of the number of collisions needed for establishing rotational equilibrium obtained from ultrasonic measurements [301], are presented in Table 15, while those obtained from the thermal conductivity measurements are given in Table 16 [547].

A direct comparison of experimental and calculated values is difficult, since in the above work again only one side of the problem, *i.e.*, the quantum-mechanical problem of calculating the transition probabilities, was considered. To determine the relaxation time or the number of collisions it is necessary to formulate the statistical problem.

In [274], the theoretical and experimental data are compared by using an expression for the relaxation time obtained for the two-level model [87]. The calculated results are in agreement with the experimentally obtained relaxation times. However, the use of the two-level model in this case is not substantiated. In particular [588] presents calculations of the dispersion curve for H_2; here in order to obtain agreement between the experimental and theoretical data it was necessary to take into account the rotational transitions between five levels according to the scheme 0–2, 1–3, 2–4.

Table **15.** Rotational Relaxation in Hydrogen Based on Ultrasonic Measurements

Gas	T°K	$\tau_{rot} \cdot 10^8$ sec	Z	References
Standard hydrogen (75% $o-H_2$, 25% $n-H_2$)	90	—	540	[584]
	207	1.2	215	[384]
	273	2.3	350	[549]
	273	2.1	320	[549]
	273	—	360	[549]
	285	1.1	160	[384]
	288	2.1	310	[352]
	298	1.9	270	[548]
	298	1.7	240	[548]
Para hydrogen	90	—	360	[584]
Deuterium	273	9.0	220	[549]
	273	1.8	190	[549]
	273	—	210	[549]

An attempt to consider rotational relaxation by the quantum-mechanical approach, taking into account all the rotational levels was made in [366]. The results obtained there for the rotational relaxation time of O_2 and N_2 differ from experimentally obtained data [384, 382, 344, 345] by approximately two orders of magnitude. This disagreement is due primarily to the fact that

Table **16.** Rotational Relaxation in Hydrogen Based on Measurements of the Thermal Conductivity Coefficient

T °K	$\tau_{rot} \cdot 10^8$ sec	Z
254.4	16.21	348
277.7	18.44	373
293.3	23.12	451
324.3	27.22	497
341.3	29.10	514

the model used was too rough. In [366] it was assumed that the excitation of rotational levels takes place in the one-quantum manner, *i.e.*, the selection rule $\Delta j = \pm 1$ is satisfied for the rotational quantum number j. In general, this assumption is invalid for heavy molecules. The excitation of rotational levels for a majority of molecules, with the exception of light molecules, takes place in the multiquantum manner, as a result of simultaneous transfer of several quanta of rotational energy (see Sect. 2). In addition, for all molecules, with the exception of very light ones, no quantum-mechanical calculations of the probability of excitation of rotational levels are available. The only available estimates of probabilities determined by Brout [274] are not satisfactory from the theoretical viewpoint. The method of distorted waves used by him (see Sect. 17) is inapplicable, as was shown in [559] to calculate the probabilities of rotational excitation on collision of heavy molecules.

[17] VIBRATIONAL RELAXATION

The establishing of equilibrium with respect to vibrational degrees of freedom (vibrational relaxation) belong among the most extensively studied non-equilibrium phenomena which take place behind the front of strong shock waves. The results obtained in this study are used extensively in different fields of physics – ultraacoustics, the theory of combustion, theory of transport phenomena, physical chemistry, etc [155]. Here it must be emphasized that the experimental study of vibrational relaxation in shock waves is, to a certain extent, a further development of the methods of study of nonequilibrium phenomena, since the state of a gas behind a shock front differs appreciably from, for example, conditions in ultrasonic fields.

The theoretical study of vibrational relaxation is made appreciably easier by the fact that in the majority of practical cases this process takes place so slowly that at each time instant reference can be had to the existence of an equilibrium distribution of energy between the translational and rotational degrees of molecular freedom. On the other hand, vibrational relaxation takes place so rapidly that the concentrations of dissociated and ionized molecules practically do not succeed in changing. This means that in the study of vibrational relaxation, it is possible, generally speaking, not to consider the dissociation and ionization processes. However, at high temperatures (for example, higher than 10,000 °K for O_2), the characteristic time of establishing translational, rotational and vibrational degrees of freedom, as well as the time for establishing equilibrium dissociation are found to be of the same order. In this case the process of establishing the complete statistical equilibrium can no longer be broken up into individual stages. However, with the exception of certain particular cases, which will be considered separately, no theoretical analysis of these overlapping processes is available at present.

The structure of a shock wave in the region where vibrational equilibrium is established is described by a system of gasdynamic equations, which express the laws of conservation of the mass, momentum and energy fluxes, as well as by the equation of state (see Sect. 9)

$$\left.\begin{aligned} \varrho v &= \varrho_1 V, \\ p + \varrho v^2 &= p_1 + \varrho_1 V^2, \\ h + \frac{v^2}{2} &= h_1 + \frac{V^2}{2}, \\ p &= \varrho \frac{RT}{M}; \end{aligned}\right\} \tag{17–1}$$

the subscript "1" denotes the parameters of the incoming flow, $h = h_n + E_{0\mathrm{rot}}/\varrho + E_{\mathrm{vib}}/\varrho$ is the specific enthalpy of the gas, $h_{\mathrm{trans}} = \frac{5}{2}RT/\mu$ is the enthalpy of the translational degrees of freedom, $E_{0\mathrm{rot}} = \varrho RT/\mu$ and E_{vib} are, respectively, the internal energies of rotation and vibrations referred to a unit volume. Eqs. (17–1) must be supplemented by a relaxation equation which would describe the kinetics of establishing vibrational equilibrium. This equation is usually taken in the form

$$\frac{\mathrm{d}E_{\mathrm{vib}}}{\mathrm{d}t} = \frac{E_{0_{\mathrm{vib}}}(T) - E_{\mathrm{vib}}}{\tau_{\mathrm{vib}}}, \tag{17–2}$$

where $E_{0\text{rot}}(T)$ is the equilibrium vibrational energy of a unit volume and $\tau_{\text{vib}}(p, T)$ is the relaxation time. Eqs. (17–1) and (17–2) are sufficient for describing the flow parameters in the region where vibrational equilibrium is established. The simultaneous solution of Eqs. (17–1) and (17–2) on various assumptions with respect to the relaxation time τ_{vib} has been studied in detail in [394, 258, 3, 16].

From the physical viewpoint, the solutions of Eqs. (17–1) and (17–2) is not of interest, but rather the analysis of processes on which these equations are based. This is precisely the subject matter of this section.

[1] Kinetic Equations and the Transition Probabilities

a) *Kinetic equations.* Let us start with the simplest case of vibrational relaxation in a system of diatomic molecules which constitute a moderate admixture to an inert monatomic gas. The vibrational relaxation in such a system takes place by interaction of the diatomic molecules with the molecules of the inert gas. The interaction of diatomic molecules with each other can be disregarded.

This interaction results in an energy transfer between the vibrational and translational degrees of freedom of the colliding atoms and molecules. The equations which describe the vibrational relaxation are balance equations for the molecular population at each vibrational level. If the number density of molecules of the nth vibrational level is denoted by $x_n(t)$, then these equations will take on the form

$$\frac{dx_n}{dt} = Z\Big(\sum_{n' \neq n} P_{n'n}x_{n'} - x_n \sum_{n' \neq n} P_{nn'}\Big), \quad n = 0, 1, 2, \ldots, \tag{17–3}$$

where P_{ij} is the probability of transition of a molecule from the ith to the jth vibrational level on collision with an atom, Z is the collision frequency of the molecules; it is assumed that Z is independent of the vibrational state of the molecule. The last condition is not rigorously satisfied. However, it may be assumed, since in the majority of practical cases the amplitude of atomic vibrations in molecules is small in comparison with the average distance between the atoms. In writing Eq. (17–3) the relative translational motion was considered in the classical manner. This is not a limitation, since in cases of practical interest, the density and temperature of systems are such that the quantum corrections which are due to the finite length of the de Broglie wave can be disregarded. It is also assumed in Eq. (17–3) that the vibrational motion can be separated from the other forms of internal motion: rotational; electronic; etc. Here, it must be noted that a rigorous derivation of kinetic equations for particles with internal degrees of freedom from the dynamic equations of quantum or classical mechanics is, at present, not available [209].

Before solving Eq. (17–3) we must specify explicitly the form of the transition probabilities P_{nm}. The probabilities P_{nm} are defined as follows [220]:

$$ZP_{nm} = n_A d^2 (2\pi)^{\frac{1}{2}} \left(\frac{\mu}{kT}\right)^{\frac{3}{2}} \int_0^\infty p_{nm}(v)\, e^{-\frac{\mu v^2}{2kT}} v^3 \, dv , \tag{17–4}$$

where μ is the reduced mass of the colliding particles, $p_{nm}(v)$ is the probability of transition of a molecule from the nth to the mth vribational level on collision with an atom moving with a relative velocity v, $d=(d_1+d_2)/2$, where d_1 and d_2, respectively, are the atom and molecule diameters (the solid spheres model) and n_A is the number density of the atoms of the inert monatomic gas. The probability $p_{nm}(v)$ is calculated by quantum-mechanical methods by solving Schrödinger's equation describing the collision process. According to the above, the relative motion of colliding molecules in this problem is considered in the classical manner. A detailed calculation of p_{nm} by this method was performed in [621, 622, 414, 129]. In this presentation we shall restrict ourselves only to the qualitative picture of the derivation, dwelling only on those points which are of importance for subsequent discussion. The transformation of translational into vibrational energy for the majority of molecules takes place without a change in their electronic state. The last circumstance makes it possible to neglect the electronic degrees of freedom and to consider the molecule as an oscillator. As a first approximation, the effect of rotation on p_{nm} is also neglected. However, the rotational degrees of freedom must be taken into account in the study of a number of problems of thermal dissociation, vibrational relaxation in diatomic gases, etc. The time needed for establishing equilibrium with respect to the rotational degrees of freedom is of the order of times of several mean-free paths. Hence, the problem of departure from rotational equilibrium practically does not appear in considering vibrational relaxation. The last remark is also valid with respect to the translational degrees of freedom.

b) *The semiclassical theory of transition probabilities (the linear problem).* Due to the fact that the relative motion of an oscillator with an atom is considered in the classical manner, it is described as the effect of an external force on the oscillator. For the majority of molecules under study, the time of the force application to the oscillator or the collision duration τ_{col} is appreciably greater than the time of natural vibrations of the oscillator.

According to the general theory of adiabatic invariants, the probability for the transition of an oscillator from quantum state n into another quantum state m as a result of such a collision is small; hence the theory of disturbances can be used for determining $p_{nm}(v)$.

In the elementary one-dimensional case, the Schrödinger equation, which describes the collision process, has the form [97]

$$i\hbar \frac{\mathrm{d}a_k}{\mathrm{d}t} = \sum_m a_m(t)\, V_{km}(r(t))\, e^{i\omega_{kmt}}. \tag{17–5}$$

Eq. (17–5) is written in a representation which is determined by the eigenfunctions of an undisturbed Hamiltonian of the molecules. $V_{km}(r(t))$ is a matrix excitation element, which in this problem is the intermolecular interaction potential, $r(t)$ is the classical trajectory of relative motion of the centers of gravity of colliding particles, which is determined from

$$\mu\ddot{r} = -\frac{\mathrm{d}V_{nn}}{\mathrm{d}r}. \tag{17–6}$$

Eqs. (17–5) must be supplemented by initial conditions. We are interested here in the probability of transition of the oscillator from some fixed state n to some arbitrary state m. Hence as our initial conditions we should select

$$a_k(-\infty) = \delta_{kn},$$

which means that the undisturbed oscillator was in the nth steady state before collision.

In the first approximation of the excitation theory, $p_{nm}(v)$ is given by

$$p_{nm}(v) = |a_m(\infty)|^2 = \left| \frac{1}{\hbar} \int\limits_{-\infty}^{\infty} V_{nm}(r(t))\, e^{i\omega_{nmt}}\, \mathrm{d}t \right|^2. \tag{17–7}$$

Eq. (17–7) can be simplified. The amplitude of vibrations of atoms in a molecule at lower vibrational levels is small in comparison with the radius of intermolecular interaction; hence $V(r, y)$ can be expended in a series in terms of powers of $y = Y - Y_0$, where Y is the intramolecular coordinate (see Sect. 2, subsection 3), and to restrict ourselves to the linear term. In this case

$$V_{nm}(r) = (n\,|V(r, y)|\,m) \simeq \left(n \left| V(r, 0) + y\frac{\partial V}{\partial y} \right| m\right) = y_{nm}\frac{\partial V}{\partial y}. \tag{17–8}$$

The constant term $V(r, 0)\delta_{nm}$ can be disregarded, since it does not affect the magnitude of the transition probabilities.

In the one-dimensional case it is customary to take into account only the interaction of the incoming atom with the closest atom of the molecule. The interaction with the far atom of the molecule is disregarded since, due to the fact that the collision is adiabatic, only the short-range part of the intermolecular interaction potential is of substance for transition probability calculations. The radius for this interaction, which is of the order of tenths of an angstrom, is by approximately an order of magnitude smaller than the distance between atoms in the molecule. Thus, in the approximation used here, the intermolecular interaction potential has the form

$$V(r, y) = V(r - \lambda y),$$

where λY is the distance from the center of gravity of a molecule to its atom with which the collision takes place (for homonuclear molecules $\lambda = \frac{1}{2}$), r, as before, is the distance between the centers of gravity of colliding molecules. In this case, using the approximation of Eq. (17–8)

$$V_{nm}(r) = - y_{nm}\lambda \frac{\mathrm{d}V}{\mathrm{d}r}; \quad V_{nm}(r) = V(r). \tag{17–9}$$

Using Eq. (17–9) we can rewrite Eq. (17–7) in the form

$$p_{nm}(v) = |y_{nm}|^2 \cdot \frac{1}{\hbar^2} \left| \int_{-\infty}^{\infty} F(t)\, e^{i\omega_{nm}t}\, dt \right|^2, \tag{17–10}$$

where $F(t) = -\lambda \mathrm{d}V/\mathrm{d}r$ is a force with which the incoming atom interacts with the molecule. The physical meaning of Eq. (17–10) is obvious. When an oscillator is excited by an external force $F(t)$, only the Fourier component of $F(t)$ of the corresponding frequency is of substance. The integral contained in Eq. (17–7) can be evaluated in the general case without assuming any specific form for the intermolecular interaction potential, using a method suggested in [414]. We shall consider the time t as a complex variable and we shall displace the integration path from the real axis into the upper half-plane. This displacement of the integration path is limited only by the necessity of passing by the singular points of the integrand. We shall displace the integration path in a manner such that it should pass along a straight line in the upper half-plane which is parallel to the real axis, and be removed from it through a distance of the pole of function $F(t)$, which lies closest to the real axis, and which is given by the condition

$$F(t) \to \infty \quad \text{as} \quad t \to t_1. \tag{17–11}$$

It is easy to see that the integral of Eq. (17–10) is proportional to exp $(i\omega_{nm}t_1)$, and

$$p_{nm} \sim \exp(-2i\omega_{nm}t_1) \equiv \exp(-2\omega_{nm}\tau_{\mathrm{col}}), \tag{17–12}$$

where $\tau_{\mathrm{col}} = it_1$.

Approximately the magnitude of τ_{col} can be estimated by the following considerations. The collision time is a characteristic time in the problem at hand. From the two characteristic quantities which are involved in the collision theory, *i.e.*, the velocity up to collision (v) and the intermolecular interaction radius (a), it is possible to set up only one dimensional time group, *i.e.*, $\tau_{col} \sim a/v$. An exact calculation of τ_{col} in the case of $V = W_0 \exp(-\alpha r)$ yields the value $\tau_{col} = \pi/\alpha v$. In fact, it was shown in Chapter 1, Sect. 2 [see Eq. (2–25)], that

$$F(t) = -\frac{\alpha W_0 \lambda}{\mathrm{ch}^2 \dfrac{\alpha v t}{2}}.$$

The pole of this function which is closest to the real axis is determined from

$$\frac{\alpha v t_1}{2} = i\frac{\pi}{2},$$

whence

$$t_1 = \frac{i\pi}{\alpha v}, \quad \text{and} \quad \tau_{col} = \frac{\pi}{\alpha v}.$$

Using the preexponential factor, Eq. (17–12) takes on the form

$$p_{nm}(v) = \frac{16\pi^2 \mu^2 \omega^2 \lambda^2}{\alpha^2 \hbar^2} |y_{nm}|^2 e^{-\frac{2\pi\omega_{nm}}{\alpha v}}. \tag{17–13}$$

Expression (17–13) is valid for any molecular model. The specific form of the molecular model affects only the magnitude of the matrix element y_{nm}.

The following conclusion can be made on the basis of Eq. (17–13). In the adiabatic approximation being considered $(2\pi\omega_{nm}/\alpha v) \gg 1$. Since $p_{nm}(v)$ is an exponential function of the transferred energy, only transitions between neighboring vibrational levels are practically permissible. This assertion is independent of the molecular model. In the case of the harmonic oscillator model, in approximation (17–9) it is exact, since for a harmonic oscillator y_{nk} are nonzero only when $k = n + 1$:

$$\left.\begin{aligned} |y_{n,n\pm 1}|^2 &= \left(n + \frac{1}{2} \pm \frac{1}{2}\right)\frac{\hbar}{2m\omega}, \\ |y_{n,k}|^2 &= 0 \quad k \neq n \pm 1; \end{aligned}\right\} \tag{17–14}$$

where m is the reduced oscillator mass. Using Eq. (17–14), we can write Eq. (17–13) as

$$\left.\begin{aligned} p_{n,n+1} &= (n+1)\frac{8\pi^2 \mu^2 \omega \lambda^2}{\hbar m \alpha^2} \exp\left(-\frac{2\pi\omega}{\alpha v}\right), \\ p_{nk} &= 0 \quad k \neq n \pm 1. \end{aligned}\right\} \tag{17–15}$$

c) *Quantum theory of transition probabilities (the linear problem)*. Quantum-mechanical calculation of $p_{nm}(v)$ in the one-dimensional case, without resort to the assumption that the relative translational motion of the colliding particles is quasi-classical, has been performed in [620, 388, 530]. It is interesting to compare the results obtained in these calculations with Eq. (17–15) in order to determine the limits of applicability of the latter. Calculation of $p_{nm}(v)$ is based on solving the steady-state Schrödinger equation for a system of colliding particles

$$\left[-H_0+\frac{\hbar^2}{2\mu}\frac{d^2}{dr^2}+\mathscr{E}-V(r,y)\right]\psi(r,y)=0, \qquad (17\text{–}16)$$

where $H_0=-(\hbar^2/2m)\,d^2/dy^2)+U(y)$ is the Hamiltonian of the molecule, $V(r,y)=W_0\exp[-\alpha(r-\lambda y)]$ is the intermolecular interaction potential, and ε is the total energy of the system.

The solution of Eq. (17–16) is sought in the form

$$\psi=\sum_m \psi_m(y)f_m(r), \qquad (17\text{–}17)$$

where $\psi_m(y)$ are the eigenfunctions of the Hamiltonian of the molecule H_0, and $f_m(r)$ are the functions sought which, as $r\to\infty$, satisfy the conditions

$$\left.\begin{aligned} &f_n=\exp(-ik_nr)+A_n\exp(ik_nr),\\ &f_m=A_m\exp(ik_mr)\quad n\neq m,\\ &k_n=\frac{\mu v_n}{\hbar},\quad k_m=\frac{\mu v_m}{\hbar};\end{aligned}\right\} \qquad (17\text{–}18)$$

where v_n is the relative velocity of the atom before collision and v_m is the relative velocity of the atom after collision, which is accompanied by transition of the molecule from the nth to the mth vibrational state. The probability $p_{nm}(v)$ is determined by

$$p_{nm}=\frac{k_m}{k_n}|A_m|^2. \qquad (17\text{–}19)$$

The probabilities $p_{nm}(v)$ are calculated within the framework of the method of distorted lines, which is a peculiar version of the perturbation theory.

From Eq. (17–16) for $f_m(r)$, we obtain

$$\left(\frac{d^2}{dr^2}+k_n^2-\frac{2\mu}{\hbar^2}V_{nn}\right)f_n(r)=\sum_{m\neq n}\frac{2\mu}{\hbar^2}V_{nm}(r)f_m(r), \qquad (17\text{–}20)$$

where V_{nm} and V_{nn} have the same meaning as in Eq. (17–5). Eqs. (17–20) are solved by methods of the perturbation theory; here the undisturbed functions are represented by functions which satisfy the equation

$$\left(\frac{d^2}{dr^2}+k_n^2-\frac{2\mu}{\hbar^2}V_{nn}\right)f_n^{(0)}(r)=0, \qquad (17\text{–}21)$$

which describes the process of elastic collision of an atom with a molecule.

The inelastic collision is described by functions of the following approximation $f_m^{(1)}$, obtained by solving the equation:

$$\left(\frac{d^2}{dr^2} + k_m^2 - \frac{2\mu}{\hbar^2} V_{mm}\right) f_m^{(1)} = \frac{2\mu}{\hbar^2} V_{nm} f_n^{(0)} .$$

The asymptotic behavior of $f_m^{(1)}$ is precisely the factor which determines p_{nm}, for which the following expression was obtained:

$$p_{nm}(v) = \frac{4}{k_n k_m} \left| \frac{2\mu}{\hbar^2} \int_{-\infty}^{\infty} V_{nm}(r) F_n(r) F_m(r) \, dr \right|^2 , \qquad (17\text{–}22)$$

where $F_l(r)$ is solution of Eq. (17–21) which satisfies the conditions $F_l(r) \to 0$, as $r \to -\infty$ and

$$F_l \sim \cos(k_l r + \eta) \quad \text{as} \quad r \to \infty .$$

Direct calculation of $p_{nm}(v)$ in the case of an exponential interaction potential for a model of a harmonic oscillator yields the following in Eq. (17–9):

$$\left.\begin{aligned} p_{n,\,n+1} &= (n+1) \frac{8\pi^2\mu^2\omega\lambda^2}{\hbar m \alpha^2} \frac{\operatorname{sh} q_{n+1} \cdot \operatorname{sh} q_n}{(\operatorname{ch} q_{n+1} - \operatorname{ch} q_n)^2}, \\ p_{nk} &= 0 \quad k \neq n \pm 1 , \end{aligned}\right\} \qquad (17\text{–}23)$$

where $q_i = (2\pi\mu v_i/\alpha\hbar)$.

Comparing Eq. (17–23) with (17–15), one can easily determine the limits of applicability of the latter expression. Eq. (17–23) is transformed into Eq. (17–15) when

$$\frac{\operatorname{sh} q_{n+1} \cdot \operatorname{sh} q_n}{(\operatorname{ch} q_{n+1} - \operatorname{ch} q_n)^2} \approx e^{-\frac{2\pi\omega}{\alpha\bar{v}}} ,$$

$$\bar{v} = \frac{1}{2}(v_n < v_{n+1}) \approx v_n ,$$

which comes about when

$$q_{n+1}, \quad q_n \gg 1 \qquad (17\text{–}24)$$

and

$$q_n - q_{n+1} = \frac{q_n^2 - q_{n+1}^2}{q_n + q_{n+1}} = \frac{2\pi\omega}{\alpha\bar{v}} \gg 1 . \qquad (17\text{–}25)$$

Conditions (17–24) and (17–25) have a clear physical meaning. Inequality (17–25) can be rewritten in the form $\omega\tau_{\text{col}} \gg 1$, which corresponds to the condition that the collisions are adiabatic which was postulated in deriving Eqs. (17–15). Condition (17.24) means that we are considering collisions in which the de Broglie wavelength, which corresponds to the relative motion ($\lambda_0 = \hbar/\mu v$) is appreciably smaller than the radius of intermolecular interaction (α^{-1}). This condition was also taken into account in the derivation of Eq. (17–15). In fact, in Eqs. (17–5) which serve as a basis for the derivation of Eq. (17–15), it was assumed that the relative motion of the colliding particles can be considered in the classical manner. The latter condition is precisely fulfilled when Eq. (17–24) is satisfied. In the same approximations it is possible to obtain Eq. (17–10) directly from Eq. (17–22), using the fact that in the quasi-classical approximation the matrix elements are transformed into the corresponding Fourier components [97]. A detailed comparison of results of the semiclassical and quantum-mechanical method of calculations has been performed in [605, 301, 483a]. Let us also note that the correct expression for the vibrational energy received by the oscillator on the average in one collision with an atom can also be obtained from purely classical calculations [250, 214, 481, 577, 465, 249] (see also Sect. 2).

d) The transition probability as a three-dimensional problem. The entire preceding discussion pertained to the one-dimensional case. The probabilities $p_{nm}(v)$ in the three-dimensional case which corresponds to an arbitrary orientation of the colliding particles in space were determined in [554, 531, 129, 600, 369] and the calculations are partially presented in [301, 368]. The probabilities $p_{nm}(v)$ in the three-dimensional case were calculated by the semiclassical method in [129, 301] or by the successive quantum-mechanical method in the distorted waves approximation [554, 531, 600, 369]. It was assumed in the calculations that the intermolecular interaction potential is spherically symmetrical. By virtue of the same fact, the problem was actually reduced to a one-dimensional one. In fact, within the framework of the semiclassical method, the assumption to the effect that the problem is one-dimensional is reflected only in Eq. (17–6), which determines the path of the colliding particles. In the three-dimensional case Eq. (17–6) retains its form except that the potential $V(r)$ must, as is usually done [96], be replaced by the effective potential

$$V(r) + \frac{M}{2\mu r^2} = V(r) + W_0 \frac{b^2}{r^2}, \tag{17–26}$$

which takes into account the centrifugal energy. In Eq. (17–26) M is the moment, b is the radius of interaction, $W_0 = (\mu v^2/2)$ is the initial kinetic energy of the colliding particles. Similar changes appear also in the method of distorted waves. The kinetic energy operator $(\hbar^2/2\mu) \times (d^2/dr^2)$ in Eq. (17–16) must be replaced by its three-dimensional equivalent

$$\frac{\hbar^2}{2\mu}\left(-\frac{1}{r^2}\frac{\partial}{\partial r}r^2\frac{\partial}{\partial r} + \frac{I^2}{r^2}\right),$$

where

$$I^2 = -\left[\frac{1}{\sin^2\theta}\frac{\partial^2}{\partial\varphi^2} + \frac{1}{\sin\theta}\frac{\partial}{\partial\theta}\left(\sin\theta\frac{\partial}{\partial\theta}\right)\right]$$

is the operator of the square of the moment in spherical coordinates, which corresponds to the classical square of the moment which appears in the expression for the centrifugal energy in Eq. (17–26). In this replacement the wave function in Eq. (17–16) is a function of the angles. However, this functional behavior is inconsequential, since it is assumed that the intermolecular interaction potential is spherically symmetrical. The solution of Eq. (17–16) in the three-dimensional case is sought in the form

$$\psi(r, y, \theta, \varphi) = \sum_m \psi_m(y)\, \varphi_m(r, \theta, \varphi),$$

where

$$\varphi_m(r, \theta, \varphi) = \frac{1}{r} \sum_{l,k} f_{mlk}(r)\, Y_{lk}(\theta, \varphi);$$

and Y_{lk} are known functions [97].

For f_{mlk}, which turns out to be independent of the subscript k, we get an equation which is similar to Eq. (17–20), with only this difference that $V_{nn}(r) \approx V(r)$ is replaced by the effective potential energy

$$V(r) + \frac{\hbar^2}{2\mu} \frac{j(j+1)}{r^2}$$

(j is the orbital quantum number), which is the quantum mechanical counterpart of the classical expression (17–26). Eqs. (17–6) and (17–20) in the three-dimensional case are solved on the assumption that the centrifugal potential can be replaced by a constant value taken at the direction reversal point, the vicinities of which make the greatest contribution into the values of the matrix elements (17–7) and (17–22), the so-called method of the modified wave number. This approximation is justified, since the centrifugal potential varies much slower than $V(r)$ in the region of the turnabout point and, as is shown in [517], it is satisfied with a high degree of precision. The transition probabilities thus obtained must be integrated over all the interaction radii b or summed over their corresponding orbital quantum numbers j. Calculations performed by the semiclassical method and in the approximation of distorted waves, as in the one-dimensional case, yield the same results, which are close to the corresponding results obtained in the one-dimensional case [531, 567]. Slightly different expressions for the transition probabilities in the three-dimensional case were obtained in [600]. However, this difference does not follow from any new physical assumptions, but is due rather to the character of simplifications which were introduced in [600] in calculating Eq. (17–22).

The basic shortcoming of all the above works consists in the selection of the spherically symmetrical intermolecular interaction potential. This selection of the potential automatically excludes all the processes of vibrational-rotational interaction in molecular collision, which can perceptibly change the magnitude of the transition probability [134].

An attempt to take into account the fact that the molecular interaction potential is not spherically symmetrical was made in [555, 561, 306, 369], which are discussed in [368].

d) Effect of attraction forces. When the relative velocities of the colliding particles are low, attraction forces may become important for transition probabilities. These probabilities, taking into account the attraction forces, were calculated in [622, 301, 577, 249, 311, 312] and have introduced a new multiplier in Eq. (17–15) or (17–23) which increases the transition probability. The form of this factor depends on the method of calculation and on the character of the approximations made. In particular, the following expression was obtained for it in [577, 129]:

$$e^{\frac{2\pi\omega}{\alpha v}\left(\frac{2}{\pi}\sqrt{\frac{\varepsilon}{W_0}}\right)} \tag{17–27}$$

where ε is the depth of the potential hole [dissociation energy; editor], and $W_0=(\mu v^2/2)$ is the initial kinetic energy of the relative motion of the colliding particles. This increase in the transition probability on introduction of the attraction forces has a simple physical explanation. The attraction forces actually increase the velocity of relative motion, which, according to Eq. (17–15), increases the transition probability. Thus, the introduction of the multiplier of Eq. (17–27) is equivalent to increasing the initial velocity v in Eq. (17–15) to the value [577]:

$$\bar{v} = v\left[1 - \frac{2}{\pi}\operatorname{arc\,tan}\left(\frac{\varepsilon}{W_0}\right)^{\frac{1}{2}}\right]^{-1}.$$

e) Averaging the values of transition probabilities. As was pointed out in the beginning of this subsection, we are interested not only in the value of the transition probability as a function of the velocity, but also in the averaged value of the probability, which is a function of the temperature. The averaging of $p_{nm}(v)$ over thermal velocities [see Eq. (17–4)] was performed in [530, 531], where an analytic expression was obtained for P_{nm}. Without bringing here all the calculations, we wish to note the following important feature. The integrand of Eq. (17–4) after being substituted into Eq. (17–13) has, as is easy to see, a sharp maximum, the position of which is determined by the condition that the exponent be at a minimum:

$$\omega\tau_{\text{col}} + \frac{\mu v^2}{2kT}. \tag{17–28}$$

From this condition we can find the velocity v^*, which has corresponding to it a maximum of the integrand

$$v^* = \left(\frac{2\pi kT\omega_{nm}}{\mu\alpha}\right)^{\frac{1}{3}}. \tag{17–29}$$

The main contribution to the integral of Eq. (17–4) is made by a region of variables in the vicinity of v^*. Hence,

$$P_{nm} \sim |y_{nm}|^2 e^{-\omega_{nm}\frac{2\pi}{\alpha v^*} - \frac{\mu v^{*2}}{2kT}} = |y_{nm}|^2 e^{-3\chi}, \tag{17–30}$$

where

$$\chi = \frac{\mu v^{*2}}{2kT} = \left(\frac{\pi^2\mu\omega_{nm}^2}{2\alpha^2 kT}\right)^{\frac{1}{3}} = \frac{1}{2}\omega_{nm}\tau^*. \tag{17–31}$$

Eq. (17–30) is the Landau-Teller formula. A more exact evaluation of the integral of Eq. (17–4) which does not require fulfilling the condition

$$\tfrac{1}{2}(v_n + v_{n+1}) \approx v_n,$$

which is implied in Eq. (17–25), *i.e.* which does not make use of the assumption that the relative kinetic energy of the effective collisions is appreciably greater than the energy of the transferred photon, was performed in [583]. The results thus obtained are practically identical with Eq. (17–28). The only difference is that $\ln P_{nm}$ is found to be proportional not to $T^{-\frac{1}{3}}$ as in Eq. (17–30), but to $T^{-0.34}$.

In the temperature region of practical interest, $\chi = 5$–10 for various gases [531], hence of substance for P_{nm} and, consequently, also for the conversion of the translational into vibrational energy are collisions of particles the velocities of which lie in the tail of the Maxwellian distribution. It follows from this that the relaxation characteristics at ordinary temperatures yield data on high-energy collisions, *i.e.*, on collisions typical for a gas with a temperature χT.

For a harmonic oscillator we can write Eq. (17–30) in the form

$$\left.\begin{aligned} P_{n,n+1} &= (n+1)P_{01}, \\ P_{n,n-1} &= nP_{10}, \\ P_{nm} &= 0 \quad m \neq n \pm 1, \end{aligned}\right\} \tag{17–32}$$

where $\ln P_{10} \sim T^{-\frac{1}{3}}$.

The absolute value of P_{10} has been calculated in [129, 531, 368, 465, 249, 600, 562, 563]. Thus, the following expression is quoted in [368] for P_{10}:

$$P_{10} = 1.294\left(\frac{r_c}{r_0}\right)^2 \frac{1}{Z_0}\left(1 + 1.1\frac{\varepsilon}{kT}\right)^{-1} \frac{4\pi^2\mu\omega}{\alpha^2\hbar}\sqrt{\frac{4\pi}{3}}\,\chi^{\frac{1}{2}} e^{-3\chi + \frac{\hbar\omega}{2kT} + \frac{\varepsilon}{kT}}, \tag{17–33}$$

where r_0 and ε are constants of the Lennard-Jones potential

$$V(r) = 4\varepsilon\left[\left(\frac{r_0}{r}\right)^{12} - \left(\frac{r_0}{r}\right)^{6}\right],$$

r_c is the coordinate of the turnabout point, and $Z_0 = 3$ is the orientation factor. Eq. (17–33) was obtained for an intermolecular interaction potential given by

$$V(r) = W_0 e^{-\alpha(r-\lambda y)} - \varepsilon. \tag{17–34}$$

The term ε in Eq. (17–34) takes into account the attraction forces (it has here the same meaning as in the Lennard-Jones potential). The constant α in Eq. (17–34) is determined by equating (17–34) to the Lennard-Jones potential. In the first approximation it may be assumed [530], that

$$\alpha = \frac{17.5}{r_0}. \tag{17–35}$$

More exact calculations, based on equating the values of the potentials and their first derivatives in the turnabout point (Method A), yield

$$\alpha = \frac{12}{r_0}\left[\frac{1}{2}\left(1 + \sqrt{\frac{E_m}{\varepsilon} + 1}\right)\right]^{\frac{1}{6}}\left[1 + \left(\frac{E_m}{\varepsilon} + 1\right)^{-\frac{1}{2}}\right]. \tag{17–36}$$

In addition, α can be determined by equating the values of the potential in two points, r_0 and r_c (Method B). In this case the following expression is obtained for α:

$$\alpha = \frac{1}{r_0}\left[\ln\left(\frac{E_m}{\varepsilon} + 1\right)\right]\left\{1 - \left[\frac{1}{2}\left(1 + \sqrt{\frac{E_m}{\varepsilon} + 1}\right)\right]^{-\frac{1}{6}}\right\}. \tag{17–37}$$

In Eqs. (17–36) and (17–37) $E_m = \mu v^{*2}/2$, where v^* is determined from Eq. (17–29). Numerical calculation of α using the above methods was performed in [368]. The values of α thus obtained are given in Table 17. It should, however, be remembered that the parameters of the Lennard-Jones potential were selected in order to satisfactorily describe the experimental values of the viscosity and diffusion coefficients of gas mixtures the temperature of which does not exceed 1000 °K. In this temperature region the behavior of the potential is of substance at comparatively long ranges. However, in the process of vibrational relaxation, as was pointed out above, the main contribution is made by high-energy collisions which are typical of gases with temperatures $\chi T (\chi > 1)$, which is several thousands of degrees. The short-range behavior of the potential, which is of interest in this temperature region, is practically unknown at present. Approximate quantum-mechanical calculations of short-range repulsion forces were performed for only indi-

Table 17. Values of r_0/r_c and αr as Functions of E_m/ε

E_m/ε	r_0/r_c	αr_0 Method A	αr_0 Method B
0	1	24	24
1.25	1.0377	20.75	22.32
3	1.070	19.26	21.2
5.25	1.0978	18.44	20.6
8	1.1225	17.96	20.1
15	1.165	17.48	19.58
24	1.2010	17.29	19.23
35	1.2322	17.25	19.02
63	1.2849	17.25	18.76
99	1.3286	17.54	18.63
120	1.3480	17.65	18.58
168	1.3831	17.87	18.52
224	1.4142	18.10	18.49
399	1.4797	18.65	18.48
728	1.5525	19.32	18.52

vidual pairs of molecules (O_2-O_2, O_2-N_2, N_2-N_2, H_2-H_2, etc.) [580, 581, 443].

Eq. (17–33) can be used only for estimating the order of magnitude. In fact, a more rigorous accounting of the attraction forces, which was done in [129], results in replacing $\exp(\varepsilon/kT)$ in Eq. (17–33) by $\exp[(4/\pi\sqrt{\chi\varepsilon/kT}]$. In [369] a more exact expression for Z_0 was obtained. This corresponds to the collision of identical diatomic molecules

$$Z_0 = (\alpha r_e)^2 e^{-\frac{3}{8}\frac{\alpha r_e^2}{rc}},$$

where r_e is the internuclear distance in the molecule. Thus, Eq. (17–33), apparently, correctly conveys the temperature dependence of P_{10}, although the absolute values obtained from it, even when making use of all the refinements, are approximate. In addition, at present, it is difficult to evaluate the accuracy of the approximation [604].

A somewhat different temperature dependence of P_{10} was obtained in [248–250], however, this is a result of a peculiar definition of the transition probability, which differs from that ordinarily used.

f) Transition probability in nonadiabatic collisions. Up to now we have considered only adiabatic collisions, for which $\omega\tau_{col} \gg 1$. However, in rapid collisions with heavy molecules, or with molecules which are at the upper vibrational levels, the opposite is true, *i.e.*, $\omega\tau_{col} \ll 1$. In particular, the latter condition prevails in collisions of I_2 molecules with He in a gas at room temperature. The probability $p_{nm}(v)$ under this condition can be calculated from Eqs. (17–5). In this case the exponent in Eq. (17–5) can be replaced by unity. Then Eqs. (17–5), making use of Eq. (17–9), take on a simpler form

$$i\hbar \frac{\mathrm{d}a_k}{\mathrm{d}t} = -\lambda \frac{\mathrm{d}V}{\mathrm{d}r} \sum_m a_m(t)\, y_{km}, \tag{17–38}$$

for which an exact solution is available [151, 154]

$$a_m(t) = \int_{-\infty}^{\infty} \psi_m^*(y)\, e^{\frac{i}{\hbar}\left(\lambda \int_{-\infty}^{t} \frac{\mathrm{d}V}{\mathrm{d}r}\mathrm{d}t\right) y} \psi_n(y)\, \mathrm{d}y, \tag{17–39}$$

where $\psi_p(y)$ is the wave function of the molecule in the pth vibrational state. In Eq. (17–38), as well as in Eq. (17–5), it is assumed that the excitation of the vibrational degrees of freedom is not accompanied by changes in the electronic state. Eq. (17–39) can be simplified by using Eq. (17–6). In fact, according to Eq. (17–6),

$$\frac{1}{\hbar} \int_{-\infty}^{\infty} \frac{\mathrm{d}V}{\mathrm{d}r}\, \mathrm{d}t = -2k_n, \tag{17–40}$$

where k_n is the wave number of the incoming atom.

Using Eq. (17–40) it is possible to write $p_{nm}(v)$ in the form

$$p_{nm}(v) = |a_m(\infty)|^2 = \left| \int_{-\infty}^{\infty} \psi_m^*(y)\, e^{-2ik_n\lambda y} \psi_n(y)\, dy \right|^2. \tag{17–41}$$

Eq. (17–41) is also obtained by a sequential quantum-mechanical calculation [139, 140].

Analysis of Eq. (17–41) shows that molecular collisions under the condition $\omega\tau_{\mathrm{col}} \ll 1$ can excite not only the neighboring levels, but the next ones as well. In fact, Eq. (17–41) can be written explicitly, using the harmonic oscillator model and assuming that $m=0$. On these assumptions

$$p_{0n} = e^{-q^2} \frac{q^{2n}}{n!}, \tag{17–42}$$

where

$$q = 2k_0\lambda y_{01}, \quad y_{01} = \sqrt{\frac{\hbar}{2m\omega}}. \tag{17–43}$$

It can be seen from Eq. (17–42) that the transition $0 \to 1$ is most probable only for small q ($q \ll 1$), in which case

$$p_{01} = q^2 = 4\lambda^2 k_0^2 y_{01}^2 \,. \tag{17–44}$$

For large q ($q \gg 1$) the maximum probability is obtained for transition to the level $n \sim q^2$. The average vibrational energy $\Delta\varepsilon$, which is transmitted by the incoming atom to the molecule, is

$$\Delta\varepsilon = \hbar\omega \sum_{n=0}^{\infty} n e^{-q^2} \frac{q^{2n}}{n!} = q^2 \hbar\omega = 4\lambda^2 \frac{\mu}{m} W_0 \,, \quad W_0 = \frac{\mu v^2}{2} \,, \tag{17–45}$$

which is in agreement with the corresponding classical expression [214].

Eq. (17–41) can be simplified also in another particular case, assuming $k_n y_{av} \ll 1$, where y_{av} is the amplitude of vibrations of atoms in molecules. Expanding the exponent of Eq. (17–41) in a series and restricting ourselves to the first nonvanishing term, we get

$$p_{nm} = 4\lambda^2 k_n^2 y_{nm}^2 \,. \tag{17–46}$$

A harmonic oscillator $y_{nn'}$ is different from zero only when $n' = n \pm 1$ Eq. (17–14), hence under the condition being discussed

$$\omega\tau_{col} \ll 1 \quad \text{and} \quad k_n y_{av} \ll 1 \quad \left(y_{av} \sim \sqrt{\frac{\hbar}{2m\omega}} \right) \tag{17–47}$$

only transitions to the neighboring levels are practically permissible. Conditions (17–47) are quite rigorous, but they are satisfied, for example, in collisions of I_2 with He in a gas mixture at a temperature of the order of 1000 °K [146]. Eq. (17–46) averaged over the Maxwellian distribution of the incoming particles [see Eq. (17–4)], upon use of Eq. (17–14) takes on the form

$$P_{n+1,n} = (n+1) P_{10} \,, \quad P_{10} = \frac{2kT}{\hbar\omega} \cdot \frac{\mu}{m} \,. \tag{17–48}$$

Eqs. (17–41), (17–42) and (17–46), as well as those of the adiabatic theory of collision are suitable only for estimating the order of magnitude.

In closing we wish to call attention to the following. The transition probability on atom-molecule collisions have been calculated only in the two limiting cases of $\omega\tau_{col} \gg 1$ and $\omega\tau_{col} \ll 1$. The calculation of $p_{nm}(v)$ for any values of $\omega\tau_{col}$, apparently requires the use of numerical methods. An attempt to calculate the transition probability in a wide range of $\omega\tau_{col}$ was made in [152]. In this work Eqs. (17–5), consisting of 11 equations, were solved on a computer using parameters corresponding to the O_2–Ar collision for $\omega\tau_{col} \equiv (\omega/\alpha v)$ in the interval from 0.53 to 7.56. Analysis of this solution shows that Eq. (17–13) is valid approximately up to $(\omega/\alpha v) = 1.2$. Thus, when $(\omega/\alpha v) = 1.2$ the ratio of p_{01} found by the perturbation theory [Eq. (17–13)], to that obtained by solving Eq. (17.5) numerically is 1.14. When $(\omega/\alpha v)=0.9$ this ratio is already 2.2. This is approximately the region where multiphoton transition also becomes appreciably probable. Thus, when $(\omega/\alpha v)=0.9$, $p_{02}/p_{01}\approx 0.4$, when $(\omega/\alpha v)=0.7$ this ratio is 1.4, etc. Similar results for N_2 molecules were obtained in solving the system (17–5) when it consisted of four equations [483].

Expressions for $p_{nm}(v)$ for a harmonic oscillator acted upon by an exciting potential $xF(t)$ was obtained in [240] for any values of $\omega\tau_{col}$. The transition probability was determined there without directly solving Eqs. (17.5), but rather using the energy representation of Wigner's distribution function, which describes the behavior of a system of oscillators in the phase space. This problem can be solved analytically only because of a peculiar property of the harmonic oscillator, for which Ehrenfest's quantum-mechanical equations of motion coincide with the classical Newton equations. The analytical expression for $p_{n'n}(v)$ has a quite complex form with the exception of the case $n'=0$, for which

$$p_{0n}=\frac{1}{n!}\left(\frac{\bar{\varepsilon}}{\hbar\omega}\right)^n e^{-\frac{s}{\hbar\omega}},\quad \frac{\bar{\varepsilon}}{\hbar\omega}=\frac{\lambda^2}{2\hbar m\omega}\left|\int_{-\infty}^{\infty}F(t)\,e^{i\omega t}\,dt\right|^2. \tag{17–49}$$

In the limiting cases Eq. (17–49) yields Eqs. (17–10) and (17–42). In fact, when $(\bar{\varepsilon}/\hbar\omega)\ll 1$ only the $0\to 1$ transitions are practically permissible, here $p_{01}=(\bar{\varepsilon}/\hbar\omega)$ coincides with (17–10). When $\omega\tau_{col}\ll 1$

$$\frac{\bar{\varepsilon}}{\hbar\omega}=\frac{1}{2\hbar m\omega}\left|\int_{-\infty}^{\infty}F(t)\,\mathrm{d}t\right|^2=\frac{\lambda^2 4k_0^2\hbar^2}{2\hbar m\omega}=q^2,\left(F=-\lambda\frac{\mathrm{d}V}{\mathrm{d}r}\right),$$

and Eq. (17–49) becomes Eq. (17–42).

From Eq. (17–49) it is possible to obtain the average vibrational energy received by the oscillator in one collision:

$$\Delta\varepsilon=\sum_{n=1}^{\infty}\hbar\omega n p_{0n}=\bar{\varepsilon}=\frac{1}{2m}\left|\int_{-\infty}^{\infty}F(t)\,e^{i\omega t}\,dt\right|^2. \tag{17–50}$$

The above value of $\Delta\varepsilon$ is identical to the classically calculated transferred energy [see Sect. 2, Eq. (17–17)]. When $(\bar{\varepsilon}/\hbar\omega)\ll 1$

$$\Delta\bar{\varepsilon}\approx\hbar\omega p_{01}. \tag{17–51}$$

Eq. (17–51) was derived formally in Chapter 1. Now it acquires an explicit physical meaning and, in addition, it becomes clear why it is possible to obtain within the framework of classical mechanics the correct expression for p_{01}.

We note in conclusion that Eq. (17–49), as is the case with Eqs. (17–10) and (17–42), is valid on the condition that the effect of the incoming atom in the collision process can be approximated by an exciting potential of the form $yF(t)$. In the case of multiphoton transitions, the latter is permissible if

$$\alpha y_{AV} \ll 1, \quad y_{AV} \sim \sqrt{\frac{n\hbar}{2m\omega}}, \quad \frac{\mu v^2}{2} \gg n\hbar\omega.$$

[2] Vibrational Relaxation of Diatomic Molecules which Comprise a Small Admixture in a Monatomic Gas

According to the preceding discussion (see the first subsection of this section) the equations which describe vibrational relaxation of diatomic molecules which constitute a small admixture to a monatomic gas can be written in the form

$$\frac{dx_n}{dt} = Z(P_{n+1,n}x_{n+1} - P_{n,n+1}x_n + P_{n-1,n}x_{n-1} - - P_{n,n-1}x_n), \quad n = 0, 1, 2, \ldots x_n|_{t=0} = x_n(0). \qquad (17\text{–}52)$$

It is assumed in Eqs. (17–52) that vibrational relaxation proceeds by single-photon energy transfers between the translational and vibrational degrees of freedom. The latter, as was shown in the preceding subsection, is valid for adiabatic collisions ($\omega\tau_{col} \gg 1$) or for collisions which satisfy Eq. (17–47).

Eqs. (17–52) are a linear system of equations with constant coefficients. The temperature-dependent coefficients P_{nm} can be regarded as constant, since under the conditions being considered the vibrational relaxation is isothermic. Solution of Eqs. (17–52) determines functions $x_n(t)$ and makes it possible to follow changes in the molecular population of each of the vibrational levels. The stating of the problem in this manner is meaningful and will be considered in detail subsequently. However, of interest in many practical problems is not the distribution function, but rather the behavior of certain macroscopic characteristics, in particular, of the vibrational energy, This statement of the problem is characteristic of relaxation gasdynamics.

The relaxation equation for the total store of vibrational energy can be obtained from Eqs. (17–52) without solving them [414]. Let us consider the model of the harmonic oscillator. In this case [see Eqs. (17.32) and 17.48)]

$$P_{n+1,n} = (n+1)P_{10}, \qquad (17\text{–}53)$$

here $P_{01} = P_{10}e^{-\theta}$, where $\theta = \hbar\omega/kT$. The latter condition is due to the Maxwellian distribution of the velocity of the colliding particles and follows from the detailed balancing principle.

Using Eq. (17–53), Eqs. (17–52) can be written in the form

$$\frac{dx_n}{dt} = ZP_{10}\{(n+1)x_{n+1} - [(n+1)e^{-\theta} + n]x_n + ne^{-\theta}x_{n-1}\},$$
$$n = 0, 1, 2, \ldots \qquad (17\text{–}54)$$

Let us multiply Eq. (17–54) by $n\hbar\omega$ and then sum over all n. After simple transformations we obtain

$$\frac{dE_{\text{vib}}}{dt} = -ZP_{10}(1 - e^{-\theta})\left\{E - \frac{\hbar\omega e^{-\theta}N}{1 - e^{-\theta}}\right\},$$

where $E_{\text{vib}} = \hbar\omega \sum nx_n(t)$ is the vibrational energy of a unit volume, and

$$N\hbar\omega\frac{e^{-\theta}}{1 - e^{-\theta}} = N\hbar\omega(1 - e^{-\theta})\sum_n ne^{-\theta n}$$

is the equilibrium value of the vibrational energy $E_{0\text{vib}}(T)$ $(\sum_{p=0}^{\infty} x_n(t) = N)$. Thus the relaxation equation for the vibrational energy density has the form

$$\frac{dE_{\text{vib}}}{dt} = \frac{E_{0_{\text{vib}}}(T) - E_{\text{vib}}}{\tau_{\text{vib}}}, \qquad (17\text{–}55)$$

where

$$\tau_{\text{vib}} = [ZP_{10}(1 - e^{-\theta})]^{-1}. \qquad (17\text{–}56)$$

This is a general form of the relaxation equation for any physical quantity for sufficiently small deviations from equilibrium values. In the case of the harmonic oscillator model this equation is valid for any deviations from the equilibrium state. Thus, Eq. (17–55), which is formally introduced in the beginning of this section [see Eq. (17–21)], is found to be valid for the harmonic oscillator model in the case of vibrational relaxation of diatomic molecules comprising a moderate admixture to a monatomic gas. Proceeding slightly ahead, we wish to note that Eq. (17–55) retains its form also in the more general case of an arbitrary concentration of diatomic molecules.

Eq. (17–56) determines the vibrational relaxation time. This characteristic of the relaxation process is one of the most important in the theory of phenomena which are related to establishing equilibrium with respect to the vibrational degrees of freedom.

From the mathematical point of view, the energy E is one of the normal coordinates of Eqs. (17–54) [187]. The remaining normal coordinates do not have any obvious physical meaning and, as is shown in [187, 188], relax at a much higher rate.

The solution of Eq. (17–55) is exponential in form

$$E_{\text{vib}}(t) - E_{0_{\text{vib}}} = [E_{\text{vib}}(0) - E_{0_{\text{vib}}}]e^{-\frac{t}{\tau_{\text{vib}}}} \qquad (17\text{–}57)$$

and depends only on the value of the vibrational energy at the initial time.

The manner in which the molecules are distributed with respect to the vibrational levels for a given value of vibrational energy has no effect on the relaxation process.

The latter circumstance is characteristic of the harmonic oscillator model [538]. For an anharmonic oscillator the vibrational relaxation, generally speaking, depends on the manner in which the molecules are distributed with respect to the vibrational levels, as well as on the initial vibrational energy [130]. When this functional behavior applies, the vibrational energy relaxation can no longer be described by Eq. (17–55). However, in the majority of practical cases (vibrational relaxation in shock waves and on the propagation of ultrasonic vibrations), the relaxation of vibrational energy can be described with sufficient accuracy by Eq. (17–55). This is true for all the processes in which the main role is played by molecules at lower vibrational levels, since the role of anharmonicity in this case is insignificant.

Eq. (17–55) may be extended to the case of several monatomic gases. If the diatomic molecules are a small admixture in a mixture of monatomic gases, consisting of i components with the mole fractions $\xi_i (\sum \xi_i = 1)$, then, as can easily be seen, Eq. (17–55) retains its form. However, in Eqs. (17–54) and (17–56), ZP_{10} must be replaced with

$$Z(P_{10})_{\text{ef}} = \sum Z_{A_i}(P_{10})_{A_i},$$

where $Z_{\Delta i}$ is the collision frequency of the diatomic molecule A with molecules of the monatomic gas of the ith kind, and $(P_{10})_{A_i}$ is the probability of transition $1 \to 0$ on collision of molecule A with a molecule of the i*th* kind of gas. Since $(Z_{A_i}/Z) = \xi_i$, then

$$(P_{10})_{\text{ef}} = \sum \xi_i (P_{10})_{A_i}. \tag{17–58}$$

Substituting Eq. (17–58) into (17–56) we get

$$\frac{1}{\tau_{\text{vib}}} = \sum \frac{\xi_i}{\tau_{A_i}}, \tag{17–59}$$

where

$$\tau_{A_i} = [Z(P_{10})_{A_i}(1 - e^{-\theta})]^{-1} \tag{17–60}$$

is the relaxation time of diatomic molecules, which comprise a small admixture in a monatomic gas of the ith kind, the concentration of which is equal to the total concentration of all the components of the monatomic gas in the given case.

For a binary mixture of monatomic gases B and C, Eq. (17–59) takes on the form

$$\frac{1}{\tau_{\text{vib}}} = \frac{\xi_{\text{B}}}{\tau_{\text{AB}}} + \frac{\xi_{\text{C}}}{\tau_{\text{AC}}} = \frac{1 - \xi_{\text{C}}}{\tau_{\text{AB}}} + \frac{\xi_{\text{C}}}{\tau_{\text{AC}}}; \tag{17–60'}$$

(τ_{vib}, τ_{AB} and τ_{CA} are usually referred to a pressure of one atmosphere.*)

The relaxation time given by Eq. (17–55) is measured directly during the experiment.

The majority of available experimental values of the vibrational relaxation time pertains to pure gases or mixtures where the concentration of molecules undergoing relaxation is not small. A detailed consideration of these works will be given in the third subsection of this section. Here we shall restrict ourselves to the consideration of two problems, *i.e.*, vibrational relaxation of O_2 in an O_2–Ar mixture and of I_2 in an I_2–He mixture. The adiabatic parameter $\omega\tau_{col}$ for the O_2–Ar and I_2–He collisions can be evaluated in the following manner:

$$\omega = 2\pi\nu \begin{cases} = 2.97 \times 10^{14} \text{ rad/sec} & \text{for } O_2, \\ = 4.02 \times 10^{13} \text{ rad/sec} & \text{for } I_2, \end{cases}$$

$$\tau_{col} \sim \frac{1}{\alpha v},$$

where $\alpha \sim 5 \times 10 \text{ cm}^{-1}$, and $v \sim \sqrt{2kT/\mu}$. At temperatures of the order of 1000 °K $v_{O_2-Ar} \sim 10^5$ cm/sec, and $v_{I_2-He} \sim 2 \times 10^5$ cm/sec. Thus for O_2–Ar collisions under conditions being considered $\omega\tau_{col} \approx 6$, and for I_2–He $\omega\tau_{col} \approx 0.4$. In addition, for I_2–He at these temperatures still another condition is valid:

$$kx_{av} \sim \frac{\mu v}{\hbar}\sqrt{\frac{\hbar}{2m\omega}} \approx 0.3 < 1 \quad \left(k = \frac{\mu v}{\hbar}\right).$$

These estimates show that the vibrational relaxation in O_2–Ar and I_2–He mixtures takes place by single-photon transfer and is described by Eq. (17–56). However, the relaxation time of Eq. (17–56) in the O_2–Ar and I_2–He mixtures will have a different functional behavior with respect to the temperature. In fact the condition $\omega\tau_{col} \gg 1$ applies in the O_2–Ar mixture at temperatures of the order of 1000 °K, for which reason P_{10} is given by Eq. (17–33), in accordance with which

$$\ln\left[\tau_{vib} Z(1 - e^{-\theta})\right] \sim T^{-\frac{1}{3}}. \tag{17–61}$$

For the I_2–He mixture in the above temperature region it is possible to take Eq. (17–47) as the first approximation and to use Eq. (17–48) for P_{10}:

According to Eq. (17–48),

$$\tau_{vib} Z \sim \left[\frac{\mu}{m}\frac{kT}{\hbar\omega}\left(1 - e^{-\frac{\hbar\omega}{kT}}\right)\right]^{-1}. \tag{17–62}$$

* Sometimes these quantities are referred to a density under standard conditions ($p = 1$ atm and $T = 273$ °K). In this case the relaxation time is smaller (for the same temperature).

For temperatures of the order of 1000 °K

$$(kT/\hbar\omega)\,[1 - \exp(-\hbar\omega/kT)] \sim 1;$$

hence

$$\tau_{\text{vib}} Z \sim \frac{m}{\mu}. \qquad (17\text{–}63)$$

The vibrational relaxation of oxygen in the O_2–Ar mixture has been obtained experimentally on a wide temperature range. In particular, the temperature range in [277] was 1900–7000 °K. Figure 98 shows the dependence of $\tau_{\text{vib}}[1-\exp(-\theta)]$ on $T^{-\frac{1}{3}}$. On the logarithmic scale this relationship is linear in accordance with Eq. (17–61). Experimental data on vibrational relaxation of I_2 in the I_2–He mixture under conditions where Eqs. (17–62) and (17–63) apply are not available at present. Let us note that the vibrational relaxation time in the I_2–He mixture is of the order of the time required for establishing the Maxwellian distribution in the heavier gas. The fact that the I_2 molecules do not have a Maxwellian distribution will have no effect on the vibrational relaxation rate of I_2 in ultrasonic fields, since

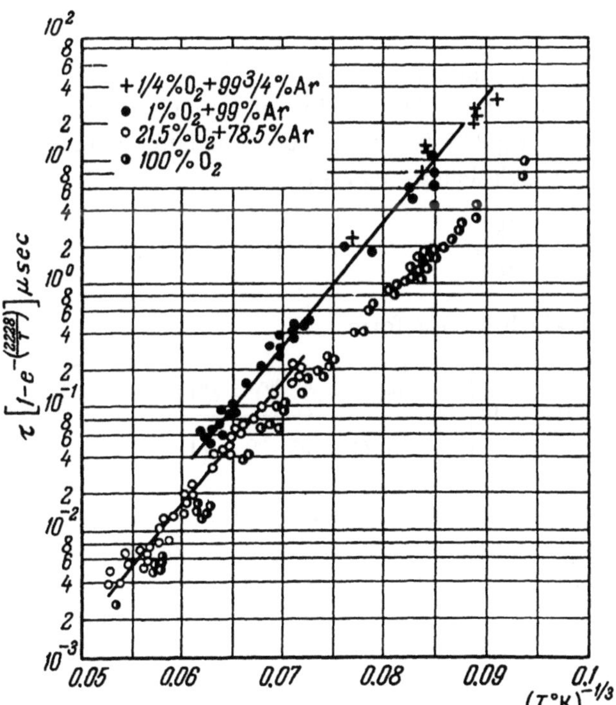

Fig. 98. The temperature dependence of $\tau_{\text{vib}}\,[1-\exp(-\theta)]$.

due to the large difference in the masses of I_2 and He, it is possible to assume that the I_2 molecules are practically at rest. A different situation arises when measuring vibrational relaxation behind a shock front. In this case it is impossible to assume that the I_2 molecules are at rest, since immediately behind the shock front they move relative to the lighter gas with an average velocity which is equal to the step increase in the flow velocity across the front. Under these conditions the vibrational relaxation time for I_2 can be found smaller than that predicted by Eqs. (17–62) and (17–63).

Studying vibrational relaxation by optical methods, in particular, by the absorption of ultraviolet radiation [113], it becomes possible to measure the molecular population of individual vibrational levels. In conjection with this it becomes important to obtain a more general solution of Eqs. (17–52), which describes the molecular distribution at individual vibrational levels.

Before we consider the methods of finding this kind of solution, let us consider one particular solution which is of great practical interest. As in the preceding case, we restrict ourselves to the harmonic oscillator model, *i.e.*, we shall study Eqs. (17–54).

Eqs. (17–54) have a solution in the form

$$x_n(t) = N(1 - e^{-\vartheta(t)})e^{-n\vartheta(t)}, \tag{17–64}$$

where $\vartheta(t)$ is known function of time which is independent of n. The latter circumstance is important. The validity of Eq. (17–64) can be checked by substituting it into Eq. (17–54). In this case the latter equations reduce to a single equation, which defines the function $\vartheta(t)$

$$\frac{d\vartheta}{dt} = ZP_{10}(1 - e^{-\vartheta})(e^{-\vartheta} - e^{-\theta})e^{\vartheta},$$

the solution of which has the form

$$\vartheta(t) = \ln\left[\frac{e^{-\frac{t}{\tau_{vib}}}(1 - e^{\theta-\vartheta_0}) - e^{\theta}(1 - e^{-\vartheta_0})}{e^{-\frac{t}{\tau_{vib}}}(1 - e^{\theta-\vartheta_0}) - (1 - e^{-\vartheta_0})}\right], \tag{17–65}$$

where τ_{vib} is given by Eq. (17–56) and $\vartheta_0 = (0)$.

The physical meaning of the above solution is obvious. Let us assume that at the initial instant the function of molecular distribution with respect to the vibrational levels has the form of a Boltzmann function for a temperature which differs from the temperature of the surrounding monatomic gas, *i.e.*, $x_n(0) = N/[1 - \exp(-\vartheta_0)\exp(-n\vartheta_0)$. In this case the vibrational relaxation will take place in a manner such that the distribution function retains its initial form, and only the vibrational temperature varies. In other words, the form of the equilibrium distribution function is retained following vibrational relaxation, if the initial distribution is a Boltzmann distribution. A similar situation is also observed in the relaxation of translational degrees of freedom in a gas consisting of Maxwellian molecules [143]. A particular solution of Eqs. (17–54) was first found by Montroll and Shuler [451] and is widely used in practical calculations [41]. It should, however, be emphasized, that the law of retention of the form of the Boltzmann distribution is valid only in a system of harmonic oscillators. In the general case the solution of Eqs. (17–54) is sought by ordinary methods. Eqs. (17–54) are a system of linear differential equations with constant coefficients, for which reason their solution can be written in the form

$$x_n(t) = \sum_m a_{nm} e^{-\mu_m t}. \tag{17–66}$$

It was shown in [451] that for Eq. (17–54)

$$a_{nm} = \alpha_m l_n(m), \quad \mu_m = m[ZP_{10}(1 - e^{-\theta})], \tag{17–67}$$

where l_n are Gotlieb's polynomials

$$l_n(m) = e^{-n\theta} \sum_{\nu=0}^{\infty} (1 - e^{\theta})^\nu \binom{n}{\nu} \binom{m}{\nu},$$

$$\binom{n}{\nu} = \begin{cases} \dfrac{n!}{\nu!(n-\nu)!} & n \geqslant \nu, \\ 0 & n < \nu \quad \text{or} \quad \nu < 0, \\ 1 & \nu = 0. \end{cases}$$

Using hypergeometric functions the Gotlieb polynomials can be written in the form

$$l_n(m) = F(-n, m+1, 1; 1 - e^{-\theta}).$$

In the particular case of $m = 0$ and 1

$$l_n(0) = e^{-n\theta},$$
$$l_n(1) = e^{-n\theta}\{1 + (1 - e^{\theta})n\}.$$

The orthogonality condition for integer and nonnegative m and n has the form

$$\sum_{\nu=0}^{\infty} e^{-\theta\nu} l_n(\nu) l_m(\nu) = \begin{cases} 0 & n \neq m, \\ e^{-n\theta}(1 - e^{-\theta})^{-1} & n = m, \end{cases}$$

$$\sum_{n=0}^{\infty} e^{n\theta} l_n(\nu) l_n(m) = \begin{cases} 0 & m \neq \nu, \\ e^{\nu\theta}(1 - e^{-\theta})^{-1} & m = \nu \end{cases}$$

Thus, the solution of Eq. (17–54) has the form

$$x_n(t) = \sum_m \alpha_m l_n(m) e^{-\frac{mt}{\tau_{\text{vib}}}}, \tag{17–68}$$

where τ_{vib} is the relaxation time [see Eq. (17–56)], and α_m are determined by initial conditions.

The peculiarity of Eq. (17–66) consists in the fact that the greatest relaxation time which actually determines the time behavior of $x_n(t)$, is identical with the relaxation time for vibrational energy.

The solution given by Eq. (17–66) is not always convenient for practical calculations, hence resort is frequently had to other forms of the solution which are obtained by the generating function method [451].

In certain cases of practical interest, it is possible to forego the use of the general Eq. (17–68) and to consider simpler particular cases. From this point of view we now consider the vibrational relaxation of electronically excited CN molecules ($B^2\Sigma^+$) in Roth's experiments [501]. Roth (see Sect. 12) has studied the time characteristics of the luminescence of the ultraviolet band of CN ($B^2\Sigma^+ \rightarrow X^2\Sigma^+$) behind a shock front. Measuring the delay time in the appearance of the (0, 0) band compared with the (1, 1) band and the intensity of the (1, 1) band at different times, Roth has calculated the vibrational relaxation time of CN ($B^2\Sigma^+$). The parentheses, as usual, shows the vibrational quantum numbers for the excited and ground electronic states. According to Roth, vibrational relaxation of CN molecules takes place according to the scheme

$$\text{CN}(B^2\Sigma^+)_{n=1} \underset{\text{M}}{\overset{k_1}{\rightarrow}} \text{CN}(B^2\Sigma^+)_{n=0} \overset{k_2}{\rightarrow} \text{CN}(X^2\Sigma^+)_{n=0} + h\nu, \tag{17–69}$$

where M are Xe atoms.

It is assumed that at the time of appearance of the peak of the (1, 1) band the population of all the higher vibrational levels with $n > 1$ is small in comparison with the population of the level $n = 1$. This time was taken as the origin for measurements. In addition, the reverse processes are disregarded in Eq. (17–69).

On these assumptions, Eq. (17–69) is, according to Roth, described by the following system of equations:

$$\left.\begin{aligned} \frac{dx_1}{dt} &= -k_1 x_1; \\ \frac{dx_0}{dt} &= k_1 x_1 - k_2 x_0. \end{aligned}\right\} \tag{17–70}$$

In Eq. (17–70), $k_1 = ZP_{10}$, x_i is the number density of CN ($B^2\Sigma^+$) molecules on the ith vibrational level.

The solution of Eq. (17–70) has the form

$$x_1(t) = x_1(0)\,e^{-k_1 t},$$
$$x_0(t) = \frac{k_1}{k_2 - k_1}\,x_1(0)\,[e^{-k_1 t} - e^{-k_2 t}]. \qquad (17\text{–}71)$$

For the time t_1 which corresponds to the appearance of the peak of the (0, 0) band ($\mathrm{d}x_0/\mathrm{d}t = 0$), Eq. (17–71) can be written in the form

$$x_1(t_1) = \frac{k_2}{k_2 - k_1}\,x_1(0)\,[e^{-k_1 t_1} - e^{-k_2 t_1}]. \qquad (17\text{–}72)$$

Remembering that $k_2 \gg k_1$, we obtain from Eq. (17–72) a formula used by Roth [501] to determine the vibrational relaxation time

$$\tau_1 \equiv \frac{1}{k_1} = \frac{t_1}{\ln\left\{\dfrac{[\mathrm{CN}(B^2\Sigma^+)_{v=1}]_0}{[\mathrm{CN}(B^2\Sigma^+)_{v=1}]_{t_1}}\right\}}. \qquad (17\text{–}73)$$

It should, however, be emphasized that the characteristic time $\tau_1 = (1/ZP_{10})$, introduced by Roth, differs from τ_{vib} defined by Eq. (17–56), by the factor $1 - e^{-\theta}$. In fact $\tau_1 = \tau_{\mathrm{E}}\,(1 - e^{-\theta})$. A similar scheme for calculations is also encountered in the study of vibrational relaxation of electronically excited NO ($A^2\Sigma^+$) molecules [503, 504] behind a shock front and of O_2 molecules which are produced by impulse photolysis of gas mixtures [155, 150, 144, 147, 149]. Equation (17–54) which takes into account only deactivation processes, was also obtained in [471].

[3] Vibrational Relaxation in Pure Gases and in Mixtures with a Monatomic Gas

In the systems considered up to now, the diatomic molecules comprised a small admixture in a monatomic gas. The study of vibrational relaxation in pure diatomic gases is also of practical interest. In such a system, in addition to the transformation of translational into vibrational energy, importance is also acquired by energy transfer between the vibrational and rotational degrees of freedom of the colliding molecules, as well as exchange of the vibrational energy.

The resonance transformation of the rotational into vibrational energy on a molecule-molecule collision is highly improbable. The physical cause for this consists in the fact that molecular collisions under ordinary conditions are adiabatic, which determines the small probability of energy transfer from rotation to vibrations. In addition, the number of molecules which have a rotational energy greater than or equal to the energy of the vibrational photon is exponentially small. Simultaneous transformation of rotational and translational into vibrational energy and conversely can be of substance. A systematic quantum-mechanical calculation of vibrational-rotational transitions is not available at present. Estimates performed in [134] show that the probability of losing or acquiring a vibrational photon, when rotational transitions are taken into account, becomes greater by several factors than the same probability calculated without considering the transformation of the rotational energy. Thus, for the O_2–O_2 and O_2–Ar collisions, it was shown in [134] that

$$P_{10}(O_2 - O_2) = p\frac{e^{2\pi\gamma}}{2\pi\gamma} P_{10}(O_2 - \mathrm{Ar}), \quad (\gamma \approx 1). \qquad (17\text{–}74)$$

The factor $p(\exp 2\pi\gamma/2\pi\gamma)$ accounts for the effect of simultaneous rotational transitions and was obtained on the following assumptions.

It is assumed that following the collision of molecules with energy of the order of $\mu v^{*2}/2$ (v^* is the relative velocity of the molecules, which corresponds to the maximum contribution to the probability P_{10} [see Eq. (17–29)]) the transitions between rotational levels with an energy difference $\Delta\varepsilon$, which satisfies the condition $(|\Delta\varepsilon|/\alpha\hbar v^*) \leqslant 1$, take place with a probability p which is determined solely by the interaction anisotropy parameter. In the region of $(|\Delta\varepsilon|/\alpha\hbar v^*) \geqslant 1$ the rotational transitions are disregarded. In addition, since P_{10} is highly dependent on the transferred energy [see Eq. (17–30)], the simultaneous deactivation of vibrational and rotational degrees of freedom (as well as the simultaneous excitation of these degrees of freedom) is disregarded. On these assumptions

$$P_{10}(O_2 - O_2) = \frac{1}{\Delta\varepsilon^*} p \int_0^{\Delta\varepsilon^*} P_{10}(\hbar\omega - \Delta\varepsilon)\, e^{-\frac{\Delta\varepsilon}{kT}}\, \mathrm{d}(\Delta\varepsilon), \qquad (17\text{–}75)$$

with the upper limit of integration determined from

$$\frac{\Delta\varepsilon^*}{\alpha\hbar v^*} \approx 1 .$$

In specific calculations the latter expression is more conveniently written in the form of an exact equality

$$\frac{\Delta\varepsilon^*}{\alpha\hbar v^*} = \gamma ,$$

assuming that $\gamma \approx 1$.

In accordance with Eq. (17–33), the closest relationship between P_{10} and T is related to the multiplier $\exp(-3\chi)$. Hence we can write in the first approximation

$$P_{10}(\hbar\omega - \Delta\varepsilon) = \mathrm{const} \cdot e^{-3\chi(\hbar\omega - \Delta\varepsilon)} . \qquad (17\text{–}76)$$

Since $(\Delta\varepsilon/\hbar\omega) \ll 1$, then

$$\chi(\hbar\omega - \Delta\varepsilon) = \chi(\hbar\omega)\left[1 - 2\frac{\Delta\varepsilon}{3\hbar\omega}\right]. \tag{17–77}$$

Substituting Eqs. (17–76) and (17–77) into (17–75) and integrating, we get Eq. (17–74).

In the above approximations the multiplying factor which accounted for simultaneous rotational transitions was found to be independent of the temperature. This circumstance is important. On the other hand, numerical evaluation of the multiplier is difficult due to the fact that it is highly dependent on p and γ, the precise values of which are unknown. According to estimates made in [134], $p(\exp 2\pi\gamma/2\pi\gamma)=5$; here $\gamma=0.8$–0.9. Hence the probability of losing or acquiring a vibrational photon becomes several-fold higher than this probability calculated without consideration of the rotational energy. Thus, the transformation of rotational into translational energy, not changing the general pattern of the relaxation phenomena considered in subsection 2 of this section, can result only in increasing the transition probability or in reducing the relaxation time.

The vibrational transfer radically changes the picture of establishing the equilibrium. This subsection is devoted to clarifying the role of these processes in establishing the equilibrium. For harmonic oscillators the translational energy is converted into vibrational energy, as was shown before, by single-photon transitions; here $P_{n,n-1}$ are determined from Eq. (17–32). On collisions of oscillators, in addition to the previously considered process, vibrational energy is also transferred from one oscillator to another. As will be shown below, these processes in a certain approximation are also of single-photon character. The probability of transfer of a vibrational photon by an oscillator in the $m+1$ state to one in the nth state will be denoted by $Q_{n,n+1}^{m+1,m}$.

a) Kinetic equations and the probability of vibrational transfer. A set of equations which describes vibrational relaxation in an isolated system of oscillators is, as in the case of an isothermal system (see first subsection of this section), a set of balance equations for the molecular number of each vibrational level.

Using the notation assumed in the beginning of this section [see Eq. (17–3)], this system of equations has the form

$$\begin{aligned}\frac{dx_n}{dt} &= Z\{P_{n+1,n}x_{n+1} - P_{n,n+1}x_n + P_{n-1,n}x_{n-1} - P_{n,n-1}x_n\} + \\ &+ \frac{Z}{N}\{(\sum_l Q_{n+1,n}^{l,l+1}x_l)\,x_{n+1} - (\sum_m Q_{n,n+1}^{m+1,m}x_{m+1})\,x_n + \\ &+ (\sum_l Q_{n-1,n}^{l+1,l}x_{l+1})\,x_{n-1} - (\sum_m Q_{n,n-1}^{m,m+1}x_m)\,x_n\}, \quad n = 0, 1, 2, \dots,\end{aligned} \tag{17–78}$$

where

$$Q_{n,n-1}^{m,m+1}(T) = 2\left(\frac{\mu}{kT}\right)^2 \int_0^\infty q_{n,n-1}^{m,m+1}(v)\, e^{-\frac{\mu v^2}{2kT}} v^3\, dv, \qquad (17\text{–}79)$$

$q_{n,n-1}^{m,m+1}(v)$ is the probability of transition of the oscillators from the state (n, m) to $(n-1, m+1)$ in collisions with a relative velocity v. The quantity $q_{n,n-1}^{m,m+1}(v)$ is calculated by quantum-mechanical methods; calculations of it are presented in [530, 154, 482].

In this presentation we shall restrict ourselves only to an outline of the derivation. As in subsection 1, we shall consider the elementary one-dimensional case. The transition probabilities are most conveniently determined by the Schrödinger equation written in a representation which is determined by the eigenfunctions of the undisturbed Hamiltonian of the molecules. In this representation the Schrödinger equation practically coincides with Eqs. (17–5). The only difference consists in the fact that the subscript l does not denote the state of one molecule, but the state of two noninteracting molecules. The subscript k is actually the double subscript (k, l). In addition, $\hbar\omega_{km}$ which corresponds to the difference in the vibrational energies prior to and after the collision, is equal to zero when vibrational energy is transferred between two identical oscillators. In this case Eq. (17–5) simplifies and takes on the form

$$i\hbar \frac{da_{kl}}{dt} = \sum_{p,q} a_{pq}(t)\, V_{(k,l)(p,q)}(r(t)), \qquad a_{kl}(-\infty) = \delta_{km} \cdot \delta_{ln}. \qquad (17\text{–}80)$$

In the first approximation of the theory of perturbations, the probability $q_{n,l}^{m,k}(v)$ $(m+n=k+l)$ is given by [see Eq. (17–7)]:

$$q_{n,l}^{m,k}(v) = |a_{kl}(\infty)|^2 = \left| \frac{1}{\hbar} \int_{-\infty}^{\infty} V_{(k,l)(m,n)}(r(t))\, dt \right|^2, \qquad (17\text{–}81)$$

where

$$V_{(kl)(mn)} \equiv (k, l\,|\,V(r, y_1, y_2)\,|\,m, n),$$

y_1 and y_2 are deviations of the intramolecular coordinates of the first and second oscillators from their equilibrium values. The potential $V(r, y_1, y_2)$ [see Eq. (17–81)] can be expanded in series in powers of $(y_1 = Y_1 - Y_0)$ and $(y_2 = Y_2 - Y_{02})$ [see Sect. 2.3]. The term responsible for the photon transfer is $\frac{1}{2} y_1 y_2 (\partial^2 V/\partial y_1 \partial y_2)$. In this case

$$V_{(k,l)(m,n)} = \frac{y_{km} y_{ln}}{2} \cdot \frac{\partial^2 V}{\partial y_1 \partial y_2}. \qquad (17\text{–}82)$$

Since $V(r, y_1, y_2) = V(r - \lambda_1 y_1 - \lambda_2 y_2)$, where λY is the distance from the colliding atom to the center of gravity of its molecule, then

$$\frac{\partial^2 V}{\partial y_1 \partial y_2} = \lambda_1 \lambda_2 \frac{\mathrm{d}^2 V}{\mathrm{d}r^2}. \qquad (17\text{–}83)$$

Substituting Eqs. (17–83) and (17–82) into (17–81), we will get an expression for the transition probability

$$q_{n,l}^{m,k}(v) = \left| \frac{y_{km} y_{ln}}{2} \lambda_1 \lambda_2 \int_{-\infty}^{\infty} \frac{\mathrm{d}^2 V}{\mathrm{d}r^2} \mathrm{d}t \right|^2 = |y_{km}|^2 \, |y_{ln}|^2 \, \beta^2, \qquad (17\text{–}84)$$

where

$$\beta = \frac{\lambda_1 \lambda_2}{2} \int_{-\infty}^{\infty} \frac{\mathrm{d}^2 V}{\mathrm{d}r^2} \mathrm{d}t.$$

Eq. (17–84) can be evaluated without specifying the form of the intermolecular interaction potential. By order of magnitude $(\mathrm{d}^2 V/\mathrm{d}r^2) \sim W_0 \alpha^2$, where α^{-1} is the intermolecular interaction radius, and $W_0 \sim \mu v^2/2$. Integration in Eq. (17–84) is limited to a time of the order of a collision duration, *i.e.*, by a quantity of the order of $1/\alpha v$. Taking into account the fact that the matrix element y_{pq} differs from zero only when $p = q \pm 1$, we get

$$\left. \begin{aligned} & q_{n,n+1}^{k,k-1}(v) = (n+1) k q_{10}, \\ \text{here} \quad & q_{10} = |y_{01}|^4 \beta^2 \sim \frac{1}{\omega (\tau_{\mathrm{col}})^2}, \quad (\omega \tau_{\mathrm{col}} \gg 1, \end{aligned} \right\} \qquad (17\text{–}85)$$

where ω is the oscillator frequency, and $\tau_{\mathrm{col}} = (1/\alpha v)$ is the collision duration. An exact calculation of q_{10} performed in [530] for the potential $V = W_0 \exp [-\alpha(r - \lambda_1 y_1 - \lambda_2 y_2)]$, taking into account the multiplying factor of 4, which was omitted there (see [567]), yields

$$q_{10} = \left(\lambda_1 \lambda_2 \frac{\mu}{m} \frac{1}{\omega \tau_{\mathrm{col}}} \right)^2; \qquad (17\text{–}86)$$

where m is the reduced mass of the oscillator, and μ is the reduced mass of the colliding particles. For diatomic homonuclear molecules

$$\lambda_1 = \lambda_2 = \tfrac{1}{2}.$$

It can be seen by comparison of Eq. (17–85) with (17–15) that the probability of transfer of vibrational photons is appreciably higher than the probability of transformation of a vibrational photon into translational energy. The latter assertion is valid for $\omega \tau_{\mathrm{col}} \gg 1$, which is the condition for applicability of the theory of perturbances in the form under consideration. When $\omega\tau_{col}\to 1$, q_{10} is of the order of p_{10}.

Averaging of q_{10} over the Maxwellian velocity distribution [see Eq. (17–79)] is not difficult and was performed in [368].

Thus,

$$Q_{10} = \lambda_1^2\lambda_2^2 \frac{\mu}{m}\frac{\alpha^2}{\omega^2}\frac{4kT}{\mu}. \qquad (17\text{–}87)$$

Equation (17–87) is identical to the corresponding expression of [530] to within a multiplier of 8 which was omitted there.

Finally, for the probability of transfer of vibrational photons in the approximation being considered, we get

$$\left.\begin{aligned} Q_{n,n+1}^{p,p-1} &= (n+1)\,pQ_{10}, \\ Q_{n,n+k}^{p,p-k} &= 0, \quad k \neq \pm 1. \end{aligned}\right\} \qquad (17\text{–}88)$$

In calculating $q_{nl}^{mk}(v)$ it is not necessary to use the perturbation theory [see Eq. (17–81)]. Equations (17–80), which define $q_{nl}^{mk}(v)$ can be solved exactly with the conditions (17–82) and (17–83), by using a method which was applied for solving Eqs. (17–88) [154]. The probabilities of vibrational photon transfer thus obtained are similar to those given by Eq. (17–39) and have the form

$$q_{nl}^{mk}(v) = |a_{kl}(\infty)|^2 = \left|\int_{-\infty}^{\infty} \psi_{kl}^*(y_1, y_2)\, e^{-iy_1y_2\beta}\, \psi_{mn}(y_1, y_2)\, dy_1 dy_2\right|^2, \qquad (17\text{–}89)$$

where $\psi_{pq}(y_1, y_2)$ are the eigenfunctions of the Hamiltonian of two non-interacting molecules. When $\omega\tau_{col} \gg 1$, which corresponds to the condition that $\beta \ll 1$, Eq. (17–89) becomes (17–84). It should, however, be emphasized that Eqs. (17–80) in the approximation (17–82) and (17–83), and, consequently, also Eq. (17–89), describe the probabilities of transfer of vibrational photon on the condition that the latter is substantially higher than the probabilities of transformation of the vibrational into translational energy.

Using Eqs. (17–88) and (17–53), we can simplify Eqs. (17–78) and write them as

$$\begin{aligned} \frac{dx_n}{dt} = {} & ZP_{10}\{(n+1)x_{n+1} - [(n+1)e^{-\theta} + n]\,x_n + ne^{-\theta}x_{n-1}\} + \\ & + ZQ_{10}\{(n+1)(1+\bar{\alpha})\,x_{n+1} - \\ & - [(n+1)\bar{\alpha} + n(1+\bar{\alpha})]\,x_n + n\bar{\alpha}x_{n-1}\}, \quad n - 0, 1, 2, \ldots. \end{aligned} \qquad (17\text{–}90)$$

where $\bar{\alpha} = \sum_{n=0}^{\infty} nx_n(t)/N$ is the average number of vibrational photons devolved upon one molecule at time t.

b) The relaxation equation for vibrational energy. Comparison with experimental results. As in the preceding subsection, we shall consider not only the distribution function, but also the behavior of certain macroscopic characteristics, in particular, of the vibrational energy.

In the general case, solution of Eqs. (17–90) is quite difficult, since they are nonlinear. However, as in subsection 2, a relaxation equation for vibrational energy can be obtained from Eq. (17–90) without solving it. We multiply Eq. (17–90) by $n\hbar\omega$ and sum over all n. It is easy to see that in such a procedure the expression which is obtained from the second braces is exactly zero, since the photon transfer does not change the total vibrational energy. The result of transforming the expression in the first braces is known [see Eqs. (17–54), (17–55) and (17–56)].

Thus, the relaxation equation for the vibrational energy density in a system consisting of diatomic molecules has the form

$$\frac{dE_{vib}}{dt} = \frac{E_{0_{vib}}(T) - E_{vib}}{\tau_{vib}}, \tag{17–91}$$

where

$$\tau_{vib} = [ZP_{10}(1 - e^{-\theta})]^{-1}, \tag{17–92}$$

and $E_{vib}(t) = \hbar\omega\bar{\alpha}(t)\,N$ is the vibrational energy density, while $E_{0vib}(T)$ is the equilibrium value of the vibrational energy. Eqs. (17–91) and (17–92) are identical to Eqs. (17–55) and (17–56). This means that the relaxation of vibrational energy in a pure diatomic gas and in a system consisting of a monatomic gas with a small addition of monatomic molecules is described by the same relaxation equation. Let us, however, stress that this simple form of the relaxation equation is obtained only for the harmonic oscillator model. The result thus obtained is physically obvious. In fact, the transfer of vibrational photons does not affect the rate of relaxation of the vibrational energy, since they result only in redistribution of vibrational photons which already exist. The rate of vibrational relaxation is determined only by the origination and disappearance of vibrational photons, which depend only on energy transfer between the translational and vibrational degrees of freedom of the colliding molecules.

Equation (17–91) can be generalized to include the case of a mixture of diatomic molecules with monatomic gases. According to subsection 2 of this section, Eq. (17–91) can describe the vibrational relaxation in a mixture of diatomic and monatomic gases of an arbitrary composition, if Eq. (17–92) for the vibrational relaxation time is replaced by Eqs. (17–59) and (17–60). Here, in the latter expression the role of molecules of the ith kind can be played by the molecules undergoing relaxation proper, if their concentration is moderate.

In the frequently encountered case of a binary mixture of a diatomic gas A with a monatomic gas B, the expression for the relaxation time takes on the form

$$\frac{1}{\tau_{\text{vib}}} = \frac{1-\xi_B}{\tau_{AA}} + \frac{\xi_B}{\tau_{AB}}, \tag{17–93}$$

where ξ_B is the mole fraction of the B molecules. Eq. (17–93) is very frequently used for calculating τ_{AA} from the known values of τ_{vib}, τ_{AB} and ξ_B. Thus, the experimentally obtained values of $\tau_{O_2-\text{Ar}}$ were used in [277] to calculate $\tau_{O_2-O_2}$. It was found that in the temperature region 1200–6000 °K

$$\tau_{O_2-\text{Ar}} = (5.0 \pm 0.5)\,\tau_{O_2-O_2}\,.$$

The difference in the effectiveness of O_2 and Ar molecules in vibrational excitation can be interpreted on the basis of simultaneous rotational-vibrational transitions [134]. However, this explanation is not the only one. The difference in the effectiveness of O_2 and Ar molecules can be also due to the small difference in the intermolecular interaction potential of O_2–Ar and O_2–O_2. In this case the ratio $\tau_{O_2-O_2}/\tau_{O_2-\text{Ar}}$ will be a function of the temperature, while the consideration of the vibrational-rotational interaction results in the first approximation in a correction which is independent of the temperature. The relative contribution of each of these effects cannot be established on the basis of experiments described in [277], since the accuracy of the experimental data quoted there is such that they cannot be used to determine the extent to which $\tau_{O_2-O_2}/\tau_{O_2-\text{Ar}}$ is temperature dependent. An explicit temperature dependence of $\tau_{O_2-O_2}/\tau_{O_2-\text{Ar}}$ was discovered in later experiments [43]. In Fig. 99, which was taken from [43], the ratio $P_{10}(O_2-O_2)/P_{10}(O_2-\text{Ar})$ is represented as a function of the temperature. As the temperature increases this ratio approaches unity, which points to the equal effectiveness of O_2 and Ar in the process of vibrational deactivation at high temperatures.

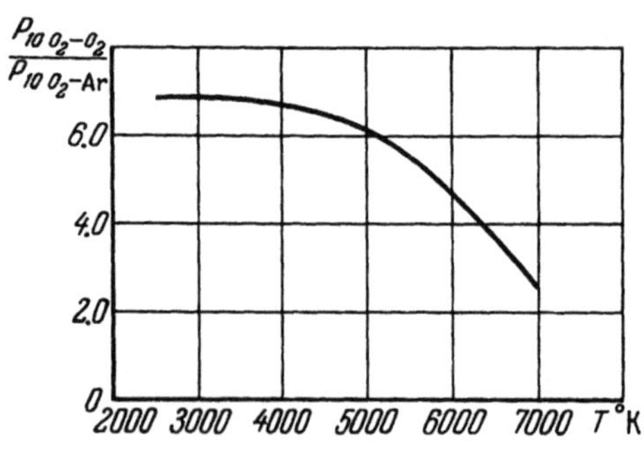

Fig. 99. Ratio of the experimentally obtained values of P_{10} for the O_2–O_2 and A–O_2r molecular collisions as a function of the temperature.

The relaxation time given by Eq. (17–92) is a quantity which is subject to direct experimental measurement. Up to now there are available tens of experimental works concerned with the study of the temperature dependence of τ_{col} for different gases (see Chapter 3). The greatest number of results pertaining to the behavior of τ_{col} at different temperatures was obtained in the study of vibrational relaxation by ultrasonic methods using shock tubes. A detailed description of the ultrasonic measurement methods and of the results obtained by them are given in [301, 368]. Lately, new methods of studying vibrational relaxation, consisting of kinetic spectroscopy and pulse photolysis, have been developed [150, 155]. The presentation of these problems is beyond the scope of this book. Here we shall restrict ourselves only to consideration of results obtained in shock tubes. Fig. 100 shows the dependence of the deactivation probability for the first vibrational level P_{10}, on collision of two identical molecules, on $T^{-1/3}$. Fig. 100 becomes more graphic if dimensionless coordinates are laid off along the axes. In particular, Fig. 101 depicts the dependence of the number of collisions needed for establishing vibrational equilibrium in a diatomic gas $Z_{vib} = \tau_{vib}/\tau_0$ (τ_0 is the time of a mean-free path traverse, and τ_{vib} is the measured relaxation time) on the adiabatic factor $\omega\tau^*_{col}$. According to Eq. (17–29),

$$\omega\tau^*_{col} = \frac{2\pi\omega}{\alpha v^*} \equiv 2\chi \equiv \left(\frac{4\pi^2\mu\omega^2}{\alpha^2 kT}\right)^{1/3}.$$

It is seen from Fig. 101, that for the majority of gases the experimental points $Z_{vib} = Z_{vib}(\omega\tau^*_{col})$ on the semilogarithmic scale situate themselves approximately on the same straight line. A similar functional behavior was found also in

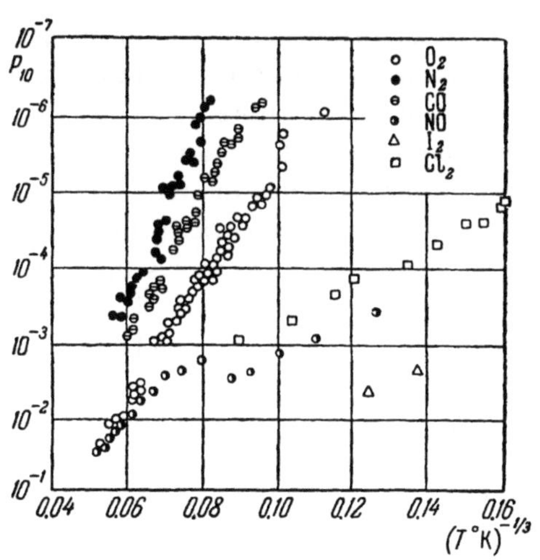

Fig. 100. Dependence of P_{10} on the temperature for different molecules.

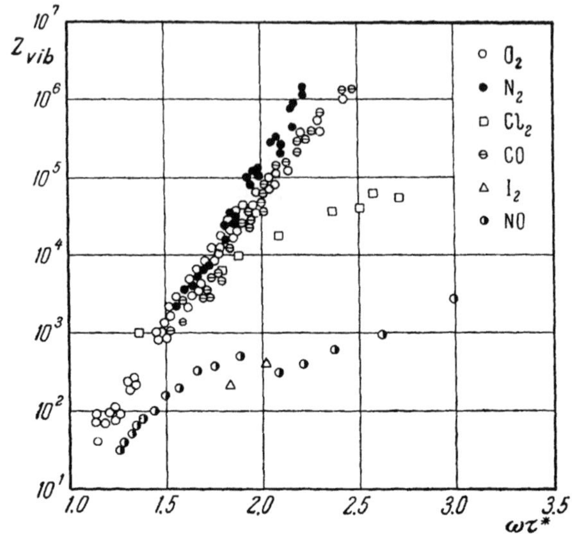

Fig. 101. Number of collisions Z_{vib}, needed for establishing vibrational equilibrium, as a function of $\omega\tau^* = (4\pi^2\mu\omega^2/\alpha^2 kT)^{1/3}$.

[447], where the following semiempirical expression was obtained for the relaxation time:

$$\ln(p\tau_{vib}) = 1.16 \cdot 10^{-3} \mu^{\frac{1}{2}} \left(\frac{\hbar\omega}{k}\right)^{\frac{4}{3}} (T^{-\frac{1}{3}} - 0.015\mu^{\frac{1}{4}}) - 18.42 .$$

In the above equation τ_{vib} is in secs, p is in atm, T and $\hbar\omega/k$ in degrees Kelvin and μ is in atomic units. This experimental result is in agreement with the conclusions of subsection 2, according to which the dependence of Z_{vib} on $T^{-\frac{1}{3}}$ is linear for an adiabatic mechanism of molecular vibrational excitation. The small deviations from linearity can be accounted for, for example, by taking into account the temperature dependence of the preexponential multiplier of P_{10} or assuming that the parameters of the intermolecular interaction potential which approximates the true molecular interaction is dependent on the energy of the arriving particles. In order to eliminate the effect of the preexponential factor, Fig. 102 shows the dependence of φ on $T^{-\frac{1}{3}}$ for an O_2 molecule [45, 46], where $\varphi = \ln Z_{vib}P$, and P is the theoretical value of the preexponential multiplier in P_{10}. It can be seen from Fig. 102 that the dependence of φ on $T^{-\frac{1}{3}}$ in a wide temperature range (1000–8000 °K) is substantially linear. At temperatures above 8000 °K the experimentally obtained vibrational relaxation times in pure oxygen lie above the straight line which corresponds to the Landau-Teller theory [47]. The latter circumstance is due to the fact that in this temperature region the concept of single-photon energy transfer following molecular collisions is found to be inapplicable,

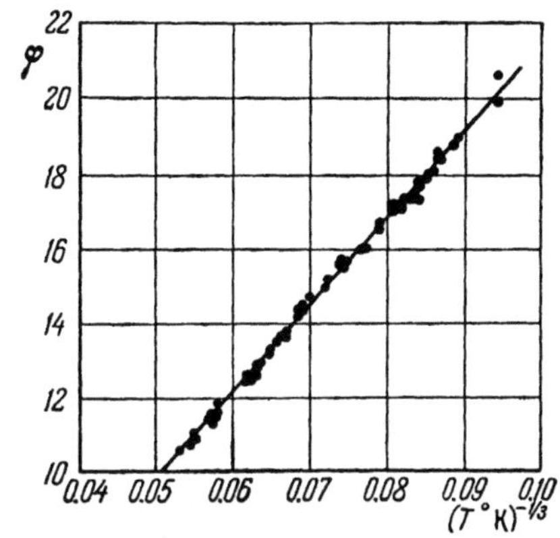

Fig. 102. The temperature dependence of $\varphi = \ln Z_{vib}P$, where P is the theoretical value of the preexponential factor in P_{10}, for O_2

and the Landau-Teller formula yields values of P_{10} on the higher side, *i.e.*, yields lower than actual vibrational relaxation times [152]. The existence of a linear relationship between $\ln Z_{vib}$ and $\omega\tau^*_{col}$ substantiates the assumption that the vibrational excitation mechanism for the majority of molecules under discussion is adiabatic. Apparently, an exception to this rule is NO (in the ground electronic state) and other radicals. The experimentally observed vibrational relaxation time for NO is found to be substantially smaller than the vibrational relaxation time in N_2 and O_2 [494, 241, 243].

Two different interpretations are offered for this effect. In [577, 494] the abnormally short time of vibrational relaxation of NO is related to the effect of attraction forces which appear in the formation of dimers. Calculations made on this assumption are in satisfactory agreement with experimental data. In [131] this difference is explained by the participation of the electronic shell of NO in the transformation of the translational into vibrational energy. It is shown, in particular, that in nonadiabatic excitation (in the sense that the electronic state of the incoming pair of NO molecules changes during collision) it is possible to have electronic transitions between states which arise from the initial state which is degenerated at infinity. When the electronic transition and nuclear vibrations frequencies coincide, there arise a resonance of a sort, which is responsible for the relatively large vibrational excitation cross sections. Calculations made on this assumption also give good agreement with experimental data. However, there is one basic difference between these two theories; they give different temperature dependences of τ_{col}, and, consequently, also of P_{10}. Thus, in the first case the temperature dependence is basically given by

$$\ln P_{10} \sim -T^{-\frac{1}{3}},$$

while in the second case it is given by

$$\ln P_{10} \sim -\frac{\varepsilon_0}{kT},$$

where $\varepsilon_0/k \sim 1000$–$2000$ °K.

It was shown by measuring the vibrational relaxation time of NO $(X^2\Pi)$ [in the ground electronic state] in a wide temperature range (1500–7000 °K) [614], that in the low-temperature region, the experimental values are in good agreement with the predictions of the theory of nonadiabatic excitation of molecular vibrations suggested by Nikitin [131]. At high temperatures (above 2000 °K) the experimental results are satisfactorily described by the ordinary adiabatic mechanism of excitation of vibrational levels. The latter result is easy to understand if we consider that at high temperatures, the probability of adiabatic excitation of molecular vibrations in NO is found to be higher than the corresponding probability calculated in the nonadiabatic approximation. Fig. 103, taken from [614], shows the experimental and theoretical values of P_{10} for NO as functions of the temperature.

It can be seen from Figs. 100–102 that the temperature curves of the experimental and theoretical values of $\tau_{vib} = \tau_{vib}(T)$ have the same shape. The absolute theoretical values differ from the experimental results by approximately an order of magnitude. A better agreement between results cannot be expected at present. In fact, the existing theoretical calculations, in addition to general shortcomings (the fact that anharmonicity is not considered, approximate consideration of rotational excitation when calculating P_{10} in the three-dimensional case, etc.), contain a practical difficulty, which is due to the unavailability of exact data on the intermolecular interaction potential and, in particular, on the value of α. The existing data can be used to deter-

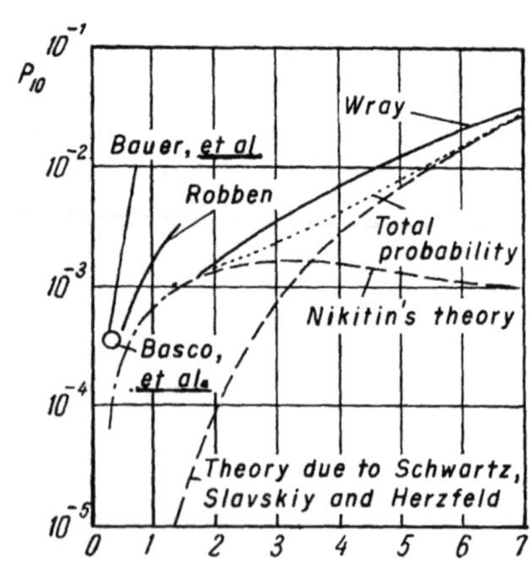

Fig. 103. The transition probability P_{10} for NO molecules as a function of the temperature. The solid curves are for the experimental values of P_{10}.

mine α to within a factor of two [368]. This accuracy is insufficient, since a change in α by 10% can, under certain conditions, change the values of τ_{vib} by an order of magnitude [531]. This sensitivity of τ_{vib} to the value of α allows us to formulate the converse problem, *i.e.*, to find α from the known experimental values of τ_{vib}. This problem is important since it suggests a method which makes it possible to determine the characteristics of the intermolecular interaction potential in an energy region which has been studied very little up to now. In fact, the intermolecular interaction potential is sufficiently well known in the energy region up to 1500 °K, where numerous data are available on the transfer coefficients, and in the energy region corresponding to several thousand degrees where data obtained by the molecular beam method are available [230]. The intermediate energy region, which is of importance for calculating the transport properties of high-temperature gases, remained unfilled. Study of vibrational relaxation makes it possible to fill this void to a certain extent. The data on the intermolecular interaction potential obtained in this manner are in good agreement with studies of the transport properties [301, 491].

The study of vibrational relaxation thus yields information not only on the energy-transfer mechanism on molecular collisions, but also makes it possible to resolve a number of related problems, such as the problem of intermolecular interaction. Thus, for example, the following was obtained for the intermolecular interaction potential for O_2–O_2 by analyzing vibrational relaxation data:

$$V(O_2 - O_2) = 20{,}300\, e^{-3.97r}\ \text{eV},\ 2.28 < r < 2.60, \qquad (17\text{–}94)$$

where r is in angstroms.

Fig. 104 shows the intermolecular interaction potentials for O_2–O_2 which are known at present. Curve 1 corresponds to Eq. (17–94), *i.e.*, it was obtained from vibrational relaxation data. Curve 2 corresponds to the theoretical potential

$$V(O_2 - O_2) = 5580 \exp(-3.355\, r)\ \text{eV}\ (1.93\ \text{A} < r < 2.70\ \text{A}),$$

calculated by the method of valency schemes in [582]. Curve 3, which is given by

$$V(O_2 - O_2) = 146.6 \exp(-2.109\, r)\ \text{eV}\ (2.5\ \text{A} < r < 2.9\ \text{A}),$$

was theoretically calculated in [443] using a model of the generalized δ-interaction. Curve 4 corresponds to the Lennard-Jones potential, the parameters of which are determined by comparison of the theoretical and experimental values of the viscosity coefficient [484]. The dash-dot line, which is a continuation of Curve 4, was obtained by extrapolating the Lennard-Jones potential to the region of small intermolecular distances.

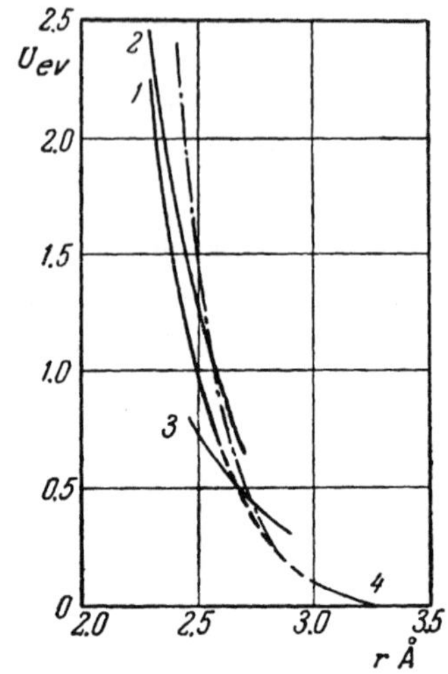

Fig. 104. Curves of the intermolecular interaction potential for $O_2 - O_2$. Curve 1 was constructed on the basis of vibrational relaxation data, Curve 2 corresponds to calculations performed in [582], Curve 3 corresponds to calculations in [443], Curve 4 is constructed on the basis of viscosity measurement data.

A similar calculation of the intermolecular interaction potential on the basis of vibrational relaxation data was performed in [582, 583], however, with a lower degree of accuracy. The accuracy of the existing methods for calculating intermolecular interaction potential parameters from experimental data on the vibrational relaxation time is determined basically by the accuracy of the theoretical expressions for the vibrational deactivation probability P_{10} which are used. The existing expressions for P_{10} differ somewhat from the temperature dependence of the preexponential factor and of the multiplier which takes into account the effect of the attraction forces (see, for example, page 239). Depending on the selection of expressions for these multipliers, different values of α are obtained [301, 441].

c) Qualitative picture of the vibrational relaxation process. As was pointed out, it is possible in the experimental study of vibrational relaxation to measure the molecular population of individual vibrational levels. This raises the problem of obtaining a more general solution of Eqs. (17–90).

In the general case the solving of a nonlinear system of equations such as (17–90) is quite difficult. However, in this problem we are aided by the fact that $Q_{10} \gg P_{10}$ [141, 145, 537]. It can be seen by inspecting Eqs. (17–90) that the process of establishing equilibrium with respect to the vibrational degrees of freedom takes place with two relaxation times, $\tau_1 \sim 1/(ZQ_{10})$ and $\tau_2 \sim 1/(ZP_{10})$; here according to the preceding discussion, $\tau_2 \gg \tau_1$. The latter inequality makes it possible to isolate two stages in the process of establishing the equilibrium. During the first, rapid stage, with a relaxation time τ_1, the vibrational degrees of freedom appear practically as an isolated system which is unrelated to the translational motion. In this stage elementary vibrational photon transfers produce a quasi-steady Boltzmann distribution. During the second, slow stage, with a relaxation time τ_2, the translational degrees of freedom interact with the vibrational degrees of freedom. The energy transfer between the translational motions and the vibrations which accompany this interaction results in establishing a single equilibrium distribution.

The establishing of the equilibrium in the rapid stage, *i.e.*, for times $t \ll \tau_2$, is described by the following equations which are obtained from (17–90) by disregarding in it terms of the order of τ_1/τ_2:

$$\frac{\mathrm{d}x_n}{\mathrm{d}t} = ZQ_{10}\{(n+1)(1+\bar{\alpha})x_{n+1} - [(n+1)\bar{\alpha} + n(1+\bar{\alpha})]x_n + \bar{\alpha}nx_{n-1}\}, \quad n = 0,1,2,\dots. \tag{17–95}$$

To within terms of the order of τ_1/τ_2, $\bar{\alpha}$ in Eq. (17–95) is a constant. In fact, the equation for $\bar{\alpha}$ which is obtained by multiplying (17–90) by n and summing over all n, has the form [see Eqs. (17–91) and (17–92)]

$$\frac{\mathrm{d}\bar{\alpha}}{\mathrm{d}t} = -ZP_{10}(1-e^{-\theta})[\bar{\alpha}(t) - \bar{\alpha}(\infty)]. \tag{17–96}$$

Disregarding in Eq. (17–96) terms of the order of τ_1/τ_2, we obtain, in the same approximation as in (17–95), the following equation:

$$\frac{\mathrm{d}\bar{\alpha}}{\mathrm{d}t} = 0 \quad \text{or} \quad \bar{\alpha} = \text{const}. \tag{17–97}$$

The last result is obvious, since during the first, rapid stage the vibrational degrees of freedom in this approximation appear practically as an isolated system. Thus, Eqs. (17–95) are a system of linear equations, the form of which coincides with that of Eq. (17–54) which was considered in subsection 2 of this section. The solution of Eq. (17–95) has the form of Eq. (17–66) and is determined from Eq. (17–68). However, of practical interest is the steady-state solution of these equations which is suitable for times $\tau_1 \ll t \ll \tau_2$. The steady-state solution of Eq. (17–95), which is determined from the condition that $(\mathrm{d}x_n/\mathrm{d}t) = 0$, as is easy to see, has the form

$$x_n = N(1-e^{-\theta_0})e^{-n\theta_0}, \tag{17–98}$$

where $\theta_0 = \ln(1+\bar{\alpha}^{-1})$.

Thus, as a result of the first stage, a quasi-steady Boltzmann distribution with a temperature determined by the initial store of vibrational photons and independent of the form of the initial function of distribution of oscillators with respect to vibrational levels, is established in the isolated system of oscillators.

The second, slow stage of establishing equilibrium is described by a system of complete equations; here according to the above, the solution is sought in the form of Eq. (17–98), where θ_0 will now be an unknown function of time. It is easy to see that for this choice of a solution, the last braces in the right-hand side of Eq. (17–90) become indentically zero, and the following equation remains for determining $x_n(t)$:

$$\frac{dx_n}{dt} = ZP_{10}\{(n+1)x_{n+1} - [(n+1)e^{-\theta} + n]x_n + ne^{-\theta}x_{n-1}\}$$

$$n = 0, 1, 2, \dots . \qquad (17\text{–}99)$$

Eqs. (17–99) differ from the analogous system (17–54) by the fact that θ, and, consequently, also P_{10} will be functions of time. The form of these functions can be found by solving Eq. (17–96) or (17–91) simultaneously with a set of equations expressing the conservation laws across the shock front. This problem must be solved numerically, and a particular solution is considered in [365].

The equilibrium with respect to the vibrational degrees of freedom in an isolated system of oscillators thus takes place in the following manner: first, during a time of the order of τ_1, there is established a quasi-steady Boltzmann distribution with a temperature which is established by the initial population of vibrational photons and which is independent of other initial conditions. The basic elementary act which takes place on oscillator collision at this stage of the process is the transfer of vibrational photons; here the total number of the latter remains practically unchanged. After the quasi-steady Boltzmann distribution has been established due to the effect of elementary processes of transformation of the translational into vibrational energy, a relatively slow process of evolving the above Boltzmann distribution into an equilibrium starts. At this stage the distribution function, remaining a Boltzmann distribution throughout, changes its temperature from θ_0 to a final value which corresponds the total equilibrium value. At temperatures below the characteristic temperature the difference between Q_{10} and P_{10} is quite appreciable. This makes it possible to observe the above stages at the time they are established. A practical example in which it is possible to follow the first stage of the establishing process is afforded by experiments due to Koshana and Pol'yana [155, 148].

In experimental installations such as a shock tube, the gas ahead of the shock front is in a state of equilibrium. In conjunction with this the initial condition of the vibrational relaxation process directly behind the shock front will have corresponding to it such a state of a gas in which there is sense in referring to two temperatures which correspond to the translational and vibrational motion. The temperature of the translational degrees of freedom is determined from the conservation laws across the shock front on the assumption that all the internal degrees of freedom of the gas are frozen. The temperature of the vibrational degrees of freedom corresponds to the temperature of the undisturbed gas ahead of the shock front. Since under these initial conditions vibrational relaxation takes place in a manner such that the tenperature of the vibrational motion retains its meaning at all times (the shape of the Boltzmann distribution is retained), the transfer of vibrational photons are found to be insubstantial in the given problem. Thus the relaxation of vibrational motion in a single-component system with an initial Boltzmann distribution of molecules with respect to the vibrational levels (with a temperature differing from that of the translational degrees of freedom) can be considered without taking into account the transfer of vibrations. The situation is different when vibrational relaxation is considered in a mixture of polyatomic gases.

[4] Vibrational Relaxation of a Mixture of Polyatomic Gases

As was shown in the preceding subsection, transfer of vibrational photons results in redistributing the molecules over the vibrational levels; here the total number of photons in the system does not change. Consequently, the transfer does not affect the rate of vibrational relaxation in pure gases. The situation differs in a mixture of polyatomic gases [142]. The vibrational relaxation time in a gas mixture is a certain total characteristic and is determined by the greatest relaxation time of individual components. If no transfer of vibrational energy between various components takes place in the system, then the relaxation time of individual components is calculated on the assumption that the molecules of the second component are deprived of the internal degrees of freedom. When the transfer processes are taken into account, this method is no longer valid. In fact, the transfer processes, while not changing the total number of vibrational photons in the system, may result in a rapid redistribution of the available vibrational energy between the different components. This can also change the relaxation times of individual components, and, consequently, also, the relaxation time of the mixture as a whole. Since the existence of transfer effect has recently been confirmed by experiment [153, 380, 446], then it is expedient to consider separately the problem of vibrational relaxation in a mixture of polyatomic gases.

Let us, for simplicity, consider a binary mixture of diatomic gases A and B. Let $x_n(t)$ and $y_n(t)$ be the number density of molecules A and B on the nth-vibrational level, in which case

$$\sum_{n=0}^{\infty} x_n(t) = N_A,$$

and

$$\sum_{n=0}^{\infty} y_n(t) = N_B .$$

The kinetic equations which describe vibrational relaxation in a binary gaseous mixture is a system of balance equations for the molecular population of A and B on each vibrational level.

If we assume that no vibrational energy transfer takes place between components A and B, then the equations for x_n and y_n will be similar to the kinetic equations for pure gases [see, for example, Eq. (17–78)]. Considering the effects of transfer between components requires the addition of new terms. Their form is similar to that of terms which describe the transfer of vibrational energy within a single component, with the only difference that all the quantities pertaining to the collision of identical particles A–A or B–B should be replaced by quantities describing collisions A–B. Assuming a harmonic oscillator model (with a frequency ω_A for A molecules and ω_B for B molecules) these equations can be written as

$$\begin{aligned}\frac{dx_n}{dt} = {} & Z_{11}\{(n+1)x_{n+1} - [(n+1)e^{-\theta_A} + n]x_n + ne^{-\theta_A}x_{n-1}\} + \\ & + Z_{12}\{(n+1)(E'_A + N_A)x_{n+1} - [(n+1)E'_A + \\ & + n(E'_A + N_A)]x_n + nE'_A x_{n-1}\} + \\ & + Z_{13}\{(n+1)(E'_A + N_B)x_{n+1} - [(n+1)E'_B e^{\theta_B - \theta_A} + \\ & + n(E'_B + N_B)]x_n + nE'_B e^{\theta_B - \theta_A}\}, \quad n = 0, 1, 2, \ldots, \end{aligned} \tag{17–100}$$

where

$$\begin{aligned} Z_{11} &= Z_{AA}P_{10}^{AA}N_A + Z_{AB}P_{10}^{AB}N_B , \\ Z_{12} &= Z_{AA}Q_{10}^{AA}, \qquad Z_{13} = Z_{AB}Q_{10}^{AB}, \\ E'_A &= \sum_{n=0}^{\infty} nx_n(t), \qquad E'_B = \sum_{n=0}^{\infty} ny_n(t), \\ \theta_A &= \frac{\hbar\omega_A}{kT}, \qquad \theta_B = \frac{\hbar\omega_B}{kT} . \end{aligned}$$

The equations for dy_n/dt are written similarly.

In Eqs. (17–100) $Z_{A_i}N_i$, where $i = A, B$ is the collision frequency of a molecule of A with molecules of i, E'_A and E'_B have the meaning of the total number of vibrational photons per unit volume (of dimensionless density of vibrational energy), respectively, for the component A and B, P_{10}^{ik} and Q_{10}^{ik} are the transitions probabilities on collisions of molecules i and k.

According to the above, Eqs. (17–100) describe the change in the number of A molecules at the nth vibrational level which takes place due to processes of mutual transformation of the vibrational energy of A molecules into translational energy of A and B molecules (the first braces) and as a result of transfer of vibrational energy on collision of A molecules among themselves (second braces) and with molecules of B (third braces).*

The relaxation equations for the average vibrational energy of each of the two components are obtained from equations for $\mathrm{d}x_n/\mathrm{d}t$ and $\mathrm{d}y_n/\mathrm{d}t$ by multiplying them by $\hbar\omega_A n$ and $\hbar\omega_B n$, respectively, and summing over all n. In this case the expression obtained from the second braces of Eq. (17–100) becomes zero. This result is obvious from the physical viewpoint (see page 257), since the transfer of vibrational photons within each component do not result in changing the average vibrational energy of the molecules. After simple transformations, the relaxation equations for the dimensional vibrational energies E'_A and E'_B take on the form

$$\left.\begin{aligned}\frac{\mathrm{d}E'_A}{\mathrm{d}t} &= -\frac{1}{\tau_A}(E'_A - E'^0_B) + \frac{1}{\tau_{AB}}[E'_B e^{\theta_B-\theta_A} \times \\ &\quad\times (E'_A + N_A) - E'_A(E'_B + N_B)],\\ \frac{\mathrm{d}E'_B}{\mathrm{d}} &= -\frac{1}{\tau_B}(E'_B = E'^0_B) + \frac{1}{\tau_{BA}} \times \\ &\quad\times [E'_A(E'_B + N_B) - E'_B e^{\theta_B-\theta_A}(E'_A + N_A)],\end{aligned}\right\} \tag{17–101}$$

where

$$E'^0_A = \frac{N_A}{e^{\theta_A} - 1}, \quad E'^0_B = \frac{N_B}{e^{\theta_B} - 1},$$

$$\tau_A = \frac{1}{(Z_{AA}P^{AA}_{10}N_A + Z_{AB}P^{AB}_{10}N_B)\ 1 - e^{-\theta_A})},$$

$$\tau_B = \frac{1}{(Z_{BB}P^{BB}_{10}N_B + Z_{BA}P^{BA}_{10}N_A)(1 - e^{-\theta_B})},$$

$$\tau_{AB} = \tau_{BA} = \frac{1}{Z_{AB}Q^{AB}_{10}}.$$

The first terms in the right-hand sides of Eqs. (17–101) describe changes in the vibrational energy of individual components due to transformation of translational into vibrational energy. The second terms, which differ only by their sign, correspond to the transfer of vibrational energy between different components.

It is obvious from a consideration of Eqs. (17–101) that vibrational relaxation in a binary mixture of diatomic gases has three characteristic times: τ_A; τ_B; and τ_{AB}. If, for some reason, it is possible to disregard the transfer

* The vibrational energy transfer on collisions between A and B molecules takes place in a manner such that the difference in the vibrational energies $\hbar\omega_A - \hbar\omega_B$ is compensated by changes in the relative translational energy.

effect, *i.e.*, if it is possible to assume that

$$\tau_{AB} \gg \tau_A, \tau_B,$$

then the system of equations (17–101) breaks up into two independent equations. On these assumptions vibrational relaxation in a binary mixture will consist of independent relaxation processes in each of the components individually. The vibrational relaxation rate in each component can be calculated without taking into account the internal degrees of freedom of the second component.

If the quantity $1/\tau_{AB}$ cannot be disregarded in comparison with $1/\tau_A$ and $1/\tau_B$, then the transfer effects must be taken into account. They manifest themselves most clearly in the cases

$$\tau_{AB} \ll \tau_A \ll \tau_B \quad \text{and} \quad \tau_A \ll \tau_{AB} \ll \tau_B.$$

A detailed discussion of these processes and of the pertinent experimental data is given in [153]. Here we restrict ourselves only to the results.

In the first case, when

$$\tau_{AB} \ll \tau_A \ll \tau_B \tag{17–102}$$

the vibrational relaxation may take place as follows. First, during a time of the order of τ_{AB} an equilibrium distribution of the available vibrational energy between components A and B is established.* The main process of this stage will be vibrational energy transfer. Then the slow process involving the transformation of the translational energy of the A and B molecules into the vibrational energy of A molecules starts. Since $\tau_A \gg \tau_{AB}$, then at each time instant during this slow process the available vibrational energy in the first approximation will be distributed in the equilibrium manner between components A and B. Thus, the establishing of vibrational equilibrium in a binary gas mixture under condition (17.102) will take place with a time τ_{vib}, which is the same for each of the components separately. The time τ_{vib} is found to be greater than τ_A and smaller than τ_B, *i.e.*,

$$\tau_A < \tau_{\text{vib}} < \tau_B.$$

The physical cause of this is clear. Under conditions being considered the collision processes which result in the formation of vibrational energy of molecules, are sources of vibrational energy not only for component A but also for component B. The appearance of additional vibrational energy in the form of component B naturally increases the time needed for establishing

* It is more correct to speak of an equilibrium distribution of the available vibrational photons, since in the process of transfer the total number of vibrational photons remains constant, while the total vibrational energy can change, since $\hbar\omega_A \neq \hbar\omega_B$.

vibrational equilibrium of component A, and, consequently, also of the entire mixture. On the other hand, the vibrational relaxation of molecules B in a mixture of diatomic gases A and B takes place more rapidly (during a relaxation time τ_{vib}) than in a mixture in which the vibrational degrees of freedom of gas A (relaxation time τ_B) are frozen, since the direct excitation of molecules of B due to direct transformation of translational into vibrational energy is found to be less effective than the indirect excitation by vibrationally excited molecules of A.

The vibrational relaxation in the opposite limiting case

$$\tau_A \ll \tau_{AB} \ll \tau_B \tag{17–103}$$

takes place as follows. First, during a time of the order of τ_A in the first approximation, an equilibrium energy distribution is established in system A. The vibrational energy of B molecules during this process practically does not change. Then, as a result of transfer of vibrational photons between A and B molecules, the slow process of establishing vibrational equilibrium in system B starts. This process proceeds with a relaxation time τ'_{vib}, which, generally speaking, is smaller than the vibrational relaxation time in the pure gas B. If the times τ_A, τ_{AB} and τ_B are found to be of the same order, then the effect of transfer processes on the rate of establishing the vibrational equilibrium is not as pronounced. Nevertheless, even in this case it can be found to be appreciable. Of interest in conjunction with this are quantitative calculations of the vibrational relaxation of individual components in air, which are considered in Sect. 22.

[18] THERMAL DISSOCIATION KINETICS

When a gas passes through the front of a strong shock wave, in addition to the excitation of translational, rotational and vibrational degrees of freedom, thermal dissociation and ionization of molecules also take place. For the majority of gases, up to temperatures of the order of characteristic temperatures (for vibrational degrees of freedom), the vibrational relaxation time is appreciably smaller than the time needed for establishing equilibrium dissociation. In conjunction with this, it is possible to assume when studying thermal dissociation, that some quasi-steady molecular distribution with respect to the vibrational levels exists at each time instant. This means that thermal dissociation can be considered by assuming that vibrational relaxation has terminated. On the other hand, there always exists a region of Mach numbers (and a temperature and pressure region corresponding to it), in which the ionization processes do not as yet play a perceptible role. Thus, in the first approximation the thermal dissociation process can be considered separately from the thermal ionization process as well as from vibrational relaxation. The relation between vibrational relaxation and thermal dissociation will be considered in subsection 3 of this section.

The change in the parameters of a diatomic gas in this nonequilibrium dissociation region is described by a system of gasdynamic equations, by the equation of state and by relaxation equations [115] (see also Sect. 9).

The relaxation equations, or the kinetic equations of the dissociation reaction are actually continuity equations and have the form

$$\left.\begin{aligned} &\frac{\partial n_i}{\partial} + \operatorname{div} n_i v = \left(\frac{\mathrm{d}n_i}{\mathrm{d}t}\right)_x, \quad i = 1,2, \\ &\left(\frac{\partial n_1}{\partial t}\right)_x = -2\left(\frac{\partial n_2}{\partial t}\right)_x = 2k'_{\mathrm{d}} n_1 n_2 + \\ &\qquad + 2k''_{\mathrm{d}} n_2^2 - 2k'_{\mathrm{rec}} n_1^3 - 2k''_{\mathrm{rec}} n_1^2 n_2 . \end{aligned}\right\} \tag{18–1}$$

Here n_1 and n_2, respectively, are the number density of atoms and molecules, k'_d, k''_d and k'_{rec}, k''_{rec} are the dissociation and recombination rate constants.

It is shown by numerical calculations that the main change in the thermodynamic parameters in the region of nonequilibrium dissociation takes place far from the equilibrium position, where it is possible to disregard the recombination. Thus, for example, in H_2, when $\mathsf{M} = 12.4$ in a region in which 90% of equilibrium dissociation has been reached, the recombination rate comprises 6×10^{-2} of the dissociation rate [286]. As the Mach number increases, this ratio is reduced. When $\mathsf{M} = 25.2$, it is only 1.7×10^{-5}. Hence in the nonequilibrium dissociation region it is possible to disregard terms with k'_{rec} and k''_{rec} in Eq. (18–1).

Methods for solving gasdynamic equations in the relaxation zone and the results thus obtained are described in many works, for example, in [115, 286]. Of interest from the physical viewpoint is not the solution of these equations, but rather the study of processes which take place in the relaxation zone and which are described by Eqs. (18–1). The unknowns in these equations are the dissociation and recombination rate constants. In a numerical solution of gasdynamic equations it is, of course, possible to use the experimental values of k_d and k_{rec}. However, these values are not always known, and extrapolation of known values of k_d and k_{rec} to a region of different experimental conditions is justified only when theoretical expressions for these two rate constants are available.

It is assumed in writing Eqs. (18–1) that the dissociation takes place as a result of double collisions (and the reverse process—recombination—takes place as a result of triple collisions). This assumption is, in the majority of cases, justified in shock waves where it is possible to neglect photodissociation and dissociation by electron collisions.

The number of collisions between molecules which would result in dissociation is, at moderate temperatures, only a small part of the total number of collisions. This raises two questions: 1) what conditions must be satisfied by the colliding molecules in order for dissociation to take place; and 2) how frequently these conditions are realized in the system in which the reaction takes place. The so-called elementary theory of collisions [210, 470] resolves these problems as follows: it is assumed that those molecules dissociate with probability P during collision, the energy of which (the energy of the internal degrees of freedom plus the kinetic energy of relative motion along the line of centers) is not smaller than D, the dissociation energy. The dissociation probability P is assumed to be independent of other characteristics of the initial condition of the colliding molecules and is subsequently determined experimentally. The second problem of the frequency with which conditions needed for dissociation are encountered is solved by the elementary collision theory by assuming that the dissociation does not disturb the equilibrium distribution of molecules with respect to energy levels.

On these assumptions, the following expression is obtained in the elementary theory of collisions for the rate constant of the bimolecular reaction [210]:

$$k_d = PZ^0 e^{-\frac{D}{kT}} \sum_{r=0}^{s} \frac{1}{r!}\left(\frac{D}{kT}\right)^r, \tag{18–2}$$

where $Z^0 = (2/\sigma_{12})\, d_{12}^2/(2\pi kT/\mu)^{\frac{1}{2}}$ is the collision frequency per unit volume when the concentration of the colliding particles is unity, μ is the reduced mass of the colliding particles, d_{12} is the collision diameter (the solid spheres model), $\sigma_{12}=2$ if both colliding molecules are identical and $\sigma_{12}=1$ in the opposite case.

It was assumed in deriving Eq. (18–2) that the probability of reaction referred to one collision has a constant value P on the condition that the sum of the relative kinetic energy along the line of centers and the energy s of the vibrational degrees of freedom (of one or both molecules), which take part in the reaction, exceeds a certain value of D and is zero in all the other cases. If the total kinetic energy of relative motion of the colliding molecules is found to be of substance in the reaction process, then s in Eq. (18–2) should be replaced by $s+1$. In the general case, when, in addition to the translational motion, the vibrational and rotational degrees of freedom participate in the reaction, the expression for the rate constant of the bimolecular reaction will have the form

$$k_d = PZ^0 e^{-\frac{D}{kT}} \sum_{r=0}^{t/2} \frac{1}{\Gamma(r+1)} \left(\frac{D}{kT}\right)^r. \tag{18–3}$$

In Eq. (18–3) the reaction probability has the constant value P if the sum of the kinetic energy along the line of centers and of the energy of t quadratic terms corresponding to the remaining degrees of freedom which participate in the reaction exceeds D. Eq. (18–3) has been derived for an even t. For an odd t

$$k_d = PZ^0 e^{-\frac{D}{kT}} \left\{ \sum_{r=\frac{1}{2}}^{t/2} \frac{1}{\Gamma(r+1)} \left(\frac{D}{kT}\right)^r + O\left(\frac{D}{kT}\right)^{-\frac{1}{2}} \right\}. \tag{18–4}$$

When performing calculations with Eqs. (18–3) and (18–4) it should be remembered that each rotational degree of freedom contributes one quadratic term corresponding to an average energy of $kT/2$ into the expression for D, while the vibrational degree of freedom contributes two quadratic terms with a total energy kT.

The degrees of freedom which correspond to translational motion can contribute only two quadratic terms to the energy expression, since the motion along the line of centers is already taken into account in Eqs. (18–3) and (18–4). Let us also note that the kinetic energy as well as the energy of the vibrational degrees of freedom can be transformed entirely into the reaction energy. This cannot be said about the energy of the rotational degrees of freedom, the transformation of which into the energy of reaction is limited by requirements imposed by the law of conservational of angular momentum.

Practical calculations are usually performed by using the formula

$$k_d = PZ^0 \frac{1}{s!} \left(\frac{D}{kT}\right)^s e^{-\frac{D}{kT}}, \tag{18–5}$$

which is obtained from Eq. (18–2) by assuming that $D \gg skT$, and in which the constants P and s are assumed to be experimentally determined empirical parameters. Eqs. (18–2)—(18–5) were obtained on the assumption that all the degrees of freedom participating in the reaction are classical. These formulas were generalized to include quantum degrees of freedom in [283, 284]. A slightly more exact calculation within the framework of classical mechanics was performed in [420, 421].

The results obtained from Eq. (18–5) are in agreement with experimental data of [283, 284]. However, this agreement is obtained only by virtue of two undetermined parameters contained in Eq. (18–5) which are the probability factor P, which is called the steric or preexponential factor, and the effective number of the degrees of freedom s which participate in the reaction.

From the theoretical viewpoint, the calculation of the reaction rate constant within the framework of the elementary theory of collisions is not consistent, since a certain arbitrariness exists in both fundamental points of the theory. The probability P which is contained in the theory of collisions should, in substance, be calculated as a function of the initial state of the colliding molecules. These probabilities, in their turn should determine the distribution of molecules with respect to the energy levels in the relaxation process. Here it is possible to know beforehand under which conditions and to what extent this distribution is an equilibrium distribution.

This raises the question of a rigorous formulation of kinetic equations describing the dissociation process. We shall assume that the main dissociation mechanism, although not the only one, is the transition of the colliding molecules from the discrete to the continuous vibrational state [193]. This is precisely the case which forms the subject matter of the next subsection.

[1] Thermal Dissociation as Molecular Transition from the Discrete to the Continuous Vibrational State

Let us start with the elementary problem of thermal dissociation in a system in which diatomic molecules are a small addition to a monatomic gas. In this case it is possible to disregard mutual collisions of the dissociating molecules and to consider only collisions of the latter with the monatomic gas.

The transition probabilities for a molecule whose intermolecular interaction potential is described by the Morse function have been calculated in [140]. The following picture of the distribution of transition probabilities as a function of the initial vibrational energy is obtained from this calculation. In the lower part of the discrete spectrum only transitions between neighboring levels are practically permissible. The probability of these transitions is given by Eq. (17–30). Due to the fact that the Morse potential curve is anharmonic, the energy difference between neighboring vibrational levels will decrease with an increase in the vibrational level number. In the upper part of the discrete spectrum, the energy levels are so close to one another that the adiabatic factor $\omega_{n,n+1}\tau_{col}$ becomes less than unity. Under these conditions (see also page 239) multiphoton transitions between discrete vibrational levels as well as transitions to the continuous spectrum are permissible. Thus, the dissociation process can be represented schematically in the form of a stochastic process of random wandering of molecules over vibrational levels in the presence of a negative source which acts on the upper levels. The wandering pitch will be different for different levels. At the lower vibrational levels the molecules can only perform single-photon transitions, *i.e.*, can move over to neighboring levels. In the region of sufficiently high vibrational energies, multiphoton transitions are also permissible; in this energy region probabilities of transition to the continuous spectrum are also nonzero.

For convenience we will make the picture even rougher. We shall assume that the molecule performs single-photon transitions up to the vibrational level with number k, from which transition to the continuous spectrum is possible. This dissociation scheme is described by kinetic equations which are similar to Eqs. (17–52), with the only difference, that a negative source describing the disintegration of the molecules into atoms will act on the upper level k.

If the number density of the dissociated molecules is denoted by $\frac{1}{2}x_d$ (x_d is the atomic number density), then these equations will take on the form

$$\frac{dx_n}{dt} = Z(P_{n+1,n}x_{n+1} - P_{n,n+1}x_n + P_{n-1,n}x_{n-1} - P_{n,n-1}x_n),$$

$$n = 0, 1, \dots, k-1. \qquad (18\text{–}6)$$

$$\frac{dx_k}{dt} = Z(P_{dk}x_d^2 - P_{kd}x_k + P_{k-1,k}x_{k-1} - P_{k,k-1}x_k), \qquad (18\text{–}7)$$

$$\frac{1}{2}\frac{dx_d}{dt} = Z(P_{kd}x_k - P_{dk}x_d^2). \qquad (18\text{–}8)$$

Here $ZP_{dk}x_d^2$ is the number of molecules which are formed per unit time in a unit volume as a result of dissociation. It is assumed that recombination produces molecules at the kth vibrational level.

In the general case solution of Eqs. (18–6)—(18–8) is quite difficult, since this system describes two processes, one of them being vibrational relaxation and the other being the establishing of equilibrium dissociation. The relaxation times of these processes, as was shown in [108, 113], are appreciably different under certain circumstances, and this makes it possible to simplify the problem solution. Since we are interested in the slow process, *i.e.*, in establishing equilibrium dissociation, then we can consider Eqs. (18–6)—(18–8) on the macroscopic time scale, assuming that the rapid process has terminated already.

In this case this latter process is not considered at all. For the subsequent discussion, only the result of this process which is synchronization of the initial distribution function with the macroscopic parameters of the problem, is of importance.

Thus, on the macroscopic time scale the solution of Eqs. (18–6)—(18–8) can be sought in the form [194]

$$x_n(t) = x_n^{(0)}(w) + \varepsilon x_n^{(1)}(w) + \cdots, \tag{18–9}$$

where $w = \frac{1}{2}(\mathrm{d}x_d/\mathrm{d}t)$ is the reaction rate, and ε is a small parameter, equal by order of magnitude to $\tau_{\mathrm{vib}}/\tau_d$, where τ_{vib} is the vibrational relaxation time and τ_d is the time needed for establishing dissociation equilibrium. This form of solution assumes that a certain quasi-steady molecular distribution exists at any given time and that it is determined by the value of the reaction rate at the same time. Thus, this distribution depends on time implicitly through the parameter w which is the dissociation rate.

To simplify the problem, we consider the case when the total number of vibrationally excited molecules is small in comparison with the number of molecules in the ground state. In this case, $x_n \ll x_0$ $(n \neq 0)$, and

$$\frac{\mathrm{d}x_n}{\mathrm{d}w}\frac{\mathrm{d}w}{\mathrm{d}t} \ll w, \quad n \neq 0, d. \tag{18–10}$$

Substituting Eq. (18–9) into (18–8) and making use of Eq. (18–10), we obtain the following system of equations for $x_n^{(0)}(w)$ [the superscript "0" is subsequently dropped]:

$$\begin{aligned} w &= Z(P_{n-1,n}x_{n-1} - P_{n,n-1}x_n), \quad n = 1, 2, \ldots, k, \\ w &= Z(P_{kd}x_k - P_{dk}x_d^2), \end{aligned} \tag{18–11}$$

where

$$w = -\frac{\mathrm{d}x_0}{\mathrm{d}t} = \frac{1}{2}\frac{\mathrm{d}x_d}{\mathrm{d}t}. \tag{18–12}$$

The results which are obtained with the above limitations are found to be valid even for more general conditions since this pertains to the behavior of the distribution function on the upper levels and to the dissociation rate. The difference between the more general case and that considered here consists only in the behavior of the distribution function at the lower levels. However, the latter circumstance is inconsequential, since in both cases this distribution is a Boltzmann distribution to a high degree of accuracy.

Assuming that the upper limit of the reaction rate is given by a quantity of the order of $ZP_{kd}x_0 \exp(-D/kT)$, $(E_{\mathrm{vib}} \sim D)$ we obtain from the condition that $\tau_d \gg \tau_{\mathrm{vib}}$ the following condition for the validity of Eq. (18–11):

$$P_{10} \gg P_{kd} e^{-\frac{D}{kT}}. \tag{18–13}$$

Let us stress still another circumstance. If the expansion of Eq. (18.9) is replaced by the ordinary series of the theory of perturbations

$$x_n(t) = x_n^{(0)}(t) + \varepsilon x_n^{(1)}(t) + \cdots, \tag{18–14}$$

then it will be possible to analyze only the first, rapid stage of the equilibrium-establishing process. In fact, the quantity $N=\sum_k^0 x_n(t)$ in the zeroth approximation is found to be independent of time. The dependence of N on the time which characterizes the dissociation process can be obtained only upon consideration of a correction term which is proportional to ε. It can be seen from this that the series (18–14) can be used only in a time interval during which N changes little in comparison with its initial value. This means that Eq. (18–14) is not practically suitable for describing the dissociation process.

We seek the solution of Eqs. (18–11) in the form

$$x_n = x_0 e^{-\varepsilon_n}(1 + \varphi_n) \quad \left(\varepsilon_n = \frac{E_n - E_0}{kT}\right), \tag{18–15}$$

where φ_n characterizes the deviations of the quasi-steady distribution from the Boltzmann equilibrium distribution; here $\varphi = 0$, while φ_n must necessarily not be much smaller than unity. Substituting Eq. (18–15) into Eq. (18–11) and taking into account the fact that according to the detailed balancing principle

$$\frac{P_{n,n-1}}{P_{n-1,n}} = e^{-(\varepsilon_{n-1} - \varepsilon_n)}, \tag{18–16}$$

we get

$$P_{kd} e^{-\varepsilon_k} \sum^{k} \frac{e^{\varepsilon_{i-1}}}{P_{i-1,i}} \gg 1,$$

By order of magnitude $w \sim Z P_{kd} x_0 \exp(-D/kT)$, hence at the lower levels $|\varphi_1| \ll 1$, while at the upper levels $|\varphi_1| \sim 1$ (excluding the case of very low temperatures). Thus dissociation results in a final disturbance of the equilibrium distribution of molecules over the vibrational levels. This disturbance is practically concentrated in that region of vibrational energies, starting with which the molecule can go over to the continuous energy spectrum. The above disturbance has an appreciable effect on the dissociation rate. Substituting Eq. (18–15) and Eq. (18–16) into the last of Eqs. (18–11), we get

$$\frac{1}{2}\frac{\mathrm{d}x_d}{\mathrm{d}t}\left(1 + P_{kd} e^{-\varepsilon_k} \sum_{i=1}^{k} \frac{e^{\varepsilon_{i-1}}}{P_{i-1,i}}\right) = P_{kd} N Z \frac{e^{-\varepsilon_k}}{\sum_{i=0}^{k} e^{-\varepsilon_i}} - P_{dk} Z x_d^2, \tag{18–17}$$

or, disregarding the recombination process,

$$w = \frac{1}{2}\frac{\mathrm{d}x_d}{\mathrm{d}t} \equiv NZ \frac{P_{kd} e^{-\varepsilon_k}}{\sum_{i=0}^{k} e^{-\varepsilon_i}\left(1 + P_{kd} e^{-\varepsilon_k} \sum_{i=1}^{k} \frac{e^{\varepsilon_{i-1}}}{P_{i-1,i}}\right)}, \quad N = \sum_{i=0}^{k} x_i. \tag{18–18}$$

Eq. (18–18) is found to be valid also in the more general case than it is assumed in Eq. (18–10). In order for Eq. (18–18) to be valid, it is sufficient to assume that it is possible to isolate a region of large vibrational energies the population of which is relatively small, and that at the same time φ at the lower boundary of this level is negligible.

In the majority of cases of practical interest

$$P_{kd}e^{-\varepsilon_k}\sum_{i=1}^{k}\frac{e^{\varepsilon_{i-1}}}{P_{i-1,i}} \gg 1,$$

hence in the braces of Eq. (18–18) it is possible to disregard the unity, and the dissociation rate w is found to be independent of the probability P_{kd}. In physical terms this result means that the limiting stage of the dissociation process in the case under study is the rate of supply of molecules to the upper vibrational levels and not the rate of their disintegration. For comparison, we wish to point out that in the elementary theory of collisions, which does not take into account departure from the Boltzmann distribution the expression for w will have the form of Eq. (18–18) if in the latter we disregard the term

$$P_{kd}e^{-\varepsilon_k}\sum\frac{e^{\varepsilon_{i-1}}}{P_{i-1,i}}$$

in comparison with unity. In this case the limiting stage of the process is the rate of transition to the continuous spectrum which is determined by P_{kd}. The rate of vibrational excitation is found to be inconsequential, since the distribution of molecules over the vibrational levels is assumed to be an equilibrium distribution.

The chain of reasoning which yields this result is not the only one, although it is quite probable, that it is the simplest. Similar results, although using slightly different methods, were obtained by other authors in [128, 603, 478, 238, 452, 540, 135a]. In particular, [452, 540] present the most systematic consideration of the thermal dissociation rate of harmonic oscillators within the framework of the theory of random walks. The dissociation rate constant in this case is determined in terms of the average time of the first appearance, *i.e.*, of the average time which is needed by the molecule to reach a vibrational level with an energy equal to the energy of dissociation.

Departure from the equilibrium energy distribution with respect to the vibrational level which is produced by dissociation, as was established, has an appreciable effect on the dissociation rate. In conjunction with this we shall now give two examples which illustrate the possible application of results thus obtained.

1. The fact that the thermal dissociation rate is determined by the nonequilibrium distribution of molecules with respect to the vibrational levels, requires a critical approach to the use of the well-known expression of statistical thermodynamics

$$\frac{k_d}{k_{rec}} = K_{eq} \tag{18–19}$$

under nonequilibrium conditions. Here K_{eq} is the chemical equilibrium constant and k_d and k_{rec} are the dissociation and recombination rate constants, respectively. In particular, in the case of thermal dissociation of diatomic molecules constituting a small addition to a monatomic gas just considered, the rate "constant" of the forward reaction differs from its equilibrium value. This fact is due to departure from the Boltzman equilibrium at the upper levels.

The rate constant of the reverse reaction retains its equilibrium value, since this reaction, under the above conditions, takes place in triple collisions of the atoms of the dissociating gas with the atoms of the host gas, and the Maxwellian velocity distribution for these particles is practically not disturbed. Consequently, the ratio (18–19) is not satisfied for the above reaction. Eq. (18–18) has the form $w = k_1 N$, where the proportionality coefficient k_1 is only a function of the temperature T (and of the density of the host gas). Hence we can write

$$\frac{k_1}{k_2} = K^*, \tag{18–20}$$

where k_2 is the rate constant for the reverse reaction, which is equal to k_{rec}, and K^* is a quantity which is independent of the concentration of the reacting components.

Equation (18–20) has the form of (18–19) and makes it possible, in general, to determine k_1 on the basis of k_2 if K^* is known. However, the meaning of Eq. (18–20) differs from that of Eq. (18–19), since K^* is not equal to the equilibrium constant and, in addition, it must be remembered that k_1 is the dissociation rate constant at the initial time, when the recombination processes can be disregarded.

The publication of works in which the appreciable effect of disturbing the equilibrium distribution function on the process of chemical transformations has been established theoretically [193, 194, 133], has restimulated the consideration of the old problem of the applicability of Eq. (18–19) under chemically nonequilibrium conditions [136, 135, 490, 479].

2. The above expression for the dissociation rate constant can be compared with experimental data. Eq. (18–18) can be simplified by making use of any specific molecular model. In particular, for the Morse oscillator in

a temperature range above the characteristic temperatures, Eq. (18–18) takes on the form* [194, 135a]:

$$k_d = Z^0 P_{k,k-1} \left(\frac{\hbar\omega}{D}\right)^2 \left(\frac{D}{kT}\right)^{1.5} e^{-\frac{D}{kT}}. \tag{18–21}$$

When Eq. (18–21) is compared with experimental data, the effect of the rotational and electronic degrees of freedom of the dissociating molecule must be taken into account. A rigorous solution of this problem within the framework of any specific model is not available at present. Estimates made in [128] show that the consideration of the rotational and electronic degrees of freedom introduces into Eq. (18–21) a new multiplying factor which is greater than unity. For the reaction $Br_2 + Ar \rightarrow Br + Br + Ar$, which was studied in a shock tube in a temperature range up to 2000 °K [463], the theoretical value of the dissociation rate constant [128], is given by

$$k_d = 5 \cdot 10^{-2} Z^0 \left(\frac{D}{kT}\right)^{1.5} e^{-\frac{D}{kT}} \text{ cm}^3/\text{mole-sec}$$

The experimental rate constant for this reaction has the form

$$k_d = 6 \cdot 10^{-1} Z^0 \left(\frac{D}{kT}\right)^{1.97} e^{-\frac{D}{kT}} \text{ cm}^3/\text{mole-sec}.$$

The above expressions yield values which coincide by order of magnitude. It should, however, be emphasized that these theoretical calculations of the dissociation rate constant can at best be used to estimate the order of magnitude. In fact, the consideration of the rotational degrees of freedom, as was pointed out previously, introduces an additional multiplying factor g_{rot} into Eq. (18–21). A value of the order of 10 has been obtained from g_{rot} in [128], while in [601] it was found to be approximately equal to 50, finally, a third value of g_{rot} is given in [489]. Due to this, the main interest in the results presented here apparently consists in the theoretical clarification of the role of disturbing the distribution function in the kinetics of the process. The established fact to the effect that equilibrium is disturbed in the process of the chemical reaction is independent of the extent to which the numerical results conform to the experimental data.

[2] Thermal Dissociation in a Single-component System

Up to now we have considered only the dissociation of diatomic molecules which constituted a small admixture in a monatomic gas. However, the consideration of dissociation in a one-component ("pure") diatomic gas is

* In [194] the multiplying factor $\hbar\omega/D$ was omitted in Eq. (31) for k_d.

also of interest. The experimental values of rate constants of thermal dissociation of certain diatomic molecules at which temperatures obtained in shock tubes are presented in Table 19. The values of the preexponential factor P in the expression for the rate constant $k_d = PZ^0 \exp(-D/kT)$ [the so-called Arrhenius law] as a function of $(D/kT)^{\frac{3}{2}}$ are given in Fig. 105. It can be seen

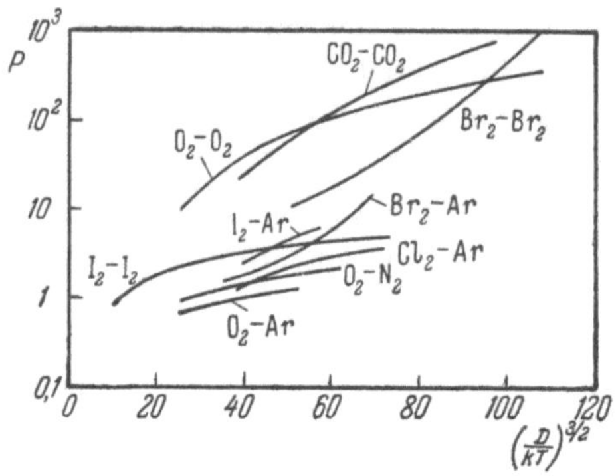

Fig. 105. The value of preexponential factor P in the expression for the dissociation rate constant $k_d = PZ \exp(-D/kT)$ as a function of $(D/kT)^{3/2}$.

from this figure that the experimental values of P form two groups. The first group corresponds to the dissociation rate constants for a "pure" diatomic gas. The second group corresponds to dissociation rate constants of a diatomic gas in a mixture with a monatomic and certain other gases. The preexponential factors in these groups differ by approximately one order of magnitude. This means that the dissociation rate in a homogeneous system is by approximately an order of magnitude higher than the dissociation rate in a monatomic gas medium.

The difference in the dissociation rates is, in all probability, related to the participation of the internal degrees of freedom of the incoming particles in the dissociation process. The main process responsible for dissociation of diatomic molecules in a monatomic host gas is the transformation of the translational energy into vibrational energy of a molecule on collision of the latter with an atom. In dissociation in a single-component system, in addition to the above process, intermolecular collision results in processes which involve transfer of vibrational and rotational energy. The first process has been studied in [144] while the second was considered in [132].

Let us analyze the first process. We consider an elementary molecular model, the so-called "cut off" harmonic oscillator [144], in which the interatomic interaction potential is approximated by a "harmonic" potential hole with depth D with the discrete vibrational levels in equidistant positions up to the dissociation energy D and with a continuous spectrum in the region of energies exceeding D. This model was used a great number of times in calculating the dissociation rate for molecules which constituted a small admixture in a monatomic gas [121, 452, 536]. The dissociation in a single-component system of these molecules is described by

$$\left.\begin{aligned}
\frac{\mathrm{d}x_n}{\mathrm{d}t} &= ZP_{10}\{(n+1)x_{n+1} - [(n+1)e^{-\theta} + n]x_n + \\
&\quad + ne^{-\theta}x_{n-1}\} + ZQ_{10}\{(n+1)(1+\bar{\alpha})x_{n+1} - \\
&\quad - [(n+1)\bar{\alpha} + n(1+\bar{\alpha})]x_n + n\bar{\alpha}x_{n-1}\}, \\
n &= 0, 1, \ldots, k-1. \\
\frac{\mathrm{d}x_k}{\mathrm{d}t} &= ZP_{10}\{ke^{-\theta}x_{k-1} - kx_k\} + \\
&\quad + ZQ_{10}\{k\bar{\alpha}x_{k-1} - k(1+\bar{\alpha})x_k\} - ZP_{kd}x_k.
\end{aligned}\right\} \qquad (18\text{–}22)$$

Here $\bar{\alpha} = (1/N)\sum_{n=0}^{k} nx_n(t)$, $N(t)$ is the oscillator number density at time t.

Eqs. (18–22) are similar to Eqs. (17–85) which describe the vibrational relaxation in an isolated system of harmonic oscillators and differ only by the last equation which describes the transition of the molecules into the dissociated state. As before, we assume that molecular dissociation takes place as a result of transition of molecules from some level k to the continuous energy spectrum. We are considering only the initial dissociation stage, for which reason we disregard the recombination processes.

The solution of Eqs. (18–22) is quite difficult. In the general case these equations describe complex relaxation processes. First, on energy transfer between the translational and vibrational degrees of freedom and also on transfer of vibrational photons, a Boltzmann distribution is very rapidly established in the first approximation, which is likely to be appreciably disturbed only in the region of large vibrational energies. The second, slow stage of the process consists in establishing dissociation equilibrium. In kinetic problems of the type under discussion, it is precisely this second, slow stage which is of interest.

It is much easier to obtain the quasi-steady function of a distribution which is established as a result of a process which is first rapid and then becomes slowly varying with time. The quasi-steady distribution, as was shown in the first subsection of this section, differs appreciably from the Boltzmann distribution only at the upper vibrational levels. This assertion is valid when only the transformation of translational into vibrational energy is considered; hence, it will be even more valid if we consider, in addition, the transfer of vibrational photons. On the basis of this we can simplify Eqs. (18–22). The value of $\bar{\alpha}$ in Eq. (18–22) is determined basically by molecules at the first vibrational levels. At these levels the Boltzmann distribution is obtained with considerable accuracy in the quasi-steady process. Hence,

$$\bar{\alpha} = \frac{1}{N}\sum_{n=0}^{k} n x_n(t) = (1 - e^{-\theta})\sum n e^{-n\theta} = \frac{1}{e^{\theta} - 1}. \tag{18–23}$$

Making use of Eq. (18–23), we can write Eqs. (18–22) in the form

$$\left.\begin{aligned} \frac{dx_n}{dt} &= ZP_{10}^{*}\{(n+1)x_{n+1} - [(n+1)e^{-\theta} + n]x_n + \\ &\quad + ne^{-\theta}x_{n-1}\}, \quad n = 0, 1, \ldots, k-1, \\ \frac{dx_k}{dt} &= ZP_{10}^{*}\{ke^{-\theta}x_{k-1} - kx_k\} - ZP_{kd}x_k, \end{aligned}\right\} \tag{18–24}$$

where

$$P_{10}^{*} = P_{10}\left[1 + (1 - e^{-\theta})^{-1}\frac{Q_{10}}{P_{10}}\right] \equiv \beta P_{10}) \tag{18–25}$$

Thus, the nonlinear system (18–22) is reduced to the linear system (18–24). Formally Eqs. (18–24) are obtained from Eq. (18–22) by substituting into the latter x_n in the form given by Eq. (18–9) and by retaining only zeroth-order terms.

The form of Eqs. (18–24) is similar to an analogous system of equations derived in the first subsection of this section [see Eq. (18–8)]. The only difference consists in the fact that P_{10} in the given case is replaced by $P_{10}^{*} = \beta P_{10}$. Eqs. (18–24) have been solved in the first section of this subsection. The form of the solution is

$$x_n = x_0 e^{-\varepsilon_n}(1 + \varphi_n).$$

For the molecular model under study in a range of temperatures smaller than or equal to the order of characteristic temperatures (for vibrational degrees of freedom), φ_n is noticeably different from zero only when $n \approx k$. In particular,

$$x_k = x_0 e^{-\varepsilon_k}\frac{1}{P_{kd}e^{-\varepsilon_k}\sum_{n=1}^{k}\dfrac{e^{\varepsilon_{n-1}}}{nP_{10}}}.$$

The dissociation rate has the form

$$w = ZP_{kd}x_k.$$

Replacement of P_{10} by P_{10}^{*} produces an increase in x_k and, consequently, also in w by a factor of β. At temperatures below the characteristic temperatures, $Q_{10} \gg P_{10}$, hence, $\beta \gg 1$, at temperatures of the order of characteristic temperatures $Q_{10} \approx P_{10}$ and $\beta \sim 1$. For sufficiently large values of β the expression $P_{kd} \exp(-\varepsilon_k) \sum_{n=1}^{k} [\exp(\varepsilon_{n-1})/n\beta P_{10}]$ may be found to be smaller than or of the order of unity. In this case w is determined by the more general equation (18–18).

The transfer of vibrational photons on molecular collisions in a dissociating gas thus increases the quasi-steady population of the upper vibrational levels (and, consequently, also the dissociation rate), bringing it closer to the equilibrium value. This increase is appreciable at temperatures below characteristic temperatures and does not make a perceptible contribution at temperatures above or of the order of characteristic temperatures. The physical cause for the increase in the number of molecules on the upper vibrational levels is obvious. At low temperature the energy transfer to highly excited molecules by transfer of vibrational photons takes place more rapidly than energy transfer produced by transformation of translational into vibrational energy. An increase in the rate of supply of molecules to the upper vibrational levels with an unchanged probability of transition into the continuous spectrum ($P_{kd} \approx 1$) is precisely the factor responsible for the increase in the quasi-steady population.

In comparing the results thus obtained with experimental data it should be remembered that the model we have considered does not take into account anharmonicity. Hence in the case of diatomic molecules this solution is only qualitative and will give values on the higher side. However, in the case of polyatomic molecules, the thermal dissociation of which takes place following the change in the multiplicity, the potential of interaction along the bond being broken up is sufficiently well approximated by the cut-off harmonic oscillator model. This, in particular, is true for N_2O following dissociation into N_2 and O. In the latter case one can expect an increase in the dissociation rate in pure N_2O in comparison with the dissociation rate of N_2O in a mixture with a monatomic gas.

The effect of vibrational energy transfer on the dissociation rate constant was studied theoretically in [132], using the dissociation of the O_2 molecule in the reactions

$$O_2 + Ar = 2O + Ar, \tag{18–26}$$

$$O_2 + O_2 = 2O + O_2. \tag{18–27}$$

Estimates made in the above reference show that in collision of two molecules, in addition to the transformation of translational into vibrational energy, it is also necessary to consider the simultaneous transformation of a part of the rotational energy of the nondissociating molecule into vibrational energy of the dissociating molecule (a process which is similar to that considered on page 251. When this phenomenon is taken into account, the dissociation rate becomes higher. Estimates made in [132] show that the rate

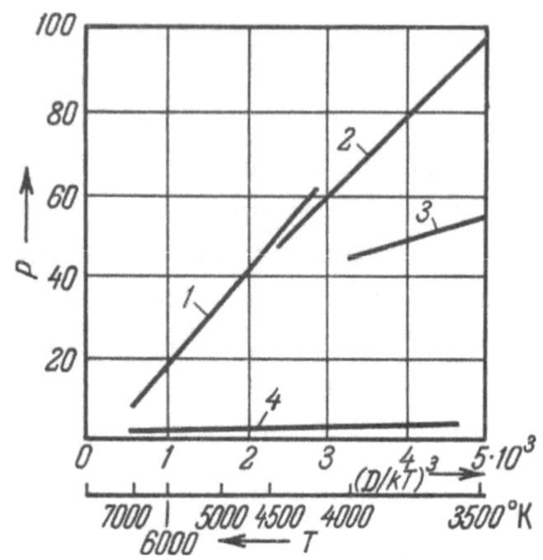

Fig. 106. The factor P in the expression for the rate constant for the dissociation of O_2 $k_d = PZ \exp(D/kT)$ as a function of $(D/kT)^{3/2}$ in collisions $O_2 - O_2$ (1, 2, 3) and O_2–Ar (4). Curve 1 represents results due to Losev [105], Curve 2 gives the results due to Matthews [437], Curve 3 was obtained by Byron [275], and Curve 4 gives the results due to Camac and Vaughan [276].

constant for reaction (18–27) should exceed the rate constant of reaction (18–26) approximately by a factor of β_1, where

$$\beta_1 = 3\gamma e^{2\gamma^2} \approx 20, \quad (\gamma \sim 1).$$

It is important that β_1 be independent of the temperature. The quantity γ has the same meaning as in Eq. (17–14).

At the present time, the reactions (18–26) and (18–27) have been studied extensively. In particular, the rate constant of reaction (18–27) in a temperature range of 3500–7000 °K was determined in [282], and in [275] the rate constant of reaction (18–26) has been found in the same temperature range. Fig. 106 shows the experimental values of P in ($k_d = PZ^0 \exp(-D/kT)$ as a function of $(D/kT)^{\frac{3}{2}}$ for collisions O_2–O_2 and O_2–Ar according to data obtained in [105, 282, 275, 437]. It follows from this figure that, as opposed to [132], $\beta_1 = P\,(O_2 - O_2)\,/\,P\,(O_2 - \mathrm{Ar})$ is a function of the temperature and decreases with a temperature increase from $\beta_1 \sim 30$–40 at $T = 3500$ °K to $\beta_1 \sim 5$–10 at $T = 7000$ °K. Thus, at present, no satisfactory explanation exists for the experimentally detected difference in the dissociation rates in a pure diatomic gas and when the latter is an admixture in an inert monatomic host gas.

[3] Concurrent Consideration of the Thermal Dissociation and Vibrational Relaxation of Diatomic Molecules

The theory of thermal dissociation which was developed in the preceding

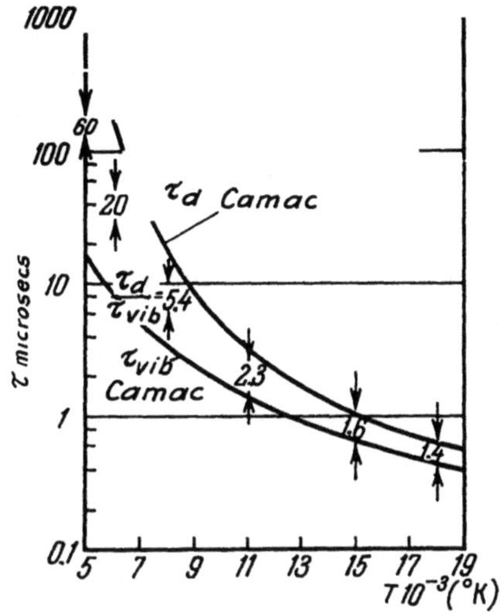

Fig. 107. The times τ_d and τ_{vib} for O_2 in the O_2–Ar mixture as a function of the temperature.

subsections is based on the assumption that τ_d, the time needed for establishing a dissociation equilibrium is appreciably greater than τ_{vib}. An experimental study of τ_d and τ_{vib} under identical conditions was performed in [108, 277, 282, 615]. In particular, it was shown in [277, 282] that at 5000°K in an O_2–Ar mixture the ratio $(\tau_d/\tau_{vib})=60$. As the temperature increases this ratio drops rapidly and at 7500°K it becomes 5.4. Results of the comparative study of τ_d and τ_{vib} which was performed in [277, 282] are presented in Fig. 107 [615]. The data for temperatures above 7500°K have been obtained by extrapolation of the experimental results which are valid in the temperature range 5000–7500°K and are suitable only for estimating the order of magni-

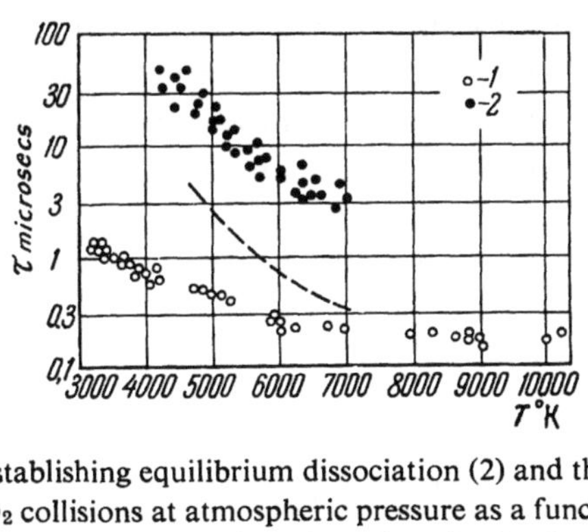

Fig. 108. Time of establishing equilibrium dissociation (2) and the vibrational relaxation time (1) for the O_2–O_2 collisions at atmospheric pressure as a function of the temperature. The dashed line denotes the average value of τ_d when the O_2–O collisions are considered.

tude. Similar relationships were obtained for τ_d and τ_{vib} in pure oxygen (Fig. 108). It can be seen from this figure that as the temperature increases, τ_d/τ_{vib} decreases; at temperatures of the order of 6000 °K, this ratio slightly exceeds an order of magnitude and at even higher temperatures, it becomes even smaller.

An experimental study of the ratio of the initial time Δt during which no appreciable dissociation takes place, to the time τ_{vib} in the O_2–Ar mixture at temperatures 5000–18,000 °K was performed in [615]. The results obtained there are presented in Table 18. This study shows that the basic assumption

Table **18.** Delay Time of Dissociation Δt in the O_2–Ar Mixture

T_{trans}	$\Delta t/\tau_{vib}$	T_{vib}/T_{trans} (at time Δt)
5500 °K	2.4	0.93
8000	1.1	0.71
11,000	0.55	0.48
14,000	0.45	0.41
18,000 °K	0.40	0.37

of the thermal dissociation theory according to which it is possible at each given time of the vibrational relaxation process to assume that the temperature of the translational and vibrational degrees of freedom are equal (the condition $\tau_d \gg \tau_{vib}$) is not valid at high temperatures. For oxygen this limit is given by temperatures of the order of 8000 °K.

Thus, the question is raised of the joint consideration of thermal dissociation and vibrational relaxation. References [355, 574, 108] are devoted to this problem. Reference [355] presents the study of the effect of vibrational relaxation on the rate of thermal dissociation of diatomic gases. The following dissociation mechanism was considered there. It was assumed that the molecule, with equal probability, can dissociate from any vibrational level if the translational energy of the colliding particles along the line of centers is sufficient for dissociation. In accordance with this model, the rate of dissociation from any level is proportional to the product of the population of this level by the number of collisions, the kinetic energy of which along the line of centers exceeds the energy $D-E_v$, where D is the energy of dissociation and E_v is the vibrational energy of the vth level. For the dissociation rate constant we can write

$$k_d \sim \frac{\sum_{v=0}^{\bar{N}-1} x_v e^{-\frac{D-Ev}{kT}}}{\sum_{v=0}^{\bar{N}-1} x_v}, \tag{18–28}$$

where x_v is the molecular number density of the vth vibrational level and $\tilde{N}$ is the total number of vibrational state, including also the continuous spectrum. It was assumed that x_v is governed by a Boltzmann distribution, the vibrational temperature of which at the given time is given by

$$\frac{d\varepsilon}{dt} = -\frac{1}{\tau_{vib}}(\varepsilon - \varepsilon_0), \tag{18–29}$$

where

$$\varepsilon = \frac{\hbar\omega}{e^{\theta} - 1}\left(\theta = \frac{\hbar\omega}{kT_{vib}}\right),$$

and

$$\varepsilon_0 = \frac{\hbar\omega}{e^{\vartheta} - 1}\left(\vartheta = \frac{\hbar\omega}{kT_{trans}}\right),$$

ε is the average value of vibrational energy devolved upon one molecule, T_{vib} and T_{trans} are the temperatures of the vibrational and translational degrees of freedom. Assuming a form of x_v which would be valid for the harmonic oscillator model, the dissociation rate constant $k_d(t)$ can be written in the form

$$k_d(t) = \frac{(k_d)_{eq}}{\tilde{N}} \frac{1 - e^{-N(\theta-\vartheta)}}{e^{\theta-\vartheta} - 1} \frac{e^{\theta} - 1}{e^{\vartheta} - 1}.$$

Figure 109, which was taken from [355], shows the theoretical values of the translational temperature of molecules behind the front of a strong shock wave in pure nitrogen, calculated without taking into account dissociation for the case of independent consideration of vibrational relaxation and of

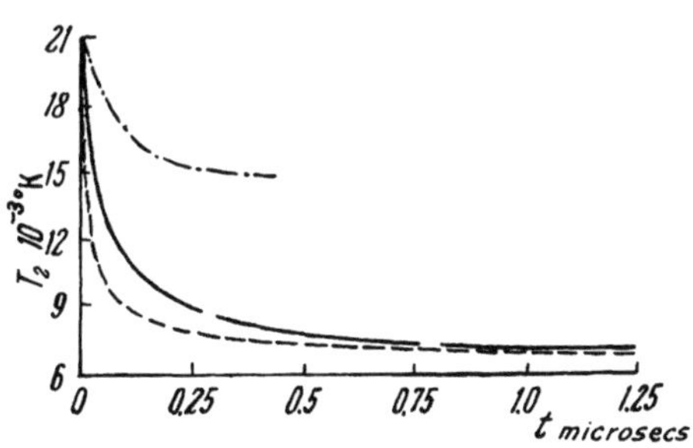

Fig. 109. Calculated values of the translational temperature of molecules behind the front of a strong shock wave in nitrogen as a function of t_l. M = 19, the initial pressure is 1 mm Hg, the equilibrium temperature behind the shock front is 6681 °K). The top curve was obtained without considering dissociation, the bottom curve corresponds to independent consideration of vibrational relaxation and of dissociation, while the middle curve was obtained by joint consideration of these two quantities.

dissociation and on joint consideration of these two quantities. As can be seen from this figure, the incompleteness of the vibrational relaxation process toward the time when dissociation starts results in a smoother change in the temperature behind the shock front.

The model considered above has a substantial shortcoming consisting in the fact that it considers only the effect of vibrational relaxation and the dissociation rate and does not take into account the effect that the dissociation process exerts on the vibrational relaxation rate. The mutual effect of these processes is considered in [574].

The relaxation equation for the vibrational energy is written in [574] as

$$\frac{d\varepsilon}{dt} = -\frac{1}{\tau_{vib}}(\varepsilon - \varepsilon_0) - \frac{\bar{E}(T_{trans}, T_{vib})}{N}\left(\frac{dN}{dt}\right)_d + \frac{\bar{G}(T_{trans})}{N}\left(\frac{dN}{dt}\right)_{rec} + \frac{\varepsilon}{N}\left\{\left(\frac{dN}{dt}\right)_d - \left(\frac{dN}{d}\right)_{rec}\right\}. \tag{18–30}$$

The last three terms in this equation, by which Eq. (18–30) differs from (18–29), take into account the effect of dissociation on the vibrational relaxation rate. The meaning of these terms is sufficiently simple. The second term describes the loss of vibrational energy due to the dissociation process; here N is the molecular number density, $\bar{E}$ $T_{trans}, T_{vib})$ is the average vibrational energy lost in each dissociation act, $(dN/dt)_d$ is the dissociation rate, *i.e.*, the number of dissociations which take place per unit volume. The third term corresponds to the vibrational relaxation which is produced by the recombination process; here $\bar{G}(T_{trans})$ is the average energy acquired by the vibrational degrees of freedom in each recombination and $(dN/dt)_{rec}$ is the recombination rate. The last term describes the change in the vibrational energy due to a change in the total number of molecules during dissociation.

Under equilibrium conditions, the right-hand side of Eq. (18–30) becomes identically zero. Hence,

$$\bar{G}(T_{trans}) = \bar{E}(T_{trans}, T_{vib}).$$

The specific form of $\bar{E}(T_{trans}, T_{vib})$ depends on the dissociation mechanism. The dissociation mechanism assumed in [574] is the same as that used in [355].

According to this, the probability of dissociation from the jth vibrational level is

$$p_j = Cx_j e^{-\frac{D-E_j}{kT_{trans}}} \tag{18–31}$$

where C is a norming constant which is determined from the condition that $\sum_{p_j} = 1$. (It is assumed that the probability of dissociation is unity if the sum of the vibrational energy and the energy of translational motion along the line of centers is greater than or equal to the dissociation energy.) Subsequently, the same authors have considered a model with predominance of dissociation from the upper vibrational levels [433]. In this case it was assumed that p_j is given by

$$p_j = CF(j)\,x_j e^{-\frac{D-E_j}{kT_{\text{trans}}}},$$

where

$$F(j) = e^{-\frac{D-E_j}{kU}}.$$

The constant U was determined from experimental data; here by order of magnitude, $U = D/6k$ (k is the Boltzmann constant). Substituting $x_j \sim \exp \times (-E_j/kT_{\text{vib}})$ into Eq. (18–31) and taking into account the norming, we get

$$p_j = \frac{1}{Q(T_m)} e^{-\frac{E_j}{kT_m}}, \tag{18–32}$$

where

$$\frac{1}{T_m} = \frac{1}{T_{\text{vib}}} - \frac{1}{T_{\text{trans}}}, \quad Q(T_m) = \sum e^{-\frac{E_j}{kT_m}}.$$

Knowing Eq. (18–32) it is easy to find $\bar{E}$ and $\bar{G}$

$$\bar{E} = \sum E_j p_j = \frac{1}{Q(T_m)} \sum E_j e^{-\frac{E_j}{kT_m}} = kT_m^2 \frac{\mathrm{d}}{\mathrm{d}T_m} \ln Q(T_m).$$

Consequently,

$$\bar{E}(T_n, T_{\text{vib}}) = \bar{E}(T_m), \quad \text{and} \quad \bar{G}(T_{\text{trans}}) = \bar{E}(\infty).$$

Putting Eq. (18–30) into an even more specific form depends on the choice of the molecular model. Reference [574] contains a consideration of a cut-off harmonic oscillator model, which has $\tilde{N}-1$ discrete levels. In this case

$$\left.\begin{aligned} Q &= \frac{1 - e^{-\frac{\tilde{N}\hbar\omega}{kT}}}{1 - e^{-\frac{\hbar\omega}{kT}}}, \\ \bar{E}(T_m) &= \frac{\hbar\omega}{e^{\frac{\hbar\omega}{kT_m}} - 1} - \frac{\tilde{N}\hbar\omega}{e^{\frac{\tilde{N}\hbar\omega}{kT_m}} - 1}, \\ \bar{G}(T_{\text{trans}}) &= \lim_{T_m \to \infty} E(T_m) = \tfrac{1}{2}\hbar\omega(\tilde{N} - 1), \end{aligned}\right\} \tag{18–33}$$

and Eq. (18–30) takes on the form

$$\frac{\mathrm{d}\varepsilon}{\mathrm{d}t} = -\frac{1}{\tau_{\text{vib}}}(\varepsilon - \varepsilon_0) - \left[\frac{\hbar\omega}{e^{\frac{\hbar\omega}{kT_m}} - 1} - \frac{\tilde{N}\hbar\omega}{e^{\frac{\tilde{N}\hbar\omega}{kT_m}} - 1} - \varepsilon\right] \frac{1}{N}\left(\frac{\mathrm{d}N}{\mathrm{d}t}\right)_d + \\ + \left[\tfrac{1}{2}\hbar\omega(\tilde{N} - 1) - \varepsilon\right] \frac{1}{N}\left(\frac{\mathrm{d}N}{\mathrm{d}t}\right)_{\text{eq}}. \tag{18–34}$$

Numerical calculation of the nonequilibrium zone behind the shock front with the use of Eq. (18–34) shows [574] that recombination [the last term in Eq. (18–34)] is found to be appreciable only when the equilibrium fraction of the dissociated molecules exceeds 10–20%. The temperature distribution behind the front of strong shock waves in nitrogen and oxygen in accordance with the considered dissociation models was also determined in [574]. The results of these calculations are presented in Figs. 110 and 111. The dashed line corresponds to the temperature distribution calculated by the model used in [355] which takes into account only the one-sided effect of vibrational relaxation on dissociation. The solid lines were obtained for the model of [574] and take into account the mutual effect of the above two processes. An interesting result of these calculations is the fact that the vibrational temperature distribution behind a shock front in oxygen is not monotonic, which results in the appearance of a maximum in this distribution. Similar calculations for the temperature distribution in nitrogen show that the T_{vib} curve is monotonic. On the whole, however, the consideration of the interaction between vibrational relaxation and dissociation yields a smoother change in the parameters behind a shock front, than would be obtained by methods of [355]. The latter assertion, generally speaking, is independent of the selection of the dissociation mechanism and of the molecular model. On the other hand, the dissociation mechanism considered in [355, 574] is highly objectionable (see page 275). As in the elementary collision theory, in these works the dissociation probability is postulated (without any quantum-mechanical substantiation whatsoever), and it is there assumed that the

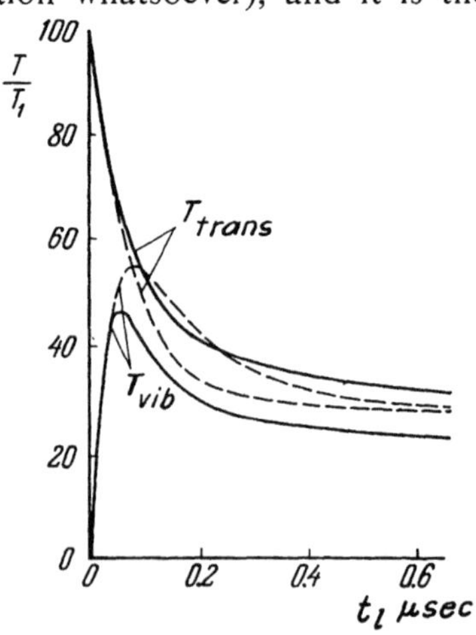

Fig. 110. The distributions of T_{trans} and T_{vib} behind a shock front in O_2($V = 7$ km/sec, $p_1 = 2.23 \times 106^4$ atm). The dashed line denotes the temperature distribution obtained by considering only the effect of vibrational relaxation on dissociation; the solid line was calculated by considering the mutual effect of these processes.

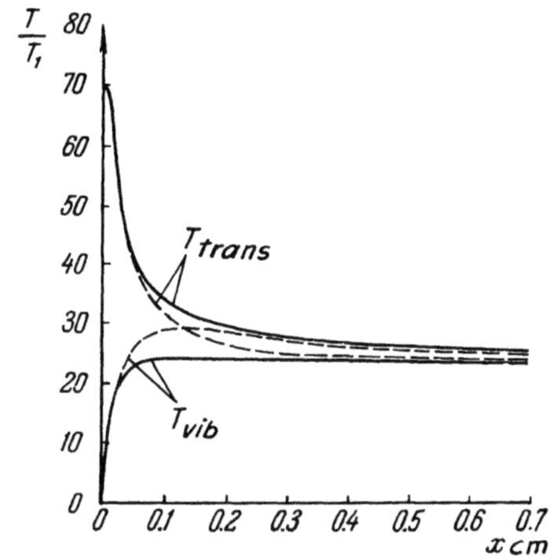

Fig. 111. The distributions of T_{trans} and T_{vib} behind a shock front in N_2($V = 7$ km/sec. $p_1 = 1$ mm Hg). The notation is the same as in Fig. 110.

dissociation does not disturb the Boltzmann distribution of molecules with respect to the vibrational levels. Due to this fact, although the solution of the joint problem of dissociation and vibrational relaxation given in [355, 574, 433] is in agreement with experimental data, it is not consistent from the theoretical viewpoint.

[19] THERMAL IONIZATION KINETICS

The propagation of a shock wave in a gas can be accompanied by appreciable ionization. Thus, for example, in the experiment of Petschek and Byron [159, 474], who have studied the ionization of argon behind a shock wave, the equilibrium degree of ionization for $\mathsf{M} \sim 18$ was 25% at temperatures of $\sim 14{,}000\,°$K. It is natural to begin the study of ionization with the simplest case of a monatomic gas, restricting the consideration to shock waves in which the temperature behind the front is appreciably smaller than I/k, where I is the ionization potential. This will be discussed in this section. The ionization kinetics of di- and polyatomic molecules and their mixtures have not been studied extensively. An exception is air, the ionization kinetics of which are considered in Chapter 5.

The structure of a shock wave in a monatomic gas is determined only by processes of establishing a Maxwellian distribution and of electronic excitation, which also includes ionization. The Maxwellian distribution is established sufficiently rapidly, so that the ionization and electronic excitation can be considered by assuming that the velocity distribution of the atoms and ions is an equilibrium one. The form of the electronic distribution function depends on the electron concentration and will be considered below.

The structure of a shock wave in a monatomic gas is described by the following system of gasdynamic equations:

$$\varrho v = \varrho_1 V; \tag{19–1}$$

$$p + \varrho v^2 = p_1 + \varrho_1 V^2; \tag{19–2}$$

$$h + \frac{v^2}{2} = h_1 + \frac{V^2}{2}; \tag{19–3}$$

by the equation of state

$$p = \frac{\varrho}{m_A} kT_A + \alpha \frac{\varrho}{m_A} kT_e; \tag{19–4}$$

and by the relaxation equation

$$\frac{d\alpha}{dt} = f(\alpha, \varrho, T_A, T_e).$$

Here h is the enthalpy per unit mass,

$$m_A h = \tfrac{5}{2}kT_A + \alpha\tfrac{5}{2}kT_e + \alpha kT_{\text{ion}}, \tag{19–5}$$

v is the velocity of the heated gas relative to the shock front, ϱ is the density, p is the pressure, T_A is the atomic temperature, T_e is the electronic temperature, T_{ion} is the ionization potential of atoms expressed in degrees Kelvin (for Ar $T_{\text{ion}} = 1.82 \times 10^5\,°\text{K}$), m_A is the mass of an atom, α is the degree of ionization, equal to the ratio of free electrons to the number of atoms and ions (see page 49). The subscript "1" denotes quantities pertaining to the cold gas ahead of the shock wave. Eqs. (19–1)—(19–3), as in all the preceding cases are written in a coordinate system moving with the shock front. The viscosity and thermal conductivity were disregarded in writing Eqs. (19–1)—(19–3). The latter is justified, since the gradients of macroscopic quantities in the relaxation zone are small. In addition, the light energy and momentum fluxes are not considered in Eqs. (19–1)—(19–8) due to their assumed smallness. The conservation equations at the shock front in which radiation is taken into account are derived in [166a] (see also [71, 434]). It should be emphasized that even if the energy devolved upon the radiation is small, it is necessary to consider the transport of resonance radiation, which can be found to be important for understanding the ionization kinetics.

Appreciable ionization is observed only in strong shock waves. In this case Eqs. (19–1)—(19–3) can be simplified [71]. It follows from the first two equations that

$$p = p_1 + \varrho_1 V^2\left(1 - \frac{\varrho_1}{\varrho}\right). \tag{19–6}$$

In the relaxation zone behind the front of a strong shock wave, the value of $1-(\varrho_1/\varrho)$ varies from $1-(\varrho_1/\varrho')$ to $1-(\varrho_1/\varrho_2)$, where ϱ' is the flow density directly behind the shock front and ϱ_2 is the density at the end of the relaxation zone. For a monatomic gas $(\varrho_1/\varrho')=\frac{1}{4}$, in addition, the relation

$$\frac{\varrho_1}{\varrho_2} < 1$$

is always valid. Hence,

$$1 > 1 - \frac{\varrho_1}{\varrho} \geqslant \frac{3}{4}. \tag{19–7}$$

Thus the pressure in the relaxation zone cannot rise by more than 25%. The behavior of the enthalpy in the relaxation zone is the same. It follows from Eqs. (19–1) and (19–3) that

$$h = h_1 + \frac{V^2}{2}\left(1 - \frac{\varrho_1^2}{\varrho^2}\right). \tag{19–8}$$

Since $(\varrho_1/\varrho)^2 > \frac{1}{16}$, the enthalpy increase in the relaxation zone in any case does not exceed 5–6%. This means that ionization behind the front of a strong shock wave takes place practically at constant pressure and enthalpy.

When the degree of ionization is small ($\alpha \ll 1$), it is possible to make future simplifications, writing Eqs. (19–4) and (19–5) in the form

$$p = \frac{\varrho}{m_A} kT_A, \tag{19–9}$$

$$m_A h = \tfrac{5}{2}kT_A + \alpha k T_{\text{ion}}. \tag{19–10}$$

From these equations, also using the condition that the enthalpy and pressure are constant, it is possible to express the temperature, velocity and density as a function of the initial conditions and of α:

$$\left.\begin{aligned} T_A &= T_{A0} - \tfrac{2}{5}\alpha T_{\text{ion}}, \\ \frac{V}{v} &= \frac{\varrho}{\varrho_1} = 4\left(1 - \tfrac{2}{5}\alpha\frac{T_{\text{ion}}}{T_{A0}}\right)^{-1}. \end{aligned}\right\} \tag{19–11}$$

Here

$$T_{A0} = \frac{3}{16}\frac{m_A V^2}{k}$$

is the atomic temperature directly behind the shock front, where $\alpha=0$.

Eqs. (19–1)—(19–5) or (19–8)—(19–11) can be used to obtain the flow parameters of the ionized gas in the nonequilibrium zone behind the shock, if the degree of ionization in the shock wave is known as a function of the distance. This means that the above equations must be supplemented by still another equation which would describe the kinetics of the ionization process.

From among all the relaxation processes which are considered in this chapter, the process of establishing equilibrium ionization has been the least studied. The available experimental data [474, 430, 462] (see also, Chapter 3) suffice for determining the time of establishing equilibrium ionization behind the shock front, however, the physical processes which are responsible for establishing the equilibrium ionization are by far not clear. The main elementary acts, which result in ionization in shock waves, *i.e.*, photoionization, ionization on electronic and atomic collisions, are known. However, the role of these processes in the thermal ionization mechanism, with the exception of the case of electron-atom collisions, has not been finally clarified.

The electron-atom collisions are most effective with respect to ionization [159, 410, 474]. By virtue of this, the ionization behind a shock front can be broken up arbitrarily into two stages. In the first stage, a result of atom-atom collisions, photoionization or some other processes, the needed initial amount of electrons is produced. At the second stage, the electron-atom collisions predominate in the ionization process. In this latter case the equation of ionization kinetics is quite easy to write [474] if it is considered that the electrons have a Maxwellian distribution (with a temperature which is generally different from that of atoms and ions).

In fact, when the electron concentration is sufficiently high (for argon, according to estimates of [474], for degrees of ionization higher than 10^{-8}) the energy transfer in electron-electron collisions will take place at a higher rate than in collisions with heavy components (atoms and ions). The latter is due to the great difference in the electron and atom masses. Thus, for example, in order for an electron to acquire its energy in argon, it must undergo about 10^5 collisions with atoms, while only several collisions with other electrons are necessary for establishing the Maxwellian distribution in the first approximation. Hence, the Maxwellian distribution and, consequently, also the electron temperature will exist in the first approximation if not less than one electron-electron collision will take place per each 10^5 electron-atom collisions. Taking into account the difference in the effective cross sections for these collisions, the degree of ionization here should be greater than 10^{-8} [159, 474]. Strictly speaking, the existence of negative and positive sources which in this case are due to the disappearance of rapid and birth of slow electrons following ionization, can result in the fact that the quasi-steady electronic distribution will differ perceptibly from the Maxwellian distribution. However, in our case this is inconsequential. The negative source, as was shown in [477], can result in disturbing the Maxwellian distribution in the case when it acts in the energy region where $E > 3kT$. In the problems being considered, the value of E which is equal to the ionization potential is appreciably greater than kT. On the other hand, the birth of slow electrons following ionization, although it can produce a local departure from the Maxwellian distribution in the region of small energies (which departure will be insignificant, since exp $(-T_{\text{ion}}/T) \ll 1$), will not be felt at the tail of the Maxwellian distribution which determines the ionization rate.

Thus, in the first approximation it is possible to refer to an equilibrium distribution of electronic velocities with a temperature which generally will be different from the temperature of the heavy components [474, 477]. The fact that the electron distribution is Maxwellian appreciably simplifies the calculations. In fact, when the electron temperature is known it is easy to calculate the number of collisions which result in ionization in electron-atom collisions (it is assumed that the ionization cross section is known), and, consequently, it is also easy to calculate the ionization rate [260, 474]. The electronic temperature is determined from the equation of electronic energy balance in accordance with which the change in the thermal energy of electrons is equal to the external energy influx, which is denoted by $Q_{el} - Q_{inel}$ where Q_{el} is the energy transferred to the electrons per unit time in elastic collisions between atoms and ions, and Q_{inel} is the energy lost by the electrons per unit due to inelastic collisions.

Under the conditions being considered, the energy influx due to elastic collisions with heavy particles compensates to a large extent energy losses in inelastic collisions. Hence, a quasi-steady state in which the electronic distribution function at each time instant is determined by the density and the electronic density at the same time and does not explicitly depend on time, is established in the electronic gas. To determine the electron temperature in the first approximation, it is possible to disregard changes in the thermal energy of electrons. In this case the electronic energy balance equation will be written in the form

$$Q_{el} = Q_{inel}. \qquad (19\text{–}12)$$

The main contribution to Q_{el} is made by electron-ion collisions, since their effectiveness is appreciably greater than the effectiveness of the electron-atom collisions. Thus, for Ar with an electronic temperature of the order of 10,000 °K energy transfer between electrons and ions plays a certain role as early as for $\alpha \sim 10^{-3}$. Since the temperature of atoms and ions is the same (due to the fact that their masses are the same, the energy transfer between the two takes place very rapidly), the ions behave as an effective intermediary in transfering energy from atoms to electrons.

The solution of Eq. (19–12) for energy transfer between ions and electrons in Ar is presented in [474] and makes it possible to determine the electronic temperature as a function of the atom temperature and the degree of ionization. Fig. 112 shows the atomic temperature as a function of the electronic temperature for different degrees of ionization. The transverse curves (starting with numbers 10–40) are constant enthalpy lines and determine the electronic and atomic temperature, as well as the degree of ionization along the particle paths when the latter move in the nonequilibrium zone. It can be seen from Fig. 112 that the electronic temperature can differ appreciably from the atomic temperature. For example, when the gas temperature behind the shock (where ionization has not as yet occurred) is about 30,000 °K, the initial electronic temperature is only 10,000 °K and reaches a maximum of 16,000 °K when ionization takes place. The physical cause of this is obvious. The electronic gas loses energy relatively easily in inelastic collisions and acquires it very slowly in elastic collisions. In fact, in each ionization the electron gas loses energy which is equal to the ionization potential *I*.

In argon $I = 15.7$ eV. Hence at an electronic temperature of the order of 10,000 °K, the energy expended for the formation of each new electron is equal to the thermal energy of approximately ten electrons. The energy transfer between ions and electrons, which replenishes the energy lost for ionization, takes place very slowly due to the large difference in the ion and

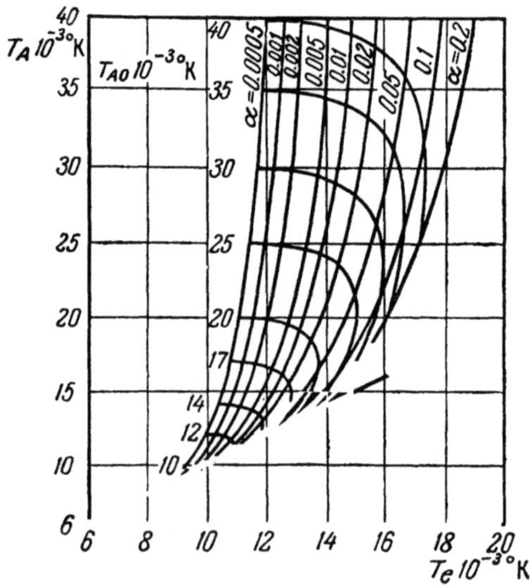

Fig. 112. The atomic temperature T_A (in the nonequilibrium zone behind a shock front in Ar) as a function of T_e for different degrees of ionization.

electron masses. Hence for small degrees of ionization, in the initial stage of development of the electron avalanche, the electronic temperature differs greatly from the ionic (atomic) temperature. As the ionic concentration is increased, the energy transfer between the ionic and electronic gases increases and the temperature difference is reduced. Thus, the limiting stage in the development of the electron avalanche, which determines the time of establishing equilibrium ionization is the energy transfer between the ions and electrons.

The ionization rate is calculated from the condition that the enthalpy is constant by dividing Q_{el} or Q_{inel} by the energy devolved upon one newly formed electron, and can be approximated (when $\alpha > 10^{-3}$) by

$$\frac{d\alpha}{dt} = n_A B(T_{A_0})\alpha^2 \ln\left(\frac{\alpha^0}{\alpha}\right), \tag{19–13}$$

where α^0 is the equilibrium degree of ionization, n_A is the atomic number density, and B is a function of enthalpy and density. In practice, B depends very little on the density.

In [474], the ionization rate is determined experimentally by using two different methods. In the first method the radiation intensity of the gas in the continuous spectral region was measured. This radiation is produced on recombination of electrons with ions and is proportional to the square of the electronic number density (see Sect. 13). Fig. 113 represents a graph

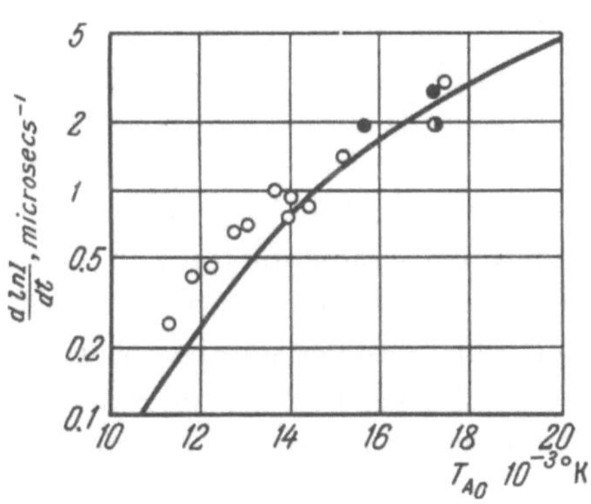

Fig. 113. The time derivative of the logarithm of the intensity (in that part of the non-equilibrium zone behind a shock front in Ar, where the intensity is about $^1/_4$ of the equilibrium intensity) as a function of the initial atomic temperature.

of the dependence of the derivative of the intensity logarithm with respect to time on the initial atomic temperature calculated by the shock-wave velocity. The derivative was measured in a point where the intensity comprised $\frac{1}{4}$ of the equilibrium intensity of light. In this point the degree of ionization was about half of its equilibrium value. As can be seen from the figure, the experimental points and the theoretical curve which was constructed on the basis of Eq. (19–13), are in good agreement with one another. The second group of experiments consisted in measuring the electric potentials in the gas. The results of these measurements give a poorer agreement with the theory than the data on the intensity of the continuous spectrum, but they also show that the rate of ionization by electrons was correctly calculated and that this process predominates in the state close to equilibrium.

It is thus possible to consider as established the fact that at the second stage, the ionization process takes place on electron-atom collisions. However, two questions can be raised: 1) what is the duration of the second stage in the total nonequilibrium process? and 2) what is the initial ionization level which is created at the first stage? In [474] these questions are resolved as follows. Eq. (19–13) is integrated from a region in which complete statistical equilibrium exists behind the shock front, to the shock front. This calculation shows that about 0.1 of the equilibrium degree of ionization should be produced by some ionization mechanism other than electron-atom collisions. In this method of calculations the main part of the nonequilibrium zone belongs to the region of electron-atom ionization. To clarify the initial stage of ionization, three possible mechanisms were considered:

1) ionization in atom-atom collisions;
2) photoionization; and
3) ionization due to admixtures.

As was pointed out in [474], not one of these processes can account for the existence of priming electrons at the early ionization stage. Due to this, attempts were made recently, to ascribe initial ionization to other processes. In particular, an attempt was made in [19, 94] to explain the initial ionization stage by diffusion of resonance radiation through the shock front; in [339], a mechanism of radiational excitation of atoms ahead of the shock front with subsequent ionization from the excited state is considered; in [594–596], a study was made of ionization of a gas by diffusing electrons, and [340, 341, 505] have considered the formation of electrons by photoeffect from the shock-tube walls. At present, it is quite difficult to determine which of these mechanisms predominates and under what conditions [410].

Specific calculations which take into account the effect of radiation on the establishing of ionization equilibrium behind a shock front have shown that under certain conditions, ionization by radiation can behave as the priming mechanism [93]. However, this process is critically dependent on the shock-wave strength. On the other hand, it was shown in [25] by direct calculations that consideration of the energy which is supplied to the relaxation zone by adsorbing the radiation of the equilibrium region (with frequencies corresponding to the wings of the absorption lines of the relaxation zone), results in a sharp reduction in the dimensions of the relaxation zone compared with the experimentally observed values. The primary ionization mechanism under these conditions is ionization of electronically excited atoms upon collision with an electron or atom. At the same time, there exists a group of experiments in which it is shown that under certain conditions the main mechanism of the early ionization stage are atom-atom ionization processes [395, 411].

However, it should be emphasized that this solution of the problem of the initial ionization level and of the duration of each stage of the ionization process is arbitrary. In fact, the theoretical calculations of the degree of ionization can be started not from the equilibrium zone far behind the shock front, as done in [474], but directly from the shock front by assuming that the first ionization stage is brought about by atom-atom collisions. In particular, this thermal ionization mechanism was considered by Weyman [593, 229], who postulated that the first ionization stage proceeds by atom-atom collisions. Here, the collisions first produce electronically excited atoms, which are ionized by subsequent collisions. When the degree of ionization becomes higher than a certain small value (which according to Bond is 10^{-3} [27, 260, 229, 261]), the electron-atom collisions (second stage) become the predominant process and subsequent ionization takes place very rapidly. Assuming that the experimental time of establishing ionization equilibrium is determined mainly by the time of the first stage, Weyman analyzed the available experimental data and obtained a value of the activation energy which is in good agreement with the ionization potential. Let us note in conjunction with this that direct measurement of the ionization rate of Xe which was performed in [395, 411, 412] shows that the activation time to the initial ionization stage corresponds to the energy of the first electronically excited states of Xe. This also provides experimental substantiation of the stepwise ionization mechanism of atom-atom collisions postulated by Weyman. Some critical remarks concerning the above theory are presented in [229].

Thus, according to existing concepts, thermal ionization in a monatomic gas takes place in two stages. Ionization by electron-atom collisions is decisive at the second stage. At present, the processes which take place at the initial ionization stage are not clear. At the same time it is entirely obvious that in analyzing these processes consideration must be given to: 1) admixtures (when the admixture level is sufficiently high, reactions with participation of the latter can become decisive); 2) radiant energy transfer; and 3) relaxation with participation of electronically excited atoms. Thermal ionization kinetics in gas mixtures is more complicated and can include the formation of intermediate, easily ionized components. For more details on this see Chapter 5.

[20] "NONEQUILIBRIUM" RADIATION OF GASES BEHIND THE FRONT OF STRONG SHOCK WAVES

A process which is very closely related to ionization and electronic excitation phenomena considered in the preceding section is the luminescence of the gas. It is known that the propagation of a strong shock wave in a gas is accompanied by intensive radiation; here the radiation from the nonequilibrium zone differs from the radiation of the gas which is at equilibrium. For convenience, we shall call the radiation from the nonequilibrium zone "nonequilibrium" radiation. The quotation marks are used in order to stress the inexactness of this terminology, since, in general, radiation even from an equilibrium zone may be thermodynamically nonequilibrium. All depends on the optical thickness of the equilibrium zone. The appearance of the "nonequilibrium" radiation reflects the existence of different relaxation processes behind the shock front and can be used for the study of these processes [474, 501, 503] (see Chapter 3). On the other hand, the radiation may affect the flow of the cold gas and the flow parameters in the relaxation zone and in the plug. The energy effect of the radiant flux on the shock-wave parameters becomes felt only at temperatures of the order of several tens of thousand degrees and above [71]. This temperature region is not considered in this book. However, even at lower temperatures, the radiation, without changing the parameters of the shock wave, can have a substantial effect on the size of the nonequilibrium zone and on the flow parameters in the plug (see Sects. 19 and 6, as well as [476]). From the viewpoint of experimental methods of studying the state of a gas in shock waves, these problems were considered in Chapter 3. On the whole, however, the problems relating to the radiation of the shock tube as a special radiation source are beyond the scope of this book and will not be considered here. In this section we wish to analyze only one typically nonequilibrium phenomenon—the so-called excess radiation effect. This effect was first observed in nitrogen heated by a strong shock wave [279] and it consists in the fact that the intensity of the radiation of the first negative system of bands of N_2^+ ($B^2\Sigma \rightarrow X^2\Sigma$) in the nonequilibrium zone exceeds appreciably the intensity of the same radiation in the equilibrium region. The appearance of excess radiation of N_2^+ $(1-)$, points to the existence of a nonequilibrium distribution of electronically excited N_2^+ molecules in the relaxation zone. Since the excess radiation effect can be quite appreciable, and apparently, typical not only of nitrogen, its consideration is of general interest.

A detailed experimental study of "nonequilibrium" radiation of the first negative system of bands of N_2^+ behind a shock wave as a function of the temperature and pressure was performed in [506, 507, 508]. Reference [507] is a study of the time characteristics of the luminescence distribution of different bands of N_2^+ behind a shock front in nitrogen, and it also contains the determination of time needed for reaching the maximum excess intensity of "nonequilibrium" radiation as a function of the temperature, and of the intensity in the maximum as a function of the temperature and the initial density of N_2. The observations were performed in a Xe–N_2 mixture in a temperature range of 7500–9850°K with partial pressures of N_2 of the order of 0.01 mm Hg.

These observations served as a basis for the following mechanism of formation of the excess population of electronically excited N_2^+ molecules in the nonequilibrium zone [507] (the asterisk denotes the electronically excited atoms and molecules):

$$N_2 + M \rightarrow N + N + M; \tag{20–1}$$

$$N + M \rightarrow N^* + M; \tag{20–2}$$

$$N + N^* \rightleftarrows N_2^+ + e; \tag{20–3}$$

$$N_2^+ + M \rightarrow N_2^{+*} + M; \tag{20–4}$$

$$N_2^{+*} \rightarrow N_2^+ + h\nu(1-)\ \text{band}; \tag{20–5}$$

$$N_2^{+*} + N + M \rightarrow N_3^+ + M. \tag{20–6}$$

In reactions (20–1) and (20–4) the third particle M can be an electron. The sequence with which the bands appear in the nonequilibrium zone [band (0,1) appears before band (1,2)] shows that the electronically excited N_2^+ ions ($B^2\Sigma$) arise in the ground vibrational state. Had the N_2^{+*} ions been formed directly in the recombination of the N^* and N^+ atoms, then due to the fact that the vibrational relaxation time is finite, transitions from the higher vibrational levels which are accompanied by radiation would have been observed earlier than transitions from the lower levels. Since reactions (20–5) and (20–6) proceed without expenditure of activation energy, then the temperature dependence of the inequilibrium radiation intensity in the maximum should show an activation energy which is equivalent to electronic excitation of N_2^+ in accordance with reaction (20–4). This excitation energy is well known and is equal to 3.2 eV. The experimental temperature dependence of the nonequilibrium radiation intensity in the maximum corresponds to an activation energy of 3.5 eV. Reaction (20–6) has been introduced to obtain agreement with data of mass spectrometer studies [519].

An attempt to calculate the excess radiation effect theoretically was made in [135], where the following mechanism of formation of excess population of N_2^+ ($B^2\Sigma$) in the nonequilibrium zone was considered:

$$N_2 + M \rightleftarrows 2N + M; \tag{20–7}$$

$$N + N \rightarrow N_2^+ + e; \tag{20–8}$$

$$N_2^+ + N \rightleftarrows N^+ + N_2; \tag{20–9}$$

$$N_2^+ + N_2^k \to N_2^{+*} + N_2; \tag{20–10}$$

$$\left.\begin{aligned} N_2^{+*} &\to N_2^+ + h\nu(1-)\ \text{band}; \\ N_2^{+*} + N_2 &\to N_2^+ + N_2^*. \end{aligned}\right\} \tag{20–11}$$

In the reaction (20–10) N_2^k denotes a vibrationally excited molecule of N_2.

Solution of equations expressing the conservation laws across the shock front together with the kinetic reaction equations (20–7)—(20–11) determines the concentration of N_2^{+*} (and, consequently, also the radiation intensity of N_2^+) in the nonequilibrium zone. In solving the problem the thermal effect of reactions (20–8)—(20–11) was neglected. This has made it possible to break up the problem of determining the nonequilibrium concentration of N_2^{+*} into two simpler ones. In the first of them, the temperature and concentration distributions of N_2 atoms and molecules were determined in the nonequilibrium zone behind a shock front neglecting the ionization and recombination processes. This problem was solved on a computer on the assumption that the dissociation rate constant has the form (see subsection 3 of Sect. 18)

$$k_d \sim \frac{\sum_{v=0}^{N-1} x_v e^{-\frac{D-E_v}{kT}}}{\sum_{v=0}^{N-1} x_v}, \tag{20–12}$$

where x_v is the fraction of molecules at the vth vibrational level, N is the total number of vibrational states, E_v is the energy of the vth vibrational level, and T is the translational temperature. It was also assumed that x_v is a Boltzmann distribution function with a temperature which differed from the translational temperature; here the vibrational energy relaxation was described by (see Sect. 2)

$$\frac{\mathrm{d}E}{\mathrm{d}t} = \frac{E_0 - E}{\tau_{\text{vib}}}.$$

The Landau-Teller formula was assumed for τ_{vib}. The results obtained by solving the first problem were used to determine the nonequilibrium concentration of N_2^{+*} (the second problem). The upper boundary of the nonequilibrium concentration of N_2^{+*} can be found by assuming that the electronically excited ions of N_2^{+*} are in equilibrium with N_2^+ ions in the ground state. In this case

$$[N_2^{+*}] = [N_2^+] e^{-\frac{3.2\text{eV}}{kT}}.$$

The concentration of $[N_2^+]$ ions is determined by considering the reactions (20–8) and (20–9); here experimental data were used for the reaction rate constants. The nonequilibrium concentration of N_2^{+*} can be found more accurately by assuming that ionization of N_2^+ ions is achieved on interaction of the latter with the vibrationally excited N_2^k molecules [see reaction (20–10)], and the deactivation process follows reactions (20–11). The kinetic equation for the concentration of $[N_2^{+*}]$ was written in the form

$$\frac{d[N_2^{+*}]}{dt} = k_1[N_2^+][N_2^k] - k_2[N_2^{+*}][N_2] - \frac{[N_2^{+*}]}{\tau^*}, \qquad (20\text{–}13)$$

where τ^* is the lifetime of the excited state. It was assumed in solving Eq. (20–13) that

$$[N_2^k] = [N_2]e^{-\frac{T_i}{T}},$$

where $[N_2^k]$ is the concentration of N_2 molecules which have sufficient vibrational energy for electronic excitation of N_2^+, and kT_i is the energy of electronic excitation of N_2^+ (3.2 eV). For reaction rate constants in Eq. (20–13) use was made of experimental data. The results of calculating N_2^{+*} predict the formation of an excess radiation peak in the nonequilibrium zone; here the intensity and position of the radiation maximum is found to be very sensitive to the choice of k_d. For k_d use was made of Eq. (20–12), according to which dissociation occurs as a result of transition of molecules from any vibrational levels to the continuous spectrum; here it is assumed that the dissociation does not disturb the existing molecular distribution with respect to vibrational levels. In light of the presentation given in subsection 3, Sect. 18, this assumption requires additional substantiation. In addition, the mechanism of formation of $N_2^+(B^2\Sigma)$ [reactions (20–7)—(20–11)] differs from that postulated in [507] (reactions (20–1)—(20–6)). Thus the results of [135] can be used only for a qualitative explanation of the excess radiation effect; in the sense they are of undoubted interest.

In conclusion we wish to note the following. The process of electronic excitation and ionization even in the elementary case of a monatomic gas (see Sect. 19) is far from being clear. Analysis of this process in the case of a diatomic gas involves (as was shown in this section) even greater difficulties, which arise on transition to more complex systems [356].

5

Nonequilibrium Phenomena in Shock Waves in Air

[21] HIGH-TEMPERATURE THERMODYNAMIC AND OPTICAL PROPERTIES OF AIR

The study of nonequilibrium phenomena in air is of interest by itself due to the importance of these processes in such practical problems as supersonic flow past bodies in the Earth's atmosphere, strong explosions, etc. The initial composition of air at room temperature is quite simple; its main components are molecular nitrogen and oxygen.* The composition of air behind the front of a strong shock wave is quite complicated. An increase in the temperature results in exciting internal degrees of molecular freedom, dissociation and ionization take place, nitrogen oxides are formed, etc. Before we consider nonequilibrium phenomena in a shock wave in air it is expedient to describe briefly the thermodynamic and optical properties of heated air which is at complete statistical equilibrium.

At present, experimental methods for direct study of thermodynamic properties of gases at high temperatures are by far not sufficiently developed. Hence the theoretical approach acquires great importance. Methods of statistical thermodynamics in their quantum-mechanical form, in addition to spectroscopic data, serve as a reliable basis for theoretical calculation of thermodynamic properties of gas systems and, in particular, of gas mixtures which can react chemically with one another. Air at high temperatures is an example of such a mixture. The thermodynamic properties of air in a wide temperature (from 200° to 20,000 °K) and pressure (from 0.00001 to 1000 atm) range were sufficiently and completely studied, for example, in [160–165, 195, 198, 551]. It is precisely this region of variations in physical conditions which is of interest for problems considered in this book. In certain temperature and pressure regions, the known spectoscopic data are insufficient for de-

* The composition (by volume) of dry air at sea level is: N_2 – 78.09%; O_2 – 20.95%; Ar – 0.93%; CO_2 – 0.03%; Ne – 1.8×10^{-3}%; He – 5.2×10^{-4}%; CH_4 – 1.5×10^{-4}%; Kr – 1×10^{-4}%; N_2O – 5×10^{-5}%; H_2 – 5×10^{-5}%; O_3 – 4×10^{-5}%; Xe – 8×10^{-6}%.

termining the thermodynamic variables. In this case resort must be had to additional theoretical estimates of the possible energy levels of molecules and ions, which are sometimes based on empirically determined relationships.

Under conditions which make it possible to disregard intermolecular interactions (a mixture of reacting ideal gases), the knowing of the energy levels makes it possible to determine thermodynamic properties with a high degree of accuracy which exceeds the practical needs. Departures from ideal behavior are considered only within the framework of certain approximations. We can point out three causes for departure from ideal behavior: 1) interaction between neutral atoms and molecules in ground or close to

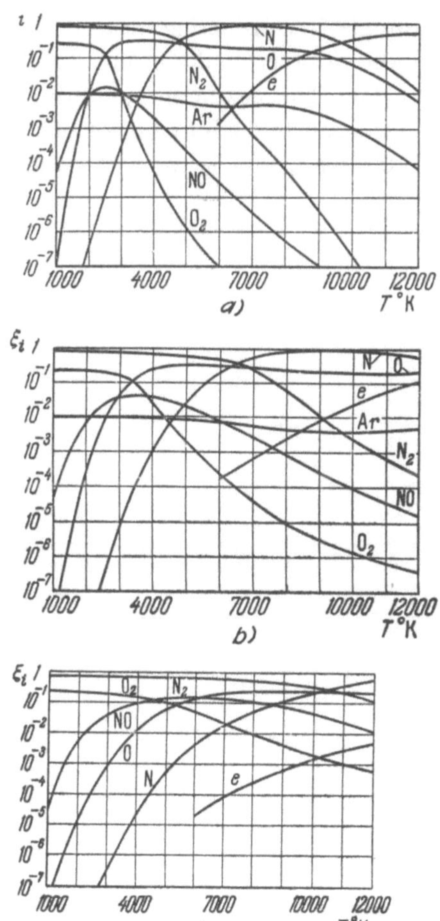

Fig. 114. The mole fractions of the molecular and atomic components of air and of electrons as a function of temperature in the thermodynamic equilibrium state at pressures 0.001 atm (a), 1 atm (b) and 1000 atm (c).

ground excited states (for example, Van der Waals-type forces); 2) peculiar departure from ideal behavior which is due to the fact that the highly excited states of atoms and molecules which have relatively large linear dimensions are not achieved, in practice, due to the exciting effect of the surrounding particles; 3) Coulomb interactions in the ionized gases. The Coulomb interactions are usually considered in the Debye-Hückel approximation.

The results obtained from these calculations are illustrated by the graphs shown in Fig. 114.

It was assumed in making these calculations that the contribution to the thermodynamics of hot air made by all nitrogen oxides, with the exception of NO, is negligible. This assertion is confirmed by direct estimates. However, in certain cases (for example, in considering the optical properties of heated air), the consideration also of other oxides of nitrogen, and in particular of NO_2, may become important. The maximum equilibrium ratio of the number density of NO_2 molecules in air to the initial molecular number density in a temperature range of $2000° \leqslant T \leqslant 5000\,°K$ and with standard density does not

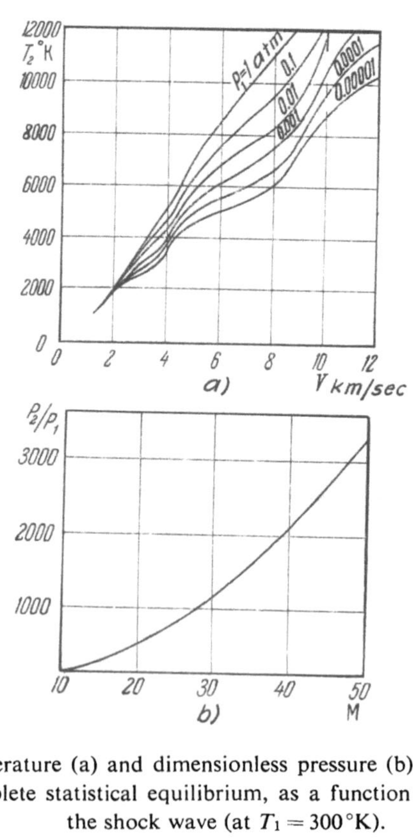

Fig. 115. The temperature (a) and dimensionless pressure (b) behind a shock front in air which is in complete statistical equilibrium, as a function of the Mach number of the shock wave (at $T_1 = 300\,°K$).

exceed 8×10^{-5} [71]. Let us note that the concentration of NO_2 in air decreases with a reduction in density.

Knowing the thermodynamic properties and the composition of air as a function of the temperature and pressure it is possible to resolve the problem of the equilibrium state of air behind a shock front. Such a problem (without taking into account radiant heat transfer) was solved, for example, in [162, 165, 172]. The solution was found from the equations of conservation of the mass, momentum and energy fluxes, and of the state of the gas ahead of and behind the shock front. The calculations were performed on a computer. For illustration, Fig. 115 shows the equilibrium values of temperature, density and pressure behind a shock front in air obtained in [165, 172]. A detailed description of results, which include other thermodynamic parameters, can be found in [162, 165, 163, 71].

When studying phenomena in shock waves propagating in air with very high velocities (more than 8–9 km/sec), and the gas temperature behind the front exceeds 10,000 °K, the radiant energy transfer must be considered. A number of works (for example [20, 71, 64]), contain a detailed analysis of these phenomena in shock waves under complete statistical equilibrium conditions. Biberman, *et al.* [20], have shown that the main processes which make the greatest contribution to the air emissivity at temperatures above 10,000 °K are photoionization of nitrogen and oxygen atoms, light absorption following the motion of electrons in ion fields (continuous spectrum) and light absorption in the spectral lines of nitrogen and oxygen atoms and ions (line spectrum) [Fig. 116]. The dependence of the absorption coefficient of

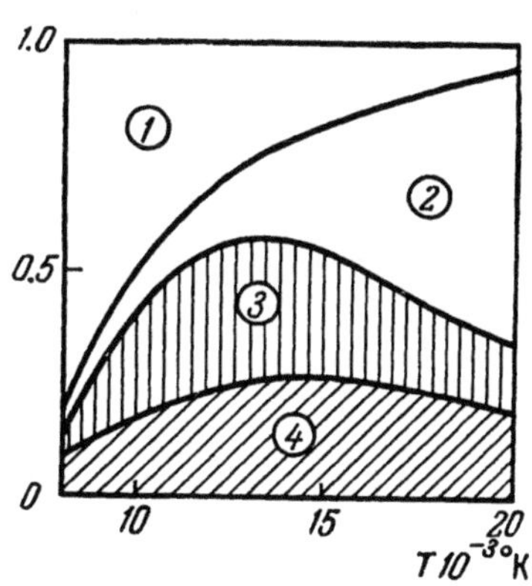

Fig. 116. Relative contribution of different processes to the radiant energy of an air layer 10 cm thick at atmospheric pressure. The unshaded region denotes the contribution of the continuous spectrum (1 is for $\nu' < 80{,}000$ cm^{-1} and 2 is for $\nu' > 80{,}000$ cm^{-1}), and the shaded region represents the contribution of spectral lines (3 of an ensemble of weak lines, considered together, 4 of several strong lines considered separately) [20].

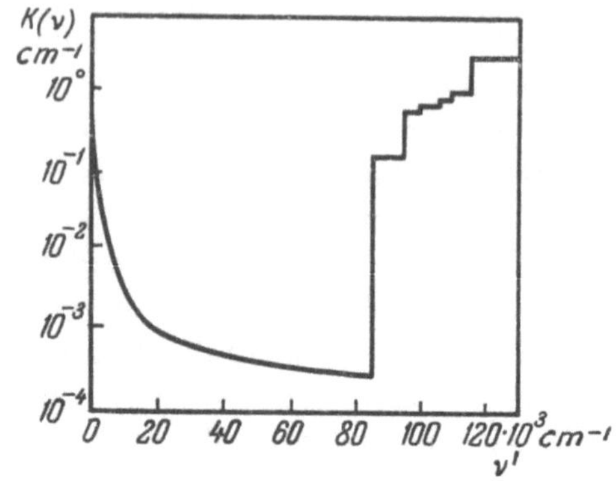

Fig. 117. The theoretical relationship between the absorption coefficient $k(\nu)$ for air at $T = 14{,}000\,°K$ and under atmospheric pressure and the wave number ν' [20].

air on the wave number in the continuous spectrum is illustrated by the graph shown in Fig. 117. At $T \sim 14{,}000\,°K$ for wave numbers up to $\sim 80{,}000$ cm^{-1}, the main contribution is made by photoionization of the upper excited atomic levels, and for moderate wave numbers it is made by free-free transitions in ion fields. The absorption coefficient of air undergoes a step increase in the region $\nu' \sim 88{,}000\ cm^{-1}$ which is the threshold frequency for phototransition from the ^2P state of the nitrogen atom. The practical significance of considering radiant heat transfer can be shown by comparing the thermal flux due to radiant and convective heat transfer in the lead part [region in the vicinity of the stagnation point; editor] of a body flying at a

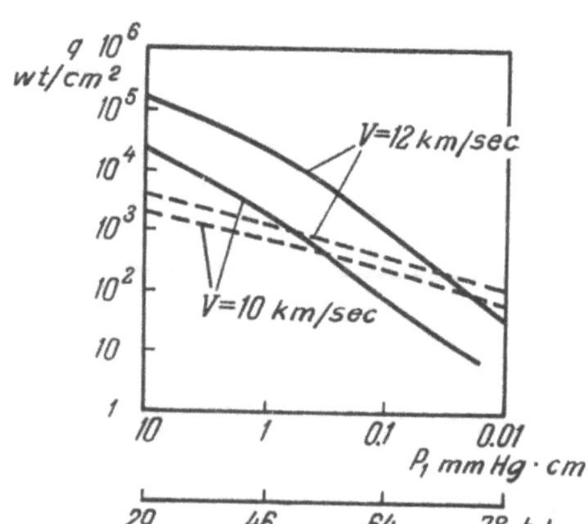

Fig. 118. The theoretical values of the radiant (solid lines) and convective (dashed lines) thermal fluxes in the stagnation region of a sphere with a radius of one meter moving with velocities of 10 and 12 km/sec when the shock wave is at a distance of 10 cm (according to Biberman, *et al.* [20]), as a function of the initial pressure p_1 or the flight altitude h.

high speed. In accordance with estimates [20], when a spherically shaped body with a radius of 100 cm performs a flight at moderate altitudes in the Earth's atmosphere with a speed of more than 8 km/sec and the shock front is situated 10 cm from the body [at the stagnation point], the radiant heating of the body can appreciably exceed the convection heating (Fig. 118).* Under these conditions the radiation which overtakes the shock wave can result in perceptible photoionization of oxygen molecules ahead of the shock front [24].

At temperatures below 8000–10,000 °K the radiant heat transfer is a small fraction of the total energy balance. The study of air radiation in this case is of interest mainly from the point of view of study of the physicochemical transformation kinetics in the gas behind the shock front. Here the main contribution to the emissivity of the air is again made by band spectra of molecules, and among them of nitrogen molecules (the first and second

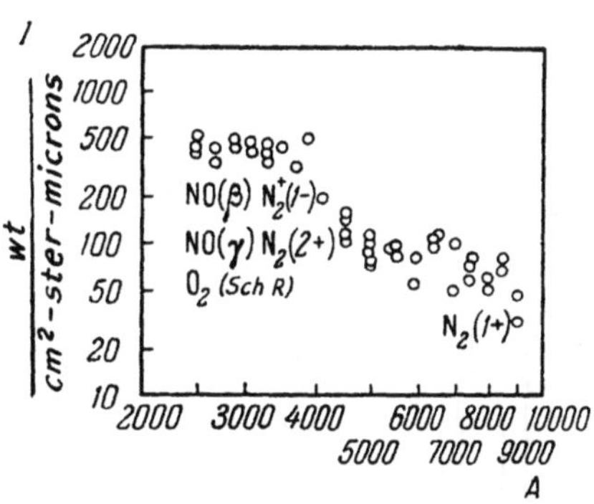

Fig. 119. The spectral intensity of the radiation of an air column about 1 cm high in a shock wave reflected from the end of a shock tube at $T = 8000\,°K$ and a density equal to 0.83 of standard, as a function of the wavelength. The figure shows the position of the main systems of the observed molecular bands.

positive band system), by ionized nitrogen N_2^+ (first negative system), NO (the β, γ, δ and ε systems), O_2 (the Schumann-Runge system). The proportional contribution of each of these bands is determined by the oscillator force of the corresponding transition and by the concentration of the corresponding molecules. The known probabilities of optical transitions for diatomic gases have been summarized in [192, 461, 185]. The further procedure for obtaining the contribution of bands to the total absorption coefficient (which is related by Kirchhoff's law to the emissivity) requires time-consuming calculation of the contour of the absorption bands which is de-

* At high altitudes importance may be acquired by the contribution of the radiation of the nonequilibrium flow behind the shock front (when $V < 9$ km/sec, see Sect. 24).

termined by the rotational and vibrational structure of the electronic transition. A number of methods for simplifying these calculations have been worked out; some information on the methods for calculating the emissivity of heated gases can be found, for example, in [158, 71]. It was calculated that some contribution to the radiation of air at temperatures up to 10,000°K is also made by the continuous spectrum which is due to the formation of a negative oxygen ion.

The radiation of heated air was experimentally studied primarily in shock tubes, using arrangements described in Sect. 12. From among the large number of experimental works on this topic, we wish to note [126, 401a, 280]. As an example, Fig. 119 shows the spectral characteristics of radiation of heated air obtained in [401a].

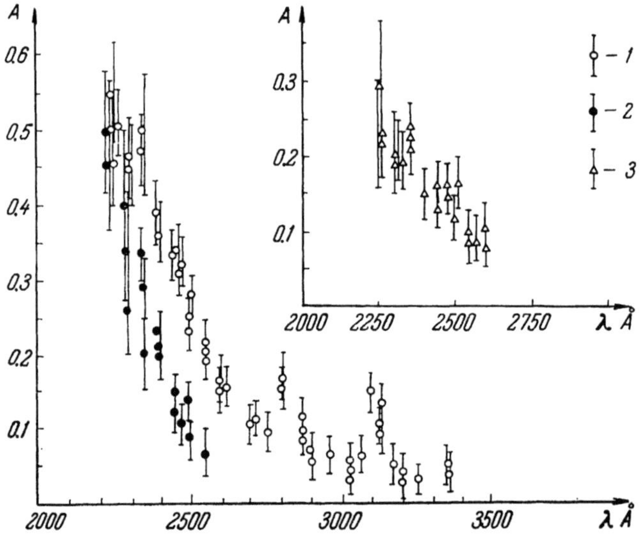

Fig. 120. Measured absorptivities of air (at a 5 cm length) in the ultraviolet spectral region as a function of the wavelength. 1) For $T \sim 2600$°K and $p \sim 6.5$ (± 0.5) atm, 2) for $T \sim 2200$°K and $p \sim 5.0$ (± 0.5) atm, 3) for $T \sim 3100$°K and $p \sim 1.2$ (± 0.2) atm.

Using the arrangement described in Sect. 11, it was possible, for example, to study the absorptivity of air under equilibrium conditions in the ultraviolet spectral region. These results are described in [112] and are represented in Fig. 120. Analysis and comparison of the results thus obtained with experimental data obtained in pure oxygen show that the main contribution to the absorption which is observed is made by the Schumann-Runge band system for O_2.

A more detailed discussion of problems which relate to the character of optical radiation in shock waves (particularly, in air) can be found in the aforementioned works, as well as in [71, 158].

As a whole, the problem of the composition and characteristics of air behind the shock front in the state of total statistical equilibrium at present can be regarded as basically solved, while only the first steps have been made in the study of processes of establishing the equilibrium.

[22] VIBRATIONAL RELAXATION

It is expedient to begin the study of nonequilibrium phenomena in a shock wave propagating in air with the excitation of the vibrational degrees of freedom of the air molecules. The most rapid relaxation processes which are due to the establishment of an equilibrium energy distribution with respect to the translational and rotational degrees of freedom, take place in air in the same manner as in pure gases, *i.e.*, during a time of the order of the time [of traverse] of a mean-free path. A perceptible increase in the relaxation time of the translational and rotational degrees of freedom can be obtained only in mixtures having main components of highly differing masses. Air, however, is not such a mixture. The situation is different in the study of slow relaxation processes, such as vibrational relaxation, dissociation and ionization. Transfer of vibrational photons, recharging and chemical reactions between the different air components can change appreciably the relaxation times of individual components in comparison with the case of a pure gas.

The simplest process of this type is vibrational relaxation of molecules in the air. To consider it we picture the problem in the following manner. The air will be regarded as a binary mixture of O_2 and N_2 gases undergoing relaxation. In addition, we assume that during the vibrational relaxation the energy distribution with respect to the translational and rotational degrees of freedom is an equilibrium one, and no dissociation or ionization take place.*

At moderate temperatures (for air not above 10,000 °K), the molecules can be approximated with a sufficient degree of accuracy by the harmonic oscillator model which makes it possible to consider only single-photon vibrational energy transitions. The relaxation equations which describe the change in the vibrational energy of components of a binary mixture of diatomic

* The last assumption is not sufficiently justified, since the vibrational relaxation time for nitrogen molecules is of the same order of magnitude as the dissociation time of oxygen molecules.

gases have been derived in Sect. 17 [see Eq. (17–96)] and have the form

$$\left.\begin{aligned}\frac{dE'_A}{dt} &= -\frac{1}{\tau_A}(E'_A - E^{0\prime}_A) + \frac{1}{\tau_{AB}} \times \\ &\quad \times [E'_B(E'_A + N_A)e^{\theta_B - \theta_A} - E'_A(E'_B + N_B)], \\ \frac{dE'_B}{dt} &= -\frac{1}{\tau_B}(E'_B - E'^{0}_B) + \frac{1}{\tau_{AB}} \times \\ &\quad \times [E'_A(E'_B + N_B) - E'_B(E'_A + N_A)e^{\theta_B - \theta_A}].\end{aligned}\right\} \tag{22–1}$$

Here E'_A is the total number density of the vibrational photons of oxygen or the dimensionless vibrational energy density ($E'_A = E_A/h\nu_A$, where E_A is the vibrational energy of the oxygen molecules referred to a unit volume), and E'_B is the same, but for nitrogen molecules. The quantities τ_A, τ_B, τ_{AB}, N_A, N_B have been defined in Sect. 17. The transition probabilities P^{AB}_{10}, P^{AA}_{10}, P^{BB}_{10}, Q^{AB}_{10} which are contained in τ_A, τ_B and τ_{AB} can be calculated from expressions obtained in [368]; thus, for the probability of vibrational deactivation of A molecules P^{AB}_{10} in A–B collisions we have

$$P^{AB}_{10} = \frac{a}{Z_0 Z_+ Z^{AB}_{vib} Z^{AB}_{trans}}, \tag{22–2}$$

where the multiplying factors in the denominator take into account the contribution to the probability made by the orientation of the colliding molecules (Z_0), by the attraction forces (Z_+), by molecular vibrations (Z^{AB}_{vib}), and by the translational motion (Z^{AB}_{trans}):

$$\left.\begin{aligned} Z^{AB}_{vib} &= \frac{\frac{1}{2}\left(1 + \frac{m_A}{m_B}\right)\vartheta'_A}{\pi^2 \theta'_A}, \\ Z^{AB}_{trans} &= \pi^2 \sqrt{\frac{3}{2\pi}}\left(\frac{T}{\vartheta'_A}\right)^{\frac{1}{6}} e^{\frac{3}{2}\left(\frac{\vartheta'_A}{T}\right)^{\frac{1}{3}} - \frac{\theta'_A}{2T}}, \\ \vartheta'_A &= \frac{16\pi^4 k(\theta'_A)\tilde{m}_A}{h^1 \alpha^2_{AB}}, \quad \theta'_A = \frac{h\nu_A}{k}. \\ Z_0 &= 3; \quad Z_+ = \exp\frac{4}{\pi}\sqrt{\chi\frac{\varepsilon}{kT}}. \end{aligned}\right\} \tag{22–3}^*$$

As shown in Sect. 17, Eqs. (22–2) and (22–3) probably convey the temperature dependence of the transition probabilities most accurately. However, the absolute values of experimental and theoretical probabilities can differ.

* At present, Z_0 and Z_+ have been determined only by the order of magnitude (see p. 239).

To compensate for this, Eq. (22–2) contains a correction factor a, which subsequently will be determined by comparison with experimental data.

Also, Q_{10}^{AB}, the probability of transfer of vibrational photons between A and B, is defined similarly:

$$Q_{10}^{AB} = \frac{a}{Z_0 Z_+ Z_{\text{vib}}^{AB} Z_{\text{vib}}^{BA} Z_{\text{trans}}^{AB}} \tag{22–4}$$

with the difference that the characteristic temperature contained in Z_{trans}^{AB} is replaced by the difference $\theta'_{\Delta AB} = \theta'_A - \theta'_B$ and, consequently, ϑ' is replaced by

$$\vartheta'_{\Delta AB} = \frac{16\pi^4 k \tilde{m}_{AB}}{h^2 \alpha_{AB}^2} (\theta'_A - \theta'_B)^2 . \tag{22–5}$$

In these expressions m_A and m_B are the molecular masses of A and B, $\tilde{m}_A$ and $\tilde{m}_{AB}$ are the reduced masses of the A molecule and of the colliding pair AB, α_{AB} is the effective radius of influence of repulsion forces in the exponential potential of interaction between atoms which are a part of different colliding molecules.

Without appreciably changing the results, it is possible to make some further simplifications. Thus, it may be assumed that the factors Z_0, Z_+ and a, as well as the gaskinetic parameters for nitrogen and oxygen, are the same.

Then, after all the substitutions, simplification and corresponding transformations, we obtain equations for the change in the dimensionless vibrational energy of oxygen and nitrogen per unit volume per unit time in the form

$$\left.\begin{aligned}
\frac{dE'_A}{dt} &= CT^{\frac{1}{3}} \left[(\bar{\varphi}_{AA} + \bar{\varphi}_{AB}) \frac{2\,\mathrm{sh}\dfrac{\theta'_A}{2T}}{\theta'_A} \left(\frac{N_A}{e^{\frac{\theta'_A}{T}} - 1} - E'_A \right) + \bar{\psi}_{AB} \right], \\
\frac{dE'_B}{dt} &= CT^{\frac{1}{3}} \left[(\bar{\varphi}_{BA} + \bar{\varphi}_{BB}) \frac{2\,\mathrm{sh}\dfrac{\theta'_B}{2T}}{\theta'_B} \left(\frac{N_B}{e^{\frac{\theta'_B}{T}} - 1} - E'_B \right) + \bar{\psi}_{BA} \right],
\end{aligned}\right\} \tag{22–6}$$

where

$$\left.\begin{aligned}
\bar{\varphi}_{ik} &= \frac{N_k}{\sqrt{\tilde{\mu}_{ik}}} \frac{(\vartheta'_{ik})^{\frac{7}{6}}}{e^{\frac{3}{2}\left(\frac{\vartheta'_{ik}}{T}\right)^{\frac{1}{3}}}}, \\
\bar{\psi}_{ik} &= \frac{\pi^2}{\sqrt{\tilde{\mu}_{ik}}\,(\theta'_{\Delta ik})^2} \frac{\theta'_i \theta'_k}{\vartheta'_{ik} \vartheta'_{ki}} (\vartheta'_{\Delta ik})^{\frac{13}{6}} \times \\
&\times \frac{\left[(E'_i + N_i) E'_k e^{-\frac{\theta'_{\Delta ik}}{2T}} - (E'_k + N_k) E'_i e^{\frac{\theta'_{\Delta ik}}{2T}} \right]}{e^{\frac{3}{2}\left(\frac{\vartheta'_{\Delta ik}}{T}\right)^{\frac{1}{3}}}}, \\
i, k &= A, B,
\end{aligned}\right\} \tag{22–7}$$

$$C = \frac{a}{Z_0 Z_+} \frac{4\pi r_c^2}{\sqrt{3}} \sqrt{\frac{k}{m_H}}, \tag{22-8}$$

where r_c is the average value of the coordinate of the classical turnabout point on collision, m_H is the mass of a hydrogen atom and $\tilde{\mu}_{ik} = \tilde{m}_{ik}/m_H$.

The numerical value of C should be taken so that the absolute values of the vibrational relaxation time of oxygen molecules in a gas consisting of O_2 only correspond to the experimentally obtained [41, 46] values of τ_{O_2} in the range 2000–7000 °K; in this case C is equal to 4.482×10^{-11} cm^3-sec^{-1}-degree$^{-\frac{1}{3}}$. This value of C ensures good agreement with τ_{vib} also in the second limiting case of pure nitrogen for $T = 1600$–5900 °K [446].

In a steady one-dimensional flow behind the front of a shock wave propagating at a constant velocity, it is more convenient to consider changes in the vibrational energy of a unit mass or of a mole of the fluid rather than of a unit volume. To simplify the expressions we change to mole fractions and introduce the quantities:

$$\xi_i = \frac{N_i}{N} \equiv \frac{N_i \mu}{\varrho \mathcal{N}}; \quad \beta_i = \frac{E_i' \mu}{\varrho \mathcal{N}}; \quad i = A, B \tag{22-9}$$

($\tilde{\mu}$ is the molecular weight of the mixture and $\mathcal{N}$ is the Avogadro number).

In this case β_i defines the number density of vibrational photons of the ith component, normed for the equivalent number density of particles.

Eqs. (22–6), when the change in density (specific volume) in the relaxation zone are considered, can be written in the form:

$$\left.\begin{aligned} \frac{d\beta_A}{dt} &= C \frac{p_1}{kT_1} T^{\frac{1}{3}} \frac{\varrho}{\varrho_1} \left[(\varphi_{AA} + \varphi_{AB}) \frac{2 \operatorname{sh} \frac{\theta_A'}{2T}}{\theta_A'} \times \right. \\ &\qquad \left. \times \left(\frac{\xi_A}{e^{\frac{\theta'_A}{T}} - 1} - \beta_A \right) + \psi_{AB} \right], \\ \frac{d\beta_B}{dt} &= C \frac{p_1}{kT_1} T^{\frac{1}{3}} \frac{\varrho}{\varrho_1} \left[(\varphi_{BA} + \varphi_{BB}) \frac{2 \operatorname{sh} \frac{\theta_B'}{2T}}{\theta_B'} \times \right. \\ &\qquad \left. \times \left(\frac{\xi_B}{e^{\frac{\theta'_B}{T}} - 1} - \beta_B \right) + \psi_{BA} \right], \end{aligned}\right\} \tag{22-10}$$

where $\varphi_{AB}, \ldots, \psi_{BA}$ are analogous to quantities contained in Eqs. (22–7) except that N_i was replaced by ξ_i and E_i' was replaced by β_i.

The values of T and ϱ contained in Eq. (22–10) are determined by relationships which follow from the laws of conservation of the mass, momentum and energy fluxes and from the equation of state:

$$\left.\begin{aligned} &\varrho v = \varrho_1 V; \quad p + \varrho v^2 = p_1 + \varrho_1 V^2; \\ &\frac{7}{2}\frac{RT}{\mu} + \frac{E_A}{\varrho} + \frac{E_B}{\varrho} + \frac{v^2}{2} = \frac{7}{2}\frac{RT_1}{\mu_1} + \frac{V^2}{2}; \\ &p = \frac{\varrho RT}{\mu}. \end{aligned}\right\} \tag{22-11}$$

For the propagation of schock waves in air, it is possible to obtain from Eq. (22–11) after simple transformations, an expression for T and ϱ/ϱ_1 in the

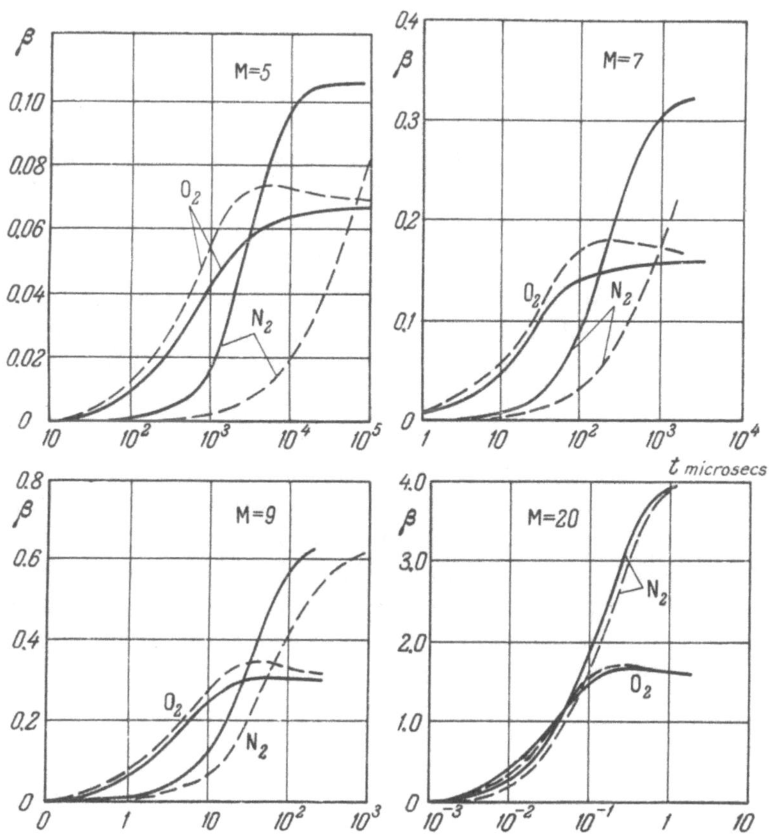

Fig. 121. The dependence of β for O_2 and N_2 on the time elapsed after the passing of a shock front in air taking into account (solid line) and not taking into account (dashed line) the transfer of vibrational photons; when the shock-wave velocities are given by M = 5, 7, 9 and 20, and the initial pressure is $p_1 = 1$ mm Hg.

explicit form, assuming that the gas A is oxygen and the gas B is nitrogen:

$$\left.\begin{aligned} T &= (97.67d_0 - 4.07c_0^2 + 4.07c_0J - 746.7\beta_A - 1118\beta_B)\,^\circ\mathrm{K}\,, \\ \frac{\varrho}{\varrho_1} &= \frac{14.2\mathsf{M}}{7c_0 - J}, \end{aligned}\right\} \qquad (22\text{–}12)$$

where

$$J = \sqrt{37c_0^2 - 48d_0 + 367\beta_A + 550\beta_B}\,,$$

$$c_0 = \frac{1.4\mathsf{M}^2 + 1}{1.183\mathsf{M}}; \quad d_0 = 0.35\mathsf{M}^2 + 3\,.$$

Here it was assumed that $\theta'_A = 2240\,^\circ$K and $\theta'_B = 3354\,^\circ$K.

Thus, Eqs. (22–10) together with (22–12) form a system of four equations with the four unknowns β_A, β_B, T and ϱ/ϱ_1. It should be assumed that initially, directly behind the shock front (at $t=0$) $\beta_A = \beta_B = 0$. Eqs. (22–10) were solved by the Runge-Kutta method using a computer. To show the role played by vibrational energy transfer between the components, the calculations were performed in two versions—with and without taking into account ($\psi_{ik}=0$) the effects of this transfer. The results obtained for shock-wave velocities corresponding to $\mathsf{M}=5$, 7, 9 and 20 are given in Figs. 121 and 122. It can be seen from Fig. 121 that the vibrational photon transfer has a different effect on the relaxation in nitrogen and oxygen. In the case of N_2 this transfer can greatly reduce the relaxation time. On the other hand, the transfer effects only insignificantly prolong the relaxation time of O_2. This result is easy to understand. The vibrational energy of O_2 molecules becomes excited under conditions when the N_2 molecules are practically not excited, for which reason there is no flow of vibrational energy from N_2 to O_2. The flow of vibrational energy in the opposite direction, *i.e.*, from O_2 to N_2, can result only in an insignificant increase in the vibrational relaxation time of O_2, since these two processes (vibrational relaxation of O_2 and N_2) are separated in time. In physical terms this means that the rate of excitation of molecules of O_2 due to transformation of the translational into the vibrational energy exceeds the rate of deactivation of O_2 molecules by transfer of vibrational energy from O_2 to N_2. On the other hand, the N_2 molecules are excited at a time when they are surrounded by the already excited O_2 molecules. In this case the rate of transfer of vibrational energy from O_2 to N_2 can appreciably exceed the rate of the direct excitation of N_2 molecules by transformation of the translational into the vibrational energy (see Fig. 122) and can result in an appreciable reduction in the relaxation time of N_2 in comparison with the case when transfer does not take place. As can be seen from Figs. 121 and 122, as the Mach number is increased, the role of the energy transfer is reduced. This is due to the fact

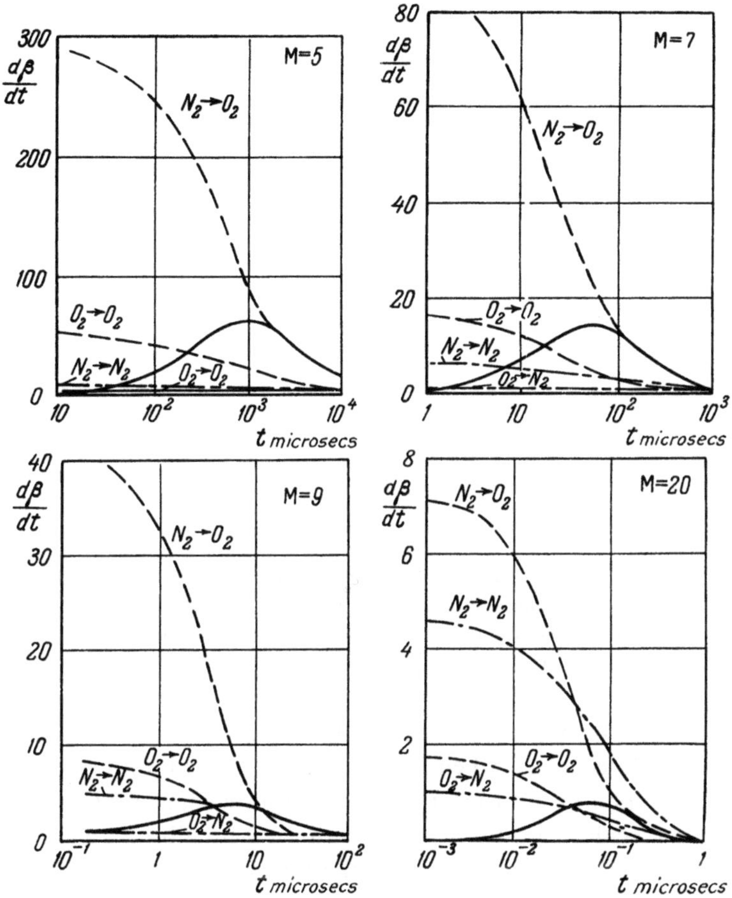

Fig. 122. Relative rate of excitation of vibrations in O_2 and N_2 molecules behind a shock front as a function of *t*. The dashed line pertains to transformation of the translational energy of N_2 and O_2 molecules into the vibrational energy of O_2 molecules; the dash-dot line is the same for excitation of N_2 molecules, while the solid line represents the rate of transfer of vibrational energy between O_2 and N_2.

that the probability of direct excitation of molecules by the energy of translational motion increases faster with an increase in the temperature than the transfer probability. At high temperatures, these probabilities differ very little and the transfer effects can be disregarded. It can be seen from the graphs presented in Fig. 121 that in the case of vibrational energy transfer, the equilibrium level of excitation of vibrations in nitrogen and oxygen is established almost simultaneously.

Consideration of the transfer of vibrational photons results in a faster reduction in the temperature and an increase in the gas density in the non-

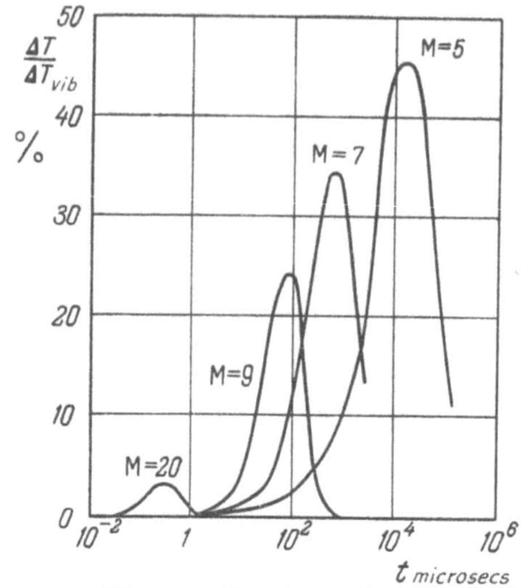

Fig. 123. The temperature difference ΔT calculated with and without considering the transfer of vibrational photons, referred to the total temperature change ΔT_{vib} behind a shock front due to vibrational excitation as a function of t. The values of T_{vib} are: 150 °K for M = 5; 400 °K for M = 7; 650 °K for M = 9; and 4550 °K for M = 20.

equilibrium zone behind the shock front. Fig. 123 shows the ratio of the temperature difference ΔT when the problem is solved with and without considering the photon transfer, to the temperature change ΔT_{vib} produced by exciting molecular vibrations in the entire nonequilibrim zone behind a shock front in air. It can be seen that the relative effect of the transfer is reduced with an increase in the Mach number of the shock wave, although the absolute value of ΔT can increase.

Thus, the effect of vibrational photon transfer in air is most appreciable for moderate Mach numbers. The above effect can be more appreciable when the effect of vibrational energy transfer from a mixture more easily excited than oxygen is considered. As an example, the latter can serve as the appreciable reduction in the vibrational relaxation time upon addition of certain gases to oxygen, which was pointed out in Sect. 7.

Since the vibrational relaxation time for nitrogen molecules in air at moderate temperatures is highly sensitive to the transfer of vibrational photons, then of the greatest interest for checking the solutions obtained are experimentally measured vibrational relaxation times of nitrogen in air (Fig. 124). The experimental results are in good agreement with calculations; the measured value of τ_{N_2} in air corresponds to the case when the transfer of vibrational energy between oxygen and nitrogen has been taken into account. Results of calculations without consideration of this transfer are

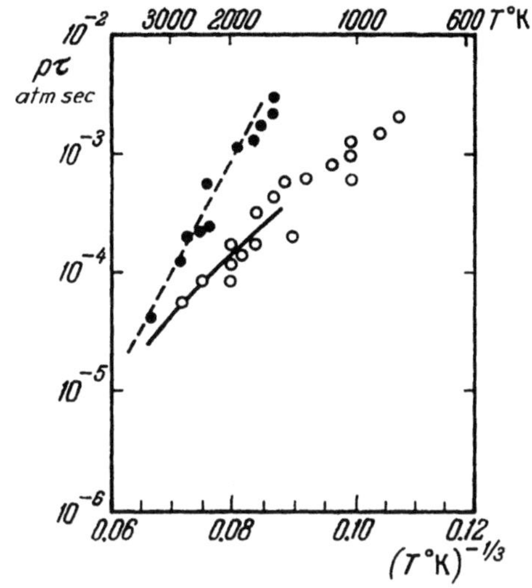

Fig. 124. Comparison of the theoretical and experimental values of the vibrational relaxation time τ_{N_2} of nitrogen molecules in air. The experimental values of τ_{N_2} are denoted by ○ in air and by ● in pure nitrogen [599a]. Calculated values of τ_{N_2} in air: the solid line was obtained when the vibrational energy transfer between oxygen and nitrogen was considered, while the dashed line is for the case where the transfer was not considered.

in agreement with experimental data on vibrational relaxation in pure nitrogen, when no such transfer takes place.

Thus, for correct description of vibrational relaxation in air it is necessary to take into account the transfer of vibrational energy between oxygen and nitrogen. Comparison with experimental data [599a] enables one to recommend for calculations of Eqs. (22–6) for the reduced vibrational energy E_A and E_B' in a unit volume. It may be assumed that the constants for $T \sim 1500$–$7000\,°$K are equal (oxygen is the A gas and nitrogen is the B gas):

$$\theta_A' = 2240; \quad \theta_B' \doteq 3354;$$
$$\theta_{\Delta AB}' = \theta_A - \theta_B; \quad \theta_{\Delta BA}' = -\,\theta_{\Delta AB}';$$
$$\sqrt{\tilde{\mu}_{AA}} = 4.0; \quad \sqrt{\tilde{\mu}_{AB}} = \sqrt{\tilde{\mu}_{BA}} = 3.742; \quad \sqrt{\tilde{\mu}_{BB}} = 3.86;$$
$$\vartheta_{AA}' = 3.76 \times 10^6; \quad \vartheta_{BA}' = 7.86 \times 10^6; \quad \vartheta_{\Delta BA}' = \vartheta_{\Delta BA}' = 8.66 \times 10^5;$$
$$\vartheta_{AB}' = 3.5 \times 10^6; \quad \vartheta_{BB}' = 7.37 \times 10^6; \quad C = 4.482 \times 10^{-11}.$$

Dissociation of O_2 and N_2 in air results in the formation of nitric oxide. The equations presented above do not take into account the vibrational relaxation of NO and the vibrational energy transfer between O_2, N_2 and NO. It is quite difficult to take NO into account due to the fact that its vibrational relaxation mechanism is very complicated (see Sect. 17). The exclusion of NO from consideration of the vibrational relaxation process in air will most probably not result in appreciable errors in calculations, since NO is not present initially in the air. It can be assumed that the NO molecules which are formed are in the excited state; in any case the vibrational relaxation time of NO is appreciably smaller than this time for O_2 and N_2. Thus, at $T=1000\,°K$ τ_{O_2} exceeds τ_{NO} by a factor of more than two hundred, at 2000 °K it exceeds it approximately by a factor of ten and at $T=5000\,°K$ it exceeds it by a factor of three.

[23] CHEMICAL REACTION KINETICS

In high-velocity shock waves, the processes taking place behind the shock front in air become much more complicated; the oxygen and nitrogen molecules undergo dissociation, nitrogen oxides are formed, and the gas components undergo electronic excitation and ionization. Here, the kinetics of dissociation of O_2 and N_2 molecules and the formation of nitrogen oxides are considered.

It is most natural to assume that dissociation of N_2 and O_2 molecules proceeds by direct disintegration into component atoms in collisions with any other molecule M. In considering the dissociation reaction of O_2, it would have been possible to take into account also the formation and disintegration of O_3. However, direct calculations of the contribution made by ozone to the dissociation kinetics of pure oxygen, on the basis of available data [419, 51, 247, 310] on the rates of reactions $O_2+O_2 \rightarrow O_3+O$ and $O_3+M \rightarrow O_2+O+M$, show that this possibility can be neglected. This also follows from experimental results [102]. When considering kinetics in air, only NO is usually taken into account from among all the oxides of nitrogen. This is due to the fact that the contribution of other oxides to the thermodynamic and composition of air at high temperatures is negligible. The study of kinetics of the formation and decomposition of nitric oxide started long ago; first, Jellinek [393] proposed a purely bimolecular mechanism of decomposition of nitric oxide. However, subsequent experiments showed this to be incorrect and a chain mechanism was assumed. This scheme for reactions in the air was suggested in [68, 70, 259] and was used extensively in the subsequent years [174, 586, 609, 400, 63, 336]. The concentration of argon (and even more so of other components) in air is so small that, in view of the fact that argon is inert it is possible to disregard its contribution to the kinetics. As a whole the chemical transformations behind a shock front in air can be described by the following ensemble of elementary reactions:

$$O_2 + M + 5.1\ \text{eV} \underset{k_{-1}}{\overset{k_1}{\rightleftarrows}} O + O + M\,, \qquad (23\text{–}1)$$

$$N_2 + M + 9.8\ \text{eV} \underset{k_{-2}}{\overset{k_2}{\rightleftarrows}} N + N + M\,, \qquad (23\text{–}2)$$

$$NO + M + 6.5\ \text{eV} \underset{k_{-3}}{\overset{k_3}{\rightleftarrows}} N + O + M\,, \qquad (23\text{–}3)$$

$$O + N_2 + 3.3\ \text{eV} \underset{k_{-4}}{\overset{k_4}{\rightleftarrows}} NO + N\,, \qquad (23\text{–}4)$$

$$N + O_2 - 1.4\ \text{eV} \underset{k_{-5}}{\overset{k_5}{\rightleftarrows}} NO + O\,, \qquad (23\text{–}5)$$

$$N_2 + O_2 + 1.9\ \text{eV} \underset{k_{-6}}{\overset{k_6}{\rightleftarrows}} NO + NO\,, \qquad (23\text{–}6)$$

where k_i and k_{-i} are the rate constants of the forward and reverse reactions ($i=1-6$), and M is one of any particles contained in the mixture.

The corresponding kinetic equations must be written on the basis of expressions usually employed in chemical reaction kinetics (see, for example [51, 87]). In the course of transformations to conform to conditions of a steady one-dimensional flow behind a shock front, different expressions can be written for the particle concentrations. Thus, in [318], the concentration is expressed in moles/cm^3; in [613] it is written as the ratio of the current number of moles to the initial number of moles; in [426] it is expressed in terms of γ_M which is the particle number density normed for the so-called equivalent molecular number density which is equal to $n_{eq}=(\varrho/\varrho_1)n_1$, where n_1 in the number density of air molecules ahead of the shock wave (see Table 4). One of the convenient representations for the concentration are the mole fractions $\xi_M=n_M/n$. In the case of complex systems with a multitude of reactions it is convenient to use as the variable characterizing the component concentration the quantity γ_M, which is equal to $\gamma_M=\xi_M(\mu_1/\mu)$, where $\mu=\Sigma\mu_M\xi_M$, and μ_1 is the initial molecular weight. Generalizing equations such as (9–23) and (9–22) for the case of air, we can represent the kinetic equations for reactions (23–1)—(23–6) for conditions behind a shock front in the following form:

$$\frac{d\gamma_{O_2}}{dt} = -\left(\sum_M k_1^M\gamma_M + k_5\gamma_N + k_6\gamma_{N_2}\right)\gamma_{O_2}\frac{\varrho n_1}{\varrho_1} + $$
$$+ (k_{-5}\gamma_{NO}\gamma_O + k_{-6}\gamma_{NO}^2)\frac{\varrho n_1}{\varrho_1} + $$
$$+ \sum_M k_{-1}^M\gamma_O^2\gamma_M\left(\frac{\varrho n_1}{\varrho_1}\right)^2, \qquad (23\text{–}7)$$

$$\frac{d\gamma_{N_2}}{dt} = -\left(\sum_M k_2^M \gamma_M + k_4\gamma_O + k_6\gamma_{O_2}\right)\gamma_{N_2}\frac{\varrho n_1}{\varrho_1} + $$
$$+ (k_{-4}\gamma_{NO}\gamma_N + k_{-6}\gamma_{NO}^2)\frac{\varrho n_1}{\varrho_1} +$$
$$+ \sum_M k_{-2}^M \gamma_N^2 \gamma_M \left(\frac{\varrho n_1}{\varrho_1}\right)^2 , \tag{23-8}$$

$$\frac{d\gamma_{NO}}{dt} = -\left(\sum_M k_3^M \gamma_M + k_{-4}\gamma_N + k_{-5}\gamma_O + k_{-6}\gamma_{NO}\right)\gamma_{NO}\frac{\varrho n_1}{\varrho_1} +$$
$$+ (k_4\gamma_O\gamma_{N_2} + k_5\gamma_N\gamma_{O_2} + 2k_6\gamma_{N_2}\gamma_{O_2})\frac{\varrho n_1}{\varrho_1} +$$
$$+ \sum k_{-3}^M \gamma_N\gamma_O\gamma_M \left(\frac{\varrho n_1}{\varrho_1}\right)^2 , \tag{23-9}$$

where k_i^M is the rate constant of the ith reaction on collision with particle M. The values of γ_O and γ_N can be determined by using the fact that the ratio of the number of nitrogen atoms to the number of oxygen atoms attendant to any chemical transformations in air is constant,

$$\frac{\gamma_N + 2\gamma_{N_2} + \gamma_{NO}}{\gamma_O + 2\gamma_{O_2} + \gamma_{NO}} = 3.72 , \tag{23-10}$$

and by using an expression which follows from Dalton's law and which relates the molecular weight of the reacting mixture with its initial value:

$$\sum_M \gamma_M \mu_M = \mu_1 \equiv 28.98 . \tag{23-11}$$

At the initial time

$$\gamma_{O_2} = 0.21 , \quad \gamma_{N_2} = 0.79 , \quad \gamma_O = \gamma_N = \gamma_{NO} = 0 . \tag{23-12}$$

Equations (23–7)—(23–9) should be supplemented by relationships which follow from the laws of conservation of mass, momentum and energy fluxes and by an equation of state which, in the given case, is most conveniently written in the form

$$\varrho v = \varrho_1 V , \tag{23-13}$$

$$p + \varrho v^2 = p_1 + \varrho_1 V^2 , \tag{23-14}$$

$$\frac{\Sigma H_M \gamma_M}{\mu_1} + \frac{v^2}{2} = \frac{H_1}{\mu_1} + \frac{V^2}{2} , \tag{23-15}$$

$$p = \frac{\varrho R T \Sigma \gamma_M}{\mu_1} . \tag{23-16}$$

To solve the above system, it is necessary to specify the reaction rate constants as a function of the gas temperature.

As follows from the discussion presented in Sect. 18, the present state of the thermal dissociation theory does not make it possible in the given case to obtain sufficiently accurate values of the reaction rate constants. Hence the main source of data on reaction rate constants are experimental results. Unfortunately, even experimental data cannot always provide the answers; the rates for a large number of reactions have not as yet been measured, and in those cases where measurements were performed, the experimental accuracy was not too high. Under these conditions only those constants can be regarded as reliable which were measured by different investigators in a sufficiently wide temperature range.

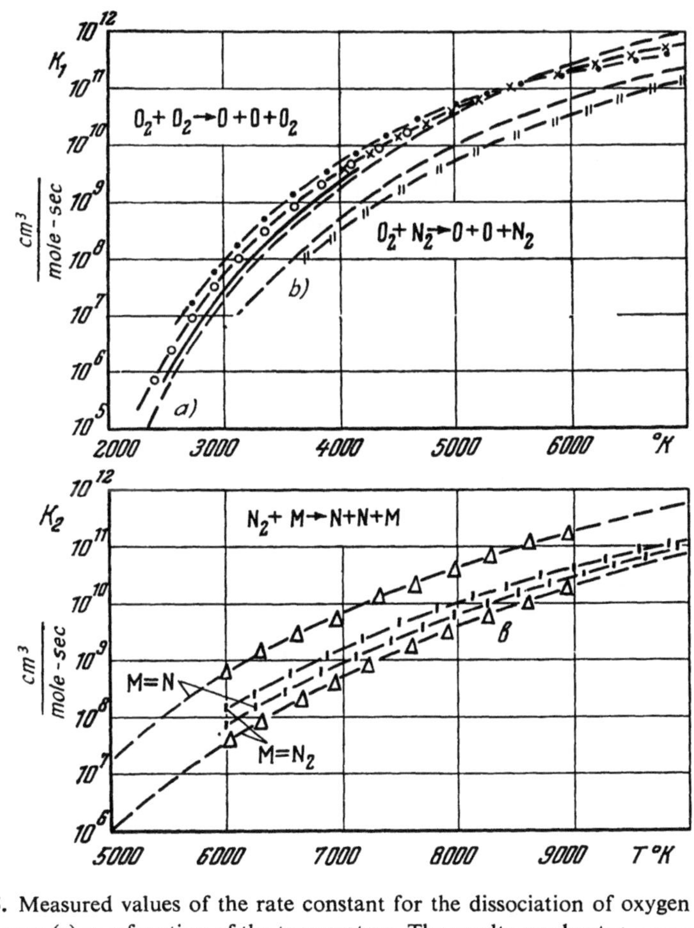

Fig. 125. Measured values of the rate constant for the dissociation of oxygen (*a* and *b*) and nitrogen (*c*) as a function of the temperature. The results are due to: — — — — Byron [275], ○-○-○-○ Matthews [437], ×-×-×-× Losev [105], -·-·-·- Generalov [43], ||-||-||-||- Generalov and Losev [49a], △-△-△-△- Byron [quoted from [613]), -|-|-|-|- Cary [282], - - - - are values recommended by Wray [613].

Let us consider each of reactions (23.1)—(23–6) separately.

1) Reaction $O_2+M \rightarrow O+O+M$ (dissociation energy $D=118$ kcal/mole. The dissociation of oxygen at high temperatures is the most extensively studied reaction. Measurements were performed with shock tubes using various methods (the interferometer method, method of absorption spectroscopy in the ultraviolet and x-ray spectral regions, etc. [43, 105, 275, 282, 437, 491]). In particular, dissociation in O_2-O_2 collisions was studied by Losev [102, 105], Matthews [437], Byron [245] and Generalov [43]. For comparison and in order to evaluate the accuracy, we have drawn the dissociation rate constants obtained by them as a function of the temperature in a graph (Fig. 125a). The same graph shows by a dashed line the values of $k_1^{O_2}$ recommended by Wray [613].* The temperature dependence of $(D/RT)^n$ was assumed by Wray to be identical for the O_2–O_2 and O_2–Ar collisions ($n=1.5$). However, experiments in pure oxygen show that in the O_2–O_2 collisions, the factor $(D/RT)^n$ is more highly temperature-dependent; the value of n obtained by various authors is two, three and even four. Cary [282] has attempted to repeat Byron's results (which are recommended by Wray). However, at $T \sim 3000\,°K$, he obtained double the value of the rate constant. Cary's results are apparently more exact, since he has used a tube with a larger cross section. Thus, it is more correct to use in calculations the expression

$$k_1^{O_2} = 5.2 \times 10^{10} \sqrt{T} \left(\frac{D}{RT}\right)^3 e^{-\frac{D}{RT}} \text{ cm}^3/\text{mole-sec}, \tag{23–17}$$

which was obtained experimentally by Matthews and Losev by different methods and under different conditions as a whole for a temperature range $T=2400$–$7000\,°K$. Generalov's experiments attest to the correctness of this recommendation.

A remarkable role in the dissociation of atmospheric oxygen can be played by O_2–N_2 collisions [44]. The value of $k_1^{N_2}$ was measured experimentally by Generalov and Losev [49a] in the range $T=3700$–$7000\,°K$. These results yield the expression

$$k_1^{N_2} = 2.5 \times 10^{11} \sqrt{T} \left(\frac{D}{RT}\right)^{1.5} e^{-\frac{D}{RT}} \text{ cm}^3/\text{mole-sec}. \tag{23–18}$$

The values which follow from this differ only slightly from Wray's recommendations, who has followed Byron in stating that $k_1^{N_2} \sim \frac{1}{4} k_1^{O_2}$.

It was discovered in experiments performed by Byron [275], Camac and Bogen [282] that the oxygen atom is highly effective in the oxygen dissoci-

* In certain calculations which were performed lately (see, for example [201]), the reaction rate constants used in the consideration of chemical processes in heated air are those recommended by Wray in 1961 [613].

ation process in the range $T=3300$–$5000\,°K$ [275] and 5000–7000 °K[282]. Wray [613] has combined these results by the relationship $k_1^{O}=25k_1^{Ar}$ where k_1^{Ar} is the rate constant for dissociation of oxygen on collision with argon atoms assumed by Wray. Thus,

$$k_1^{O} = 6.25 \times 10^{12} \sqrt{T} \left(\frac{D}{RT} \right)^{1.5} e^{-\frac{D}{RT}}\ \text{cm}^3/\text{mole-sec}. \qquad (23\text{–}19)$$

At present, no direct data are available on the effectiveness of N and NO in the oxygen dissociation process. The collision O_2+N results, with a probability practically close to unity, in the formation of NO and O, *i.e.*, results in reaction (23–5). Thus in considering the dissociation of oxygen into two atoms, it is possible to disregard the process $O_2+N \rightarrow O+O+N$. However, in the general scheme of reactions, it is necessary to consider the reverse reactions, *i.e.*, the recombination $O+O+N \rightarrow O_2+N$. The nature of the third particle here has a great influence on the rate of the process [87]. Wray assumes that the effectiveness of N (and NO), as the third particle in reaction (23–1), is equal to the effectiveness of argon. However, in the collision $O+O+N$, it is possible for complexes with relatively long lifetimes to form (unlike the $O+O+Ar$ collisions), which increases the reaction rate and disproves Wray's recommendations by equating the effectiveness of N to at least the effectiveness of O_2. It was discovered that the effectiveness of argon for the recombination $O+O+M \rightarrow O_2+M$ at room temperature is several times smaller than the effectiveness of O_2, N_2, N_2O, CO_2 and some other molecules. The effectiveness of NO in reaction (23.1) also apparently exceeds the effectiveness of argon which has a closed electronic shell. Hence it is more natural to set it equal to the effectiveness of O_2. It is obvious that these estimates require further refinement.

2) Reaction $N_2+M \rightarrow N+N+M$ (energy of dissociation $D=225$ kcal/mole). The dissociation of nitrogen has not been studied extensively. Wray [613] quotes Byron who has measured the dissociation rate constant for nitrogen in a mixture with argon at $T=6000$–$9000\,°K$. New results were obtained by Cary [282] who used a shock tube to record the gas density by an interferometer and found that his data differed substantially from those of Byron (Fig. 125c). Unlike Byron, however, Cary made control experiments in pure nitrogen which was not mixed with argon. The experimental conditions described in [284a] were such that Cary's results may be regarded as reliable. However, the problem of the value of the dissociation rate constant in nitrogen cannot be considered as resolved, since the experimental results obtained so far differ appreciably from one another.

* Similar results were obtained in the experimental study of recombination of oxygen, bromine and iodine atoms.

It is (also) difficult to solve the problem of the effectiveness of other particles in the dissociation of nitrogen. Wray assumes that the effectiveness of O_2, NO and O is equal to that of argon. However, it is known from experiments at moderate temperatures that molecular gases are more effective in the recombination of nitrogen atoms than the atoms of inert gases.* Apparently the oxygen atom with unsaturated electronic bounds also is more effective than the argon atom. Hence, it is possible to assume that the effectiveness of O_2, NO and O in the dissociation of N_2 is at least equal to the effectiveness of molecular nitrogen. Analytic expressions for the dissociation rate constant for nitrogen are presented in Table 19.

Table **19.** Summary of the Recommended Values of Reaction Rate Constants in Air

Reaction	M	k_i (cm^3/mole-sec)
(1) $O_2 + M \rightarrow O + O + M$,	O_2	$5.2 \times 10^{10} T^{1/2} (D/RT)^3 \exp(-D/RT)$
$D = 118$ kcal/mole	N_2	$2.5 \times 10^{11} T^{1/2} (D/RT)^{1.5} \exp(-D/RT)$
	O	$6.25 \times 10^{12} T^{1/2} (D/RT)^{1.5} \exp(-D/RT)$
	N, NO	$k_1{}^{O_2}$
	Ar	$4.2 \times 10^{11} T^{1/2} (D/RT) \exp(-D/RT)$
(2) $N_2 + M \rightarrow N + N + M$,	N_2	$4.25 \times 10^{11} T^{1/2} (D/RT)^{2.2} \exp(-D/RT)$
$D = 225$ kcal/mole	N	$1.85 \times 10^{12} T^{1/2} (D/RT)^{1.5} \exp(-D/RT)$
	O_2, NO, O	$k_2{}^{N}$
	Ar	$6.8 \times 10^{11} T^{1/2} (D/RT)^{1.5} \exp(-D/RT)$
(3) $NO + M \rightarrow N + O + M$,	O_2, N_2, Ar	$7.0 \times 10^{10} T^{1/2} (D/RT)^2 \exp(-D/RT)$
$D = 150$ kcal/mole	NO, O, N	$20\, k_3{}^{Ar}$
(4) $O + N_2 \rightarrow NO + N$	—	$7 \times 10^{13} \exp(-75.500/RT)$
(5) $N + O_2 \rightarrow NO + O$	—	$1.3 \times 10^{10}\, T \exp(-7100/RT)$
(6) $N_2 + O_2 \rightarrow NO + NO$	—	$9.1 \times 10^{24} T^{-5/2} \exp(-128{,}500/RT)$

3) Reaction $NO+M \rightarrow N+O+M$ (dissociation energy $D=150$ kcal/mole). The dissociation of nitric oxide in a shock wave was studied by Freedman and Daiber [324] at $T=3000$–$4300\,^\circ$K and by Wray and Teare [616] at $T=3000$–$8000\,^\circ$K by spectroscopic methods. Unfortunately, other reactions (for example, $N+NO \rightarrow N_2+O$) take place in the test gas in addition to reaction (23–3), which makes the analysis of results thus obtained difficult. Wray and Teare have performed experiments which various gas mixtures (air, mixtures of air and nitric oxide with argon) and have selected values of the dissociation rate constant for nitric oxide which would be in best agreement with experimental results. It was found that k_3^{NO} is greater by approximately a factor of 20 than k_3^{Ar}. The effectiveness of their molecules in the dissociation of nitric oxide is unknown. Since the content of nitric

oxide in air is not too high, the error produced by this indeterminacy is apparently not too great. Wray [613] assumes that $k_3^{O_2}=k_3^{N_2}=k_3^{Ar}$, $k_3^{O}=k_3^{N}=k_3^{NO}$. These values are presented in Table 19.

4) Reaction $O+N_2 \rightleftarrows NO+N$. It can be found from results obtained by Zel'dovich, *et al.* [70], that the activation energy for reaction (23–4) is equal to 68 ± 10 kcal/mole. Vetter [586] has discovered that the reverse process requires practically no activation energy. He has interpreted this result by the high affinity of NO and N in the formation of the transition complex N–O–N, which corresponds to state $^1\Sigma$ of the N_2O molecule which can dissociate into $O(^3P)+N_2/(^1\Sigma)$. Measurements of $k4$ in a shock tube performed by Glick, *et al.* [336], have given for $T=2000$–$3000\,°K$:

$$k_4 = 5 \times 10^{13} \exp(-75{,}500/RT) \text{ cm}^3\text{/mole-sec};$$

a more careful analysis of these results, which was performed by Duff and Davidson [318], has yielded values higher by 35%.

5) Reaction $N+O_2 \rightleftarrows NO+O$. It follows from [589] that the activation energy for the reverse reaction is close to 30 kcal/mole. Experiments performed by Wray and Teare [616] in a shock tube have yielded the value of k_{-5} at $T \sim 5000\,°K$ ($k_{-5}=3.4 \times 10^{11}$ cm^3/mole-sec). The constant for the forward reaction was measured at $T=394$–$516\,°K$

$$(k_5 = 2 \times 10^{12} \exp(-6200/RT) \text{ cm}^3\text{/mole-sec} \quad [404])$$

and at $T=412$–$755\,°K$

$$(k_5 = 8.3 \times 10^{12} \exp(-7100/RT) \text{ cm}^3\text{/mole-sec} \quad [299])$$

All these data can be combined by setting $k_5=1.3 \times 10^{10}\, T \exp(-7100/RT)$ cm^3/mole-sec.

6) Reaction $N_2+O_2 \rightleftarrows NO+NO$. According to measurements reported in [70], the activation energy for this reaction is about 130 kcal/mole. Subsequently, the reverse reaction was considered. Kaufman and Kelso [400] have found that at $T=1400$–$1530\,°K$,

$$k_{-6} = 2.6 \times 10^{12} \exp(-63{,}800/RT) \text{ cm}^3\text{/mole-sec}.$$

Freedman and Daiber [324] have studied this reaction at higher temperatures in a shock tube; their results are described by the expression

$$k_{-6} = 8.2 \times 10^{12} \exp(-57{,}100/RT) \text{ cm}^3\text{/mole-sec}$$

(at $T=3000$–$4300\,°K$). To obtain agreement between these data and the results of [400], Freedman and Daiber have proposed the relationship $k_{-6}=4.8 \times 10^{23}\, T^{-\frac{5}{2}} \exp(-85{,}500/RT)$ cm^3/mole-sec. This value does not contradict results obtained experimentally by Wray and Teare [616] and can be recommended for use in calculations.

A summary of the most probable expressions for reaction rate constants at high temperatures in air is given in Table 19.* These data pertain mainly to temperatures in a range of approximately 2000 to 6000–8000°K. It can be asserted that for this temperature range, the above reaction constants are known beforehand to be correct with an accuracy not worse than within the limits of an order of magnitude. The inaccuracy in the values of the constants can be estimated by the deviation between results obtained by different investigators. At present, the data on reactions (23–2) and (23–3) are least reliable; the rate constant for reaction (23–6) is not known with an accuracy higher than an order of magnitude. The greatest errors are possible at relatively moderate temperatures (1000–2000°K), when the rate constants of the main reactions are highly temperature dependent. As the temperature increases, this dependence is lessened. Extrapolation of data to very high temperatures (above 8000–10,000°K) can also result in appreciable errors due to the fact that vibrational equilibrium is not established toward the start of molecular dissociation. Fortunately, the absolute dimensions of the nonequilibrium zone in which dissociation takes place at these high temperatures are not great.

The values of the recombination rate constants are found from relationships of the type of $(k_i/k_{-i})=K$, where K is the equilibrium constant. The problem of satisfying this condition has not as yet been solved quantitatively (see Sect. 18); errors due to this may arise only in the end of the nonequilibrium zone, where the role of recombination is great. However, at the end of this zone the departures from equilibrium are insignificant.

The simultaneous solution of Eqs. (23–7)—(23–9) and (23–10)—(23–16) is obtained numerically by using a computer. We wish to note here results obtained in four works where computers were used to obtain this solution: (I) Losev [103a] in 1958; (II) Duff and Davidson [318] in 1959, (III) Wray [613] in 1961; and (IV) Lin and Teare [426] in 1963. The magnitudes of constants used in these solutions differed slightly from one another. This was because some data were not available when the earlier works were performed, and it was necessary to use only approximate expressions which were obtained from theoretical considerations. Thus, in solution (I) for rate constants of reactions (23–1)—(23–3) use was made of expressions obtained by Careri [283, 284] with a certain variated steric multiplying factor α (in the solution presented $\alpha=0.1$). For simplification, (I) and (II) did not take into account reaction (23–6), and it was also assumed that vibrational relaxation has ended toward the start of the disintegration. Among other simplifications which were assumed to reduce the computation time, it was assumed in (I) that the pressure behind the shock front remains constant, and that the

* For completness of presentation, Table 19 includes data on the constants k_1^{Ar} [615], k_2^{Ar} [282], k_3^{Ar} [616]. The values of k_1^{Ar} are valid in the range $1300°K < T < 18,000°K$.

change in the second term from the left in Eq. (23–15) can be disregarded in temperature calculations, assuming in this equation a certain average value for the flow velocity behind the shock front. Solutions (III) and (IV) did take into account all the reactions (23–1)—(23–6), and it was also assumed there that at the beginning, the vibrational degrees of freedom of nitrogen and oxygen molecules are not in equilibrium. The change in the vibrational energy of O_2 and N_2 was taken into account by using the expression

$$\frac{E_{\text{vib}} - E_{0_{\text{vib}}}}{(E_{\text{vib}})_{in} - E_{0_{\text{vib}}}} = e^{-\frac{t}{\tau_{\text{vib}}}}, \tag{23–20}$$

where E_{vib}, $(E_{\text{vib}})_{\text{in}}$ and $E_{0\text{vib}}$ are the current, initial and equilibrium values of the vibrational energy, respectively. Here the effect of dissociation on the vibrational energy and transfer of the latter between nitrogen and oxygen were not considered. In this scheme no differences are made between particles of the same chemical nature but which, however, have a different store of internal energy. On the other hand, solutions (III) and (IV) do take into account the effect of the delay in the excitation of molecular vibrations on the dissociation rate at high temperatures according to a scheme proposed by Hammerling, *et al.* [355] (see Sect. 18). All these differences between solutions (I)–(IV), however, do not lessen the interest in a comparison of results of calculations which were performed, since it is important to know the effect of some given factors (difference in the assumed rate constants, consideration of vibrational relaxation and of the effect of delay in exciting vibrations on the dissociation, etc.) on the results. Here it is natural to assume that solutions (III) and (IV) conform closer to actuality. The values of k in air which were assumed in the different solutions of the problem of molecular dissociation behind a shock front are presented in Table 20. Here the superscript of k points to the nature of the particle with which the collision which results in dissociation takes place.

Table **20.** Reaction Rate Constants in Air Assumed in Several Solutions of the Problem of Molecular Dissociation behind a Shock Front (cm³/mole-sec)

Solution	Reaction $O_2 + M \to 2O + M$, $D = 118$ kcal/mole
I	$k_1^{M} = \left(1.8 \times 10^{13}\sqrt{T} + \dfrac{3.6 \times 10^{17}}{\sqrt{T}}\right)\left(1 - e^{-\frac{2220}{T}}\right)e^{-\frac{D}{RT}}$,
II	$k_1^{M} = \dfrac{3.6 \times 10^{17}}{\sqrt{T}} e^{-\frac{D}{RT}}$,
III	$k_1^{\text{Ar,N,NO}} = 2.5 \times 10^{11}\sqrt{T}\left(\dfrac{D}{RT}\right)^{1.5} e^{-\frac{D}{RT}}$,
	$k_1^{N_2} = 2k_1^{\text{Ar}};\ k_1^{O_2} = 9k_1^{\text{Ar}};\ k_1^{O} = 25k_1^{\text{Ar}}$,

Table **20.** *Continued*

Solution	Reaction $N_2 + M \rightarrow 2N + M$, $D = 225$ kcal/mole
IV	$k_1^{O} = 2.6 \times 10^{23}\, T^{-2} e^{-\frac{D}{RT}}$,
	$k_1^{O_2} = 9.6 \times 10^{22}\, T^{-2} e^{-\frac{D}{RT}} = \frac{k_1^{O}}{3.64}$,
	$k_1^{N_2} = 7.5 \times 10^{18}\, T^{-1} e^{-\frac{D}{RT}}$,
	$k_1^{N,NO} = 3.6 \times 10^{18}\, T^{-1} e^{-\frac{D}{RT}}$.
I	$k_2^{M} = \left(2.1 \times 10^{13}\, T^{\frac{1}{2}} + 8 \times 10^{17}\, T^{-\frac{1}{2}}\right)\left(1 - e^{-\frac{3374}{T}}\right) e^{-\frac{D}{RT}}$,
II	$k_2^{M} = 5.4 \times 10^{15} e^{-\frac{D}{RT}}$,
III	$k_2^{Ar,O,O_2,NO} = 1.7 \times 10^{12}\, T^{\frac{1}{2}} \left(\frac{D}{RT}\right) e^{-\frac{D}{RT}}$,
	$k_2^{N_2} = 4.2 \times 10^{12}\, T^{\frac{1}{2}} \left(\frac{D}{RT}\right) e^{-\frac{D}{RT}}$,
	$k_2^{N} = 3.2 \times 10^{12}\, T^{\frac{1}{2}} \left(\frac{D}{RT}\right)^2 e^{-\frac{D}{RT}}$,
IV	$k_2^{O,O_2,NO} = 2 \times 10^{17}\, T^{-\frac{1}{2}} e^{-\frac{D}{RT}}$,
	$k_2^{N_2} = 5 \times 10^{17}\, T^{-\frac{1}{2}} e^{-\frac{D}{RT}}$,
	$k_2^{N} = 4.2 \times 10^{22}\, T^{-\frac{3}{2}} e^{-\frac{D}{RT}}$.
	Reaction $NO + M \rightarrow N + O + M$, $D = 150$ kcal/mole
I	$k_3^{M} = \left(3.6 \times 10^{13}\, T^{\frac{1}{2}} + 9 \times 10^{17}\, T^{-\frac{1}{2}}\right) \times \left(1 - e^{-\frac{2680}{T}}\right) e^{-\frac{D}{RT}}$,
II	$k_3^{M} = 2.4 \times 10^{15} e^{-\frac{D}{RT}}$,
III	$k_3^{Ar,O_2,N_2} = 7 \times 10^{10}\, T^{\frac{1}{2}} \left(\frac{D}{RT}\right)^2 e^{-\frac{D}{RT}}$,
	$k_3^{NO,N,O} = 20 k_3^{Ar}$,
IV	$k_3^{O,N,O_2,N_2} = 4 \times 10^{20}\, T^{-\frac{3}{2}} e^{-\frac{D}{RT}}$,
	$k_3^{NO} = 8 \times 10^{21}\, T^{-\frac{3}{2}} e^{-\frac{D}{RT}} = 20 k_3^{N_2}$.
	Reaction $O + N_2 \rightarrow NO + N$
I	$k_4 = 5 \times 10^{12}\, T^{\frac{1}{2}} e^{-\frac{78,000}{RT}}$,
II	$k_4 = 5 \times 10^{13} e^{-\frac{75,500}{RT}}$,

Table **20.** *Continued*

Solution	Reaction $N + O_2 \rightarrow NO + O$
III	$k_4 = 7 \times 10^{13} e^{-\frac{75{,}500}{RT}}$,
IV	$k_4 = 7.3 \times 10^{13} e^{-\frac{75{,}000}{RT}}$.
	Reaction $N + O_2 \rightarrow NO + O$
I	$k_5 = 5 \times 10^{12} T^{\frac{1}{2}} e^{-\frac{8000}{RT}}$,
II	$k_5 = 1 \times 10^{11} T^{\frac{1}{2}} e^{-\frac{6200}{RT}}$,
III	$k_5 = 1.3 \times 10^{10} T e^{-\frac{7080}{RT}}$,
IV	$k_5 = 1.3 \times 10^{10} T e^{-\frac{7080}{RT}}$.
	Reaction $N_2 + O_2 \rightarrow 2NO$
III	$k_6 = 9.1 \times 10^{24} T^{-\frac{5}{2}} e^{-\frac{128{,}500}{RT}}$,
IV	$k_6 = 8.6 \times 10^{13} T^{-1} e^{-\frac{148{,}300}{RT}}$.

Some results obtained in these solutions are presented in Figs. 126 and 127; all the data are for an initial pressure $p_1 = 1$ mm Hg, a temperature T_1 of about 300 °K and encompass practically almost the entire velocity range (from 3 to 7 km/sec), where in considering the vibration and dissociation effects it is possible to disregard the effect of radiant heat transfer and ionization. Analysis of solutions obtained in a wide range of other initial conditions shows that the linear dimensions of the nonequilibrium zones (and the scales on the abscissa axes in Figs. 126 and 127) practically, with the exception of regions close to equilibrium (where a large contribution is made by three-molecule recombination reactions), are inversely proportional to the pressure. It is natural that as the shock-wave velocity becomes higher, the dimensions of the nonequilibrium region are reduced and the temperature gradient across it increases. It is remarkable that in all these cases, as can be seen from solutions (III) and (IV), the nitrogen molecules are not in vibrational equilibrium almost in the entire nonequilibrium zone (curves of N_2 in Fig. 126). The vibrational relaxation of oxygen terminates comparatively rapidly (curves of O_2), practically simultaneously with the dissociation of the bulk of molecular oxygen. Consideration of departures from vibrational equilibrium (solutions III and IV), points to the possibility of values of the translational temperature in the nonequilibrium region higher than

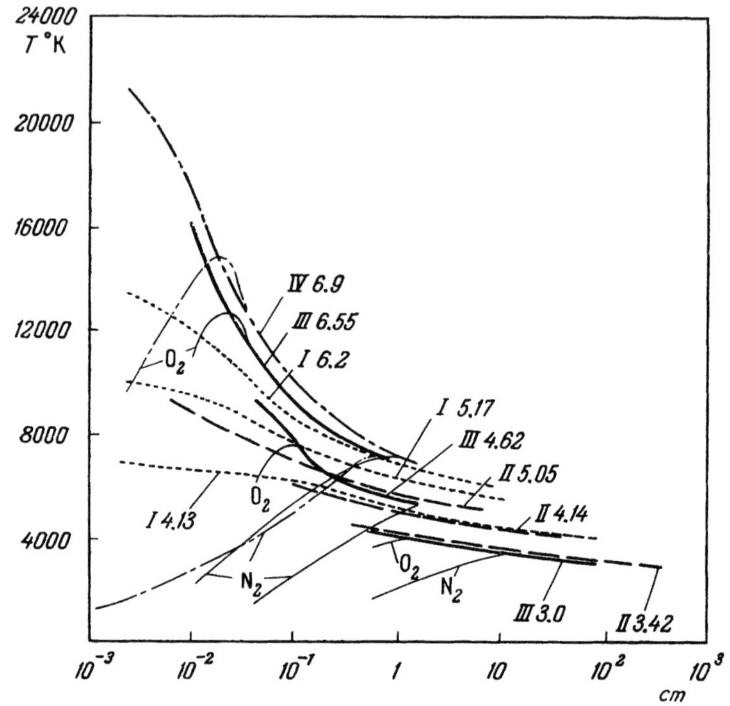

Fig. 126. Dependence of the temperature behind a shock front on the distance from the latter, obtained in solutions (I)–(IV) to the problem of molecular dissociation in air. I) Solution due to Losev, II) solution due to Duff and Davidson; III) solution due to Wray; IV) solution due to Lin and Teare; the arabic numerals denote the shock-wave velocities in km/sec. The thin lines denote changes in the vibrational temperatures of O_2 and N_2 in solutions (III) and (IV).

those which follow from solutions (I) and (II). In fact, at small distances from the forward front it is possible to have even higher temperatures, since the rotational relaxation time increases with an increase in the temperature and it can even be comparable with the time of excitation of vibrations of oxygen molecules. Thus, if it is assumed that the vibrational relaxation time for high T can comprise 100–1000 intercollision intervals, then the fact that vibrational relaxation has not been considered can introduce indeterminacy into results obtained at distances up to $10^{-2}-10^{-1}$ cm from the shock front.*

* In [426] the results of calculations are represented in the form of a function of x/l_1, where l_1 is the mean-free path ahead of the shock front. Since the ratio p_2/p_1 varies little in the nonequilibrium region, then this representation is suitable for illustration of results obtained under different initial pressures. However, for the results presented here, it makes more sense from the physical point of view (and it is more graphic) to make use of the ratio x/l_2, where l_2 is the mean-free path behind the shock front. For the solutions presented here, with $p_1 = 1$ mm Hg, l_2 is on the average close to 10^{-3} cm.

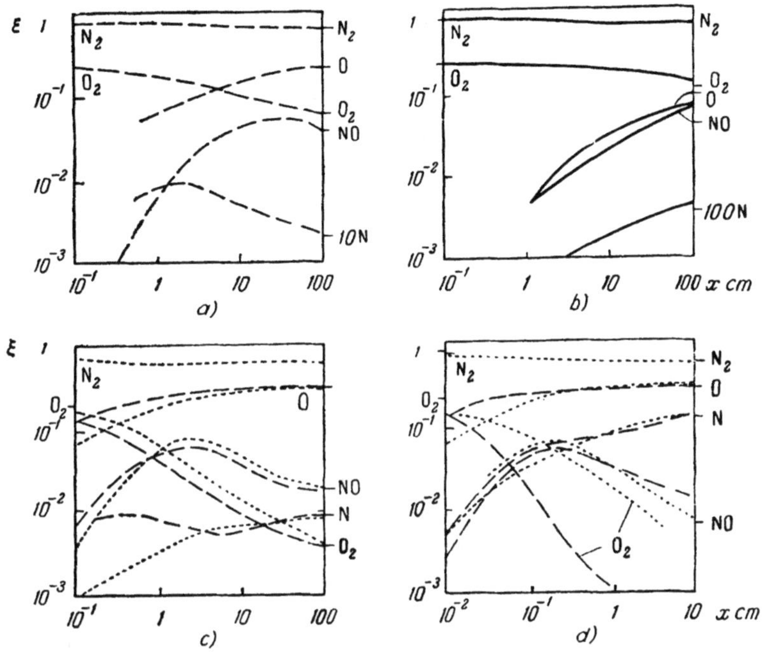

Fig. 127a, b, c, d. Mole fractions of air components as a function of the distance from the shock front: (a) according to solution II ($V = 3.42$ km/sec), according to solution III ($V = 3$ km/sec), (c) according to solutions I (dotted line), $V = 4.13$ km/sec and (II) (dashed line), $V = 4.14$ km/sec, (d) is the same for solution I with $V = 5.17$ km/sec and for solution II, $V = 5.05$ km/sec.

Graphs of the distribution of component mole fractions (Fig. 127) point to the fact that molecular oxygen dissociates rapidly (particularly for high velocities), while the concentration of N_2 does not change too significantly even when $V = 6.9$ km/sec. A distinguishing feature in all the cases is the maximum of nitric oxide concentration within the nonequilibrium zone, which was discovered by us as early as in solution (I). The maximum content of nitric oxide behind a shock front in air and the position of this maximum as a function of the wave velocity, according to the solutions being discussed are presented in Fig. 128. It can be seen from these data that the maximum concentration of NO is practically independent of the pressure; this follows from the fact that the reactions which participate in formation of NO are bimolecular [318, 92]. Comparison of the mole fraction distribution curves in Figs. 127 *b* and *d* (solutions I and II), calculated for almost the same front velocities, shows that on the whole, a refinement of the constants affects the results very little. Solution (I) for $\alpha = 0.1$, 1 and 10 points to the

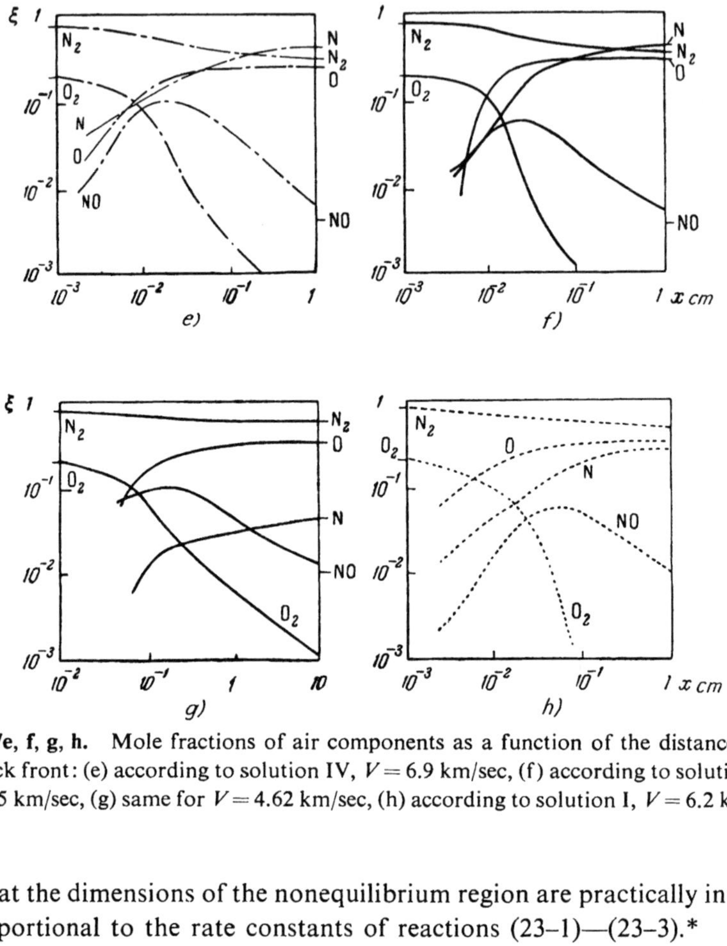

Fig. 127e, f, g, h. Mole fractions of air components as a function of the distance from the shock front: (e) according to solution IV, $V = 6.9$ km/sec, (f) according to solution III. $V = 6.55$ km/sec, (g) same for $V = 4.62$ km/sec, (h) according to solution I, $V = 6.2$ km/sec,

fact that the dimensions of the nonequilibrium region are practically inversely proportional to the rate constants of reactions (23–1)—(23–3).*

To understand the course taken by processes behind a shock front in air, it is necessary to analyze the contribution of each reaction to the total transformation mechanism. This consideration shows that the main reaction in air, which determines all the following transformations, is the dissociation of molecular oxygen. The appearance of atomic oxygen starts the chain reaction for the formation of NO and O. Initially, some NO is formed by reaction (23–6); the contribution made by this reaction is more appreciable for moderate shock-wave velocities and drops sharply with a reduction in the concentration of molecular oxygen. Calculations [318] show that equilibrium with respect to reactions (23–4) and (23–5) is established much faster than with respect to reactions (23–1)—(23–3) which is precisely the fact responsible

* For a system with a single reaction this is obvious.

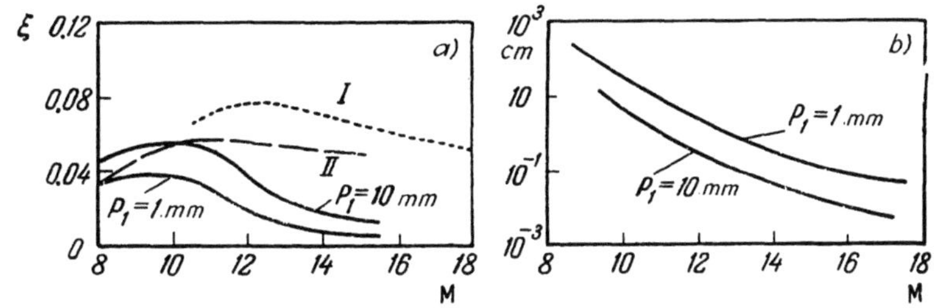

Fig. 128. Maximum mole fractions of nitric oxide behind a shock front in air (a) and the distance from the shock front to the point of maximum concentration of NO (b) as a function of the Mach number of the shock wave. (a): The solid line denotes equilibrium values, the dotted line shows results obtained by Losev for $p_1 = 1$ and 10 mm Hg (they coincide); the dashed line denotes results due to Duff and Davidson for the same p_1.

for the formation of a maximum of NO concentration. In fact, in the beginning, the formation of nitric oxide is maintained by reactions (23–4), (23–5) and (23–6). Then, after a certain amount of NO has accumulated, reaction (23–4) arrives practically to the state of quasi-steady equilibrium, which, upon further drop in the temperature, results in reducing the NO concentration. The reduction in the NO concentration is also stimulated by reaction (23–5), which proceeds in the opposite direction as a result of the large quantities of NO and O which are present in the mixture and of the reduction in the concentration of O_2. The reaction $NO + M \rightarrow N + O + M$ also acts in this direction. An additional effect on the appearance of the maximum of NO is exerted by the reduction of the contribution of reaction $O_2 + N_2 \rightarrow 2\ NO$ and also (for small shock-wave velocities) by the change in the direction in which it proceeds. The contribution of reaction (23–2) to the formation of N is almost everywhere quite small; only in high-velocity shock waves does this reaction start to play some role.

In a number of cases, attempts to obtain an approximate analytical solution of this problem are of interest. Thus, Kuznetsov [92] has noted that in comparatively weak shock waves in air the nitrogen does not dissociate at all, and in stronger waves a large part of it dissociates only after most of the oxygen has dissociated. The analytic expression for the temperature distribution behind the wave front was obtained by him in the classical approximation for the vibrational specific heat of the molecules, by assuming that $v^2 \ll V^2$ and by disregarding the contribution of NO to the total enthalpy. At comparatively low temperatures (when N_2 does not as yet dissociate) the time dependence of T is then given by

$$t \approx \frac{RT^3\left(e^{\frac{D}{RT}} - e^{\frac{D}{RT_a}}\right)}{6 \times 10^{13} \cdot p_2 (T - T'_{O_2}) D}, \tag{23–21}$$

where p_2 is a constant pressure behind the shock front (in atm), D is the dissociation energy of oxygen T'_{O_2} is the temperature corresponding to complete dissociation of O_2 and T_a is the temperature before start of dissociation. When nitrogen has partially dissociated, the relationship between t and T can be represented only in the form of two integrals, which then must be evaluated numerically. The final result practically does not differ from solution (II) [to compare his results with the numerical calculations of solution (II), Kuznetsov has used the same values of the reaction rate constants; the contribution of reaction (23–2) was disregarded].

Comparison with results obtained in shock tubes is of great importance in estimating the correctness of the above calculations. From among all the methods used to study molecular dissociation behind the shock front in air, the greatest successes were achieved by using absorption spectroscopy methods. Study of the distribution of molecular oxygen concentration by the absorption of light in the quartz and vacuum ultraviolet spectral region has made it possible to measure the half-dissociation distance L_{O_2} of molecular oxygen as a component which is most sensitive to processes in a shock-wave in air.* These measurements were performed for two groups of shock-wave velocities—for $2.7 < V < 4.25$ km/sec—by Generalov and Losev according to the scheme described in Sect. 11 (see Fig. 61) and for $4.8 < V < 7.1$ km/sec by Lin and Fyfe [422].

Generalov and Losev have determined experimentally the distribution of the absorptivity of air ($\lambda = 2245$ A), as a function of the time following the passing of the shock front, It was established by special experiments in a mixture of 21% O_2 + 79% Ar and in pure nitric oxide that the absorption which is observed in the region of $\lambda = 2245$ A is practically determined by only the molecular oxygen, since nitric oxide in this region has an absorptivity close to that of oxygen, and the concentration of NO is much smaller than that of O_2 (almost in the entire nonequilibrium region for moderate shock-wave velocities). The values of L_{O_2}, referred to atmospheric pressure behind the shock front are represented in Fig. 129 together with theoretical results of solutions (I–III). This comparison of theoretically obtained values with results of our experiments on the size of the zone of semidisintegration of molecular oxygen L_{O_2} is not ensured, however, to be free of inaccuracies which are due to the possible contribution of nitric oxide to the observed light absorption. This effect can become quite noticeable toward the end of the nonequilibrium zone (in shock waves of sufficiently high velocity), where the concentration of NO can exceed the O_2 concentration. Hence it is expedient to compare the calculations with the experiment at the very be-

* The quantity L_{O_2} can be defined as the distance at which the difference in the initial and equilibrium concentration of O_2 is reduced by a factor of two.

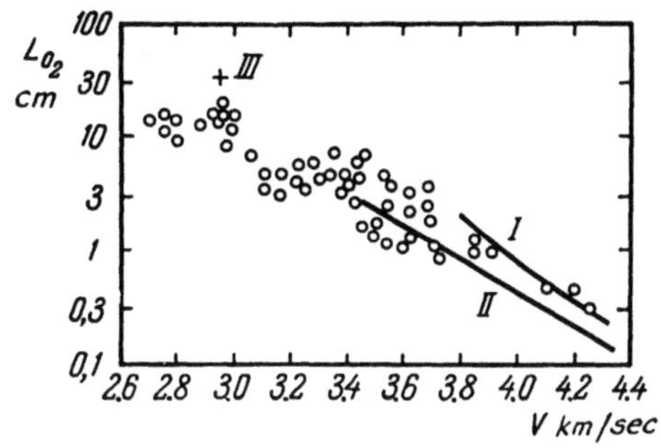

Fig. 129. Measured dimensions of the half-dissociation zone of oxygen in air as a function of the shock-wave velocity; I, II, III are the corresponding theoretical values.

ginning of the dissociation zone, where the effect of NO is negligible. The rate of dissociation of O_2 in air at the very beginning of the dissociation zone ($\xi_{O_2} \sim 0.21$) can be obtained from oscillograms using the procedure which was described in detail in Sect. 11. Here it should be remembered that the excitation of vibrations in nitrogen, which takes place simultaneously with the dissociation of O_2, affects the temperature distribution in this region behind the shock front. The experimentally obtained rates for the dissociation of O_2 in air at the beginning of the dissociation zone are presented in Fig. 130 together with the corresponding values obtained in solutions (II) and (III). Consideration of these data yields the same results as were obtained when comparing L_{O_2}.

Lin and Fyfe have performed their experiments in the same manner, except that they used a much lower initial pressure. The results of these experiments together with solutions (I)–(IV) are given in Fig. 131 for an initial pressure of 0.02 mm Hg. An examination of these figures points to

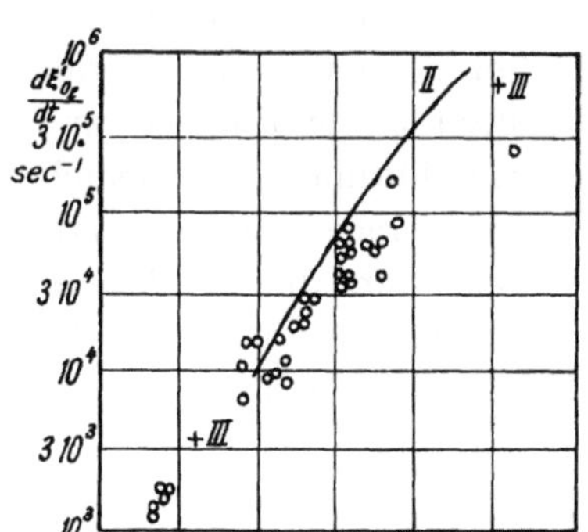

Fig. 130. Measured rates of change of the mole fraction of oxygen in the beginning of the nonequilibrium zone behind a shock front in air at $p_2 = 0.5$ atm, as a function of the temperature. II and III are values obtained theoretically by Duff, Davidson and Wray.

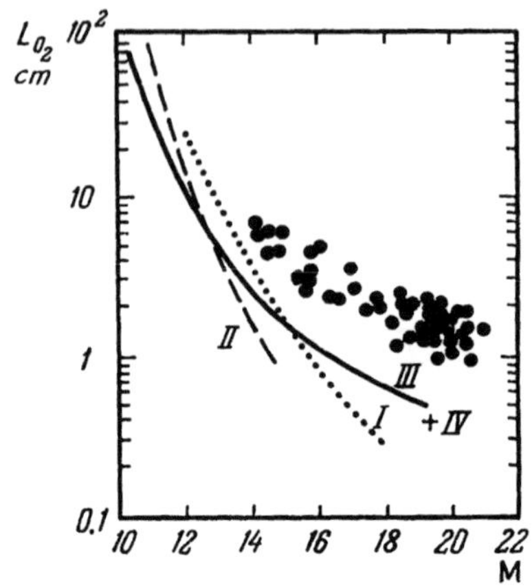

Fig. 131. Dimensions of the zone of half-dissociation of oxygen in air according to measurements made by Lin and Fyfe as a function of the Mach number. I–IV are the corresponding theoretical data.

the fact that at moderate velocities, all the solutions are in a more or less satisfactory agreement with experimental results. As the shock-wave velocity is made higher, solutions (I) and (II) show a poorer agreement with experimental data. Apparently, this is due to the assumptions made in (I) and (II) on the existence of total equilibrium and, as a result, the absence of the retarding effect of the deceleration in vibrational excitation on the rate of dissociation of O_2 and N_2. Naturally, some contribution was also made by refinements in the reaction constants which were performed in solutions (III) and (IV). At the same time, it should be noted that despite the appreciable scatter in experimental data, the results of (III) and (IV) for large Mach numbers are still below the experimental data. As one of the possible reasons for this we can point to the effect of incompleteness of rotational relaxation; consideration of retardation of rotational excitation due to reduction in the centrifugal effect can reduce the rate of molecular dissociation and incresae the dimensions of the nonequilibrium zone to match the experimental values.

Another cause may be inaccuracy of the model of interaction of molecular vibrations with their dissociation, due to Hammerling, *et al.*, which was used in calculations (III) and (IV) [see Sect. 18]. Here it should be noted that an increase in the translational temperature (following retardation of vibrational and rotational excitation), which results in reducing the zone due to acceleration of the reactions does not, apparently, have an appreciable effect on the results.*

* At very high temperatures, the exponential multiplying factors in the expression for the reaction rate constants become unity; in this case the constant is only weakly dependent on the translational temperature.

[24] THERMAL IONIZATION KINETICS AND NONEQUILIBRIUM RADIATION

The increase in the shock-wave velocity in air is, in addition to the vibrational relaxation and chemical transformations, also accompanied by thermal ionization. If the degree of ionization is sufficiently high, then the ionized particles begin to exert an appreciable effect on the rate of chemical reactions as well as on the overall energy balance. In this case, the kinetics of thermal ionization and of chemical transformations are most closely related to one another and it is impossible to separate them. At present, a joint consideration of thermal ionization and of chemical transformations in air is not available. However, in the velocity interval $V \leqslant 9$ km/sec ($\mathsf{M} \leqslant 27$), which is of practical interest, the maximum degree of ionization is of the order of 1%. In this case ionization practically does not affect the process of establishing chemical equilibrium in air. Hence the kinetics of chemical transformation behind a shock front in air can be calculated without considering the effect of ionization, which is precisely what was done in Sect. 23. In the same approximation it is possible to study the ionization kinetics by assuming that the air composition, which is determined without taking ionization into account, is known in each point of the nonequilibrium zone behind the shock front.

This simplified representation of the ionization process is found to be valid also for shock-wave velocities in excess of 9 km/sec, when the equilibrium degree of ionization is high. The ionization kinetics of air in this case are similar to those of monatomic gases (see page 296). The ionization process can be represented schematically as follows. At the first stage, in the region which extends several mean-free paths behind the shock front, chemical transformations take place and a chemical equilibrium is established in the first approximation. The contribution of charged components in this region is not too high, since their concentration is low, and they do not exert an appreciable effect on the rate of chemical reactions. However, the contribution made by this zone to the ionization process as a whole is great. The electrons which are formed here serve as primers which subsequently produce avalanche ionization by electron collisions. Avalanche ionization proceeds quite rapidly, hence the region in which equilibrium ionization is established is not much longer than the region where chemical equilibrium is established.

At present, in the overwhelming majority of theoretical and experimental works, the nonequilibrium ionization behind a shock front in air has been studied for shock-wave velocities lower than 9 km/sec. This is the region to which the subsequent discussion is confined.

The greatest difficulty encountered in the study of thermal ionization is that of selecting the main ionization mechanism. Usually, the following elementary processes, which produce ionization or which change the concentration of charged particles are considered:

1. Ionization due to electron or ion collisions.
2. Photoionization.
3. Charge exchange.
4. Electron attachment.
5. Ionization in collisions of neutral atoms and molecules.

This list of elementary processes is not exhaustive. It does not include, for example, processes in which the formation of electrons is a result of collisions with admixture atoms. Secondary effects, which are due to electron diffusion and to photoeffect from solid surfaces, are also disregarded. However, in the cases being considered, these effects are small and thus can be neglected. In practice, they must be considered only at shock-wave velocities below 4 km/sec. Under these conditions, the degree of ionization of the main components of air is very small, hence the contribution made by easily ionized admixtures (as the contribution of other side effects), can predominate.

From among the enumerated processes, charge exchange is not directly related with the change in concentration of free electrons. However, it can change the ionic concentration of individual components, which in its turn can result in changing the total rate of thermal ionization.

The concentrations of ions and electrons in the nonequilibrium zone behind a shock front in air are determined by solving the kinetic equations of reactions which describe processes 1–5. The reaction equations are solved on the assumption that the temperature distribution and the concentrations of neutral components in the nonequilibrium zone are known (see Sect. 23). The main difficulty in this formulation of the problem consists in selecting the rate constants for the different chemical reactions. At present, no reliable data on the reaction rate constants are available for the majority of the several tens of reactions which describe processes 1–5.

Reference [426] presents a detailed analysis of each of the above ionization processes and gives the most reliable data on the rate constants of the different reactions. The calculated results of the comparative contribution of each of the above processes to the total rate of thermal reaction which were obtained in [426] show that the main process which determines the thermal ionization rate in air is ionization by atomic collisions which proceeds according to the reaction $A+B \rightarrow AB^{+}+e$ (process which is the reverse of dissociative recombination). This reaction is most convenient from the energy point of view than all the others, since the energy it requires is smaller than the ionization potential by the magnitude of the dissociation energy.

At the same time, as the shock-wave velocity is increased, the contribution of electron collision ionization increases. Electronic ionization plays practically no role at velocities below 5 km/sec. On the other hand, in shock waves moving at more than 10 km/sec, electron collision ionization becomes the predominating process, just as in the case of monatomic gases. Under these conditions ionization by atom-atom collisions serves as the source of priming electrons which are needed for the development of the electron avalanche.

It was also shown in [426] that, from among all the possible ionization processes which take place following atom-atom reactions, the main role is played by the reaction which requires the smallest energy expenditure:

$$N + O + 2.8\ \text{eV} \rightleftarrows NO^{+} + e. \tag{24–1}$$

In waves with velocities in excess of 9 km/sec the above reaction has to compete with the reaction:

$$N + N + 5.8\ \text{eV} \rightleftarrows N_2^{+} + e. \tag{24–2}$$

The important role of reaction (24–1) in the thermal ionization of air was also noted in [177, 262].

The concentrations of charged components in the nonequilibrium region behind the shock front are shown in Fig. 132. It can be seen from the figure that the equilibrium concentration of charged components is not reached monotonically. In the nonequilibrium zone the concentrations of electrons of N_2^+, O_2^+ and NO^+ reach a maximum.

The experimental study of the process of establishing ionization equilibrium in air was, as a rule, limited to measuring the electron density in the nonequilibrium zone [262, 430, 462, 13, 75]. The size of the nonequilibrium

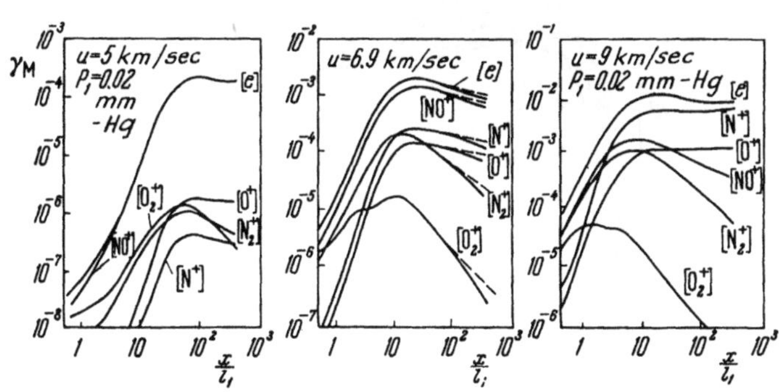

Fig. 132. The ratio γ_M of the number density of charged particles to the equivalent number density of molecules (see Table 4) in the nonequilibrium zone behind the shock front in air as a function of the reduced distance from the front. Designations: solid line corresponds to $p_1 = 0.02$ mm Hg, the dashed line is for $p_1 = 2$ mm Hg.

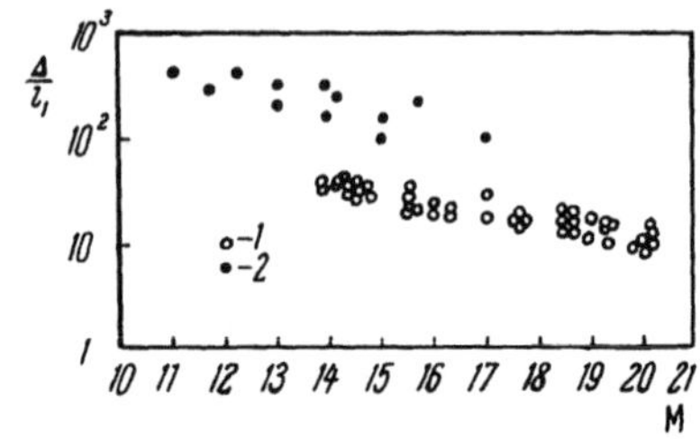

Fig. 133. Dimensions of the zone where equilibrium ionization is established (as a ratio to the mean-free path l_1 ahead of the shock wave) as a function of the Mach number of the shock wave obtained by Lin, Neal and Fyfe (1) [424] and Niblett and Blackman (2) [462].

zone as a function of the Mach number of the incoming flow is shown in Fig. 133. For convenience the ratio of the length of the nonequilibrium zone to the mean-free path l_1 ahead of the shock wave was laid off on the ordinate axis.

In [424] the electron concentration in the nonequilibrium zone behind the shock front in air was measured as a function of the distance. The results of theoretical calculation of the electron density of [424] together with experimental data are presented in Fig. 134. It can be seen from this figure that by appropriate selection of the rate constant of the main reaction (24–1), it is possible to obtain satisfactory agreement with experimental results. The moderate deviations of the experimental from the calculated data are ascribed by the authors to inaccuraties in calculations and, in particular, to the fact that the actual motion of the gas in the shock tube was approximated by one-dimensional motion.

In the same approximation in which the thermal ionization kinetics were

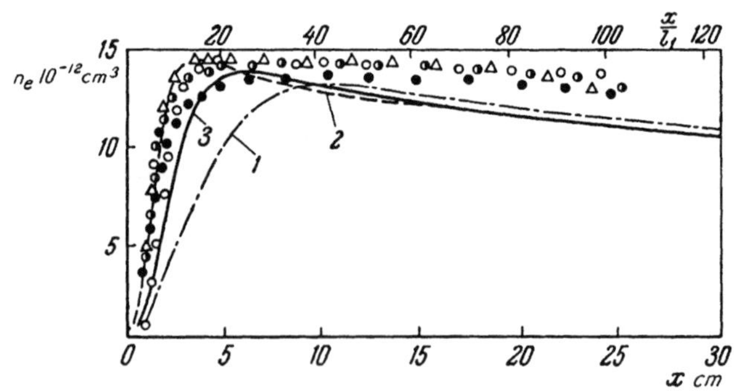

Fig. 134. The electron number density behind a shock front in air ($V = 6.9$ km/sec, $p_1 = 0.02$ mm Hg). ▲, ○, ●, ◐, are experimental values obtained in [424] from different experiments; the solid, dashed and dash-dot curves denote theoretical data. In calculating Curve 1, the rate constant for reaction (24.1) was assumed by a factor of three larger, and for Curve 2 by a factor of three smaller, than that assumed for calculating Curve 3.

calculated, it is possible to calculate also the nonequilibrium radiation in air.* If we assume that the radiation does not affect the rate at which chemical and ionization equilibrium is established in air, then calculating the nonequilibrium radiation reduces primarily to determining the concentration of electronically excited air components on the basis of the known component concentration in the ground state. To determine the intensity of the recombination continuum, it is necessary, in addition to the concentration of charged components, also to know their velocity distributions. Calculations of this kind were performed, for example, in [135, 280]. The main difficulty encountered in these calculations and their shortcoming, consist in the unavailability of reliable values of rate constants of reactions which determine the population of the excited electronic states of atoms, molecules and ions.

Thus, we shall subsequently restrict ourselves to a qualitative consideration of nonequilibrium radiation, using as our basis, well-known experimental data.

As seen from Fig. 132, the ionic concentrations (as well as electron concentrations) of individual components can reach a maximum in the nonequilibrium zone. The appearance of an ion concentration which is in excess of its equilibrium value can result in intensive luminescence of the gas in the nonequilibrium zone which in a number of cases can exceed the radiation from the equilibrium zone (the so-called "excess radiation" phenomenon; see Sect. 20). The existence of excess radiation in air is well known [279,280].

The integral intensity of radiation from the nonequilibrium region (which from now on will be called nonequilibrium radiation) is proportional to the product of the density and the length of the nonequilibrium region. Since the relaxation processes in the nonequilibrium zone are determined mainly by processes which take place in binary collisions, then the length of this zone is inversely proportional to the density (see Fig. 132). It follows from this that the integral intensity of nonequilibrium radiation is independent of the initial density of the gas. The existence of such a plateau of the luminescing front or the curve of the luminescence intensity as a function of the initial density is bound by small as well as large density values. For high densities the radiation intensity of the equilibrium zone exceeds the nonequilibrium radiation. For low densities the deactivation of electronically excited components following the radiation is not compensated by collision excitations, which reduce the stationary population of the excited electronic collisions, and, consequently, also reduce the intensity of the nonequilibrium radiation. The critical density at which the average time between two collisions is of the order of the radiation lifetime determines the boundary of the plateau

* The term nonequilibrium radiation is used here to denote the radiation from the nonequilibrium region; see also Sect. 20.

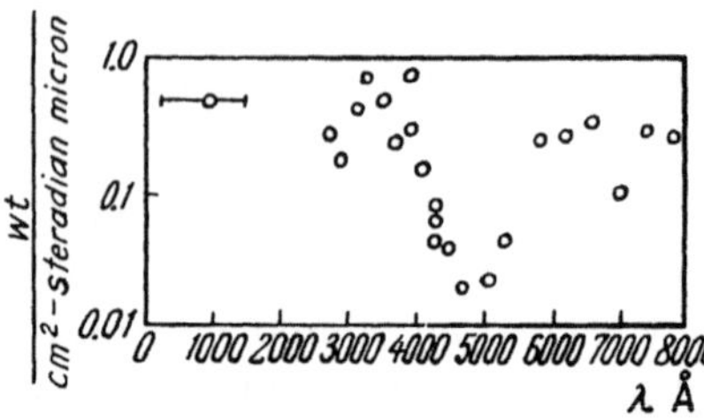

Fig. 135. Intensity of nonequilibrium radiation of air behind a shock front as a function of the wavelength ($V = 6.8$ km/sec, $p_1 = 0.02$ mm Hg).

of the luminescing front from the side of low densities. This boundary is called, for short, the "boundary of collision" (collision-limiting boundary).

In the density region which lies beyond the collision-limiting boundary, the integral intensity of nonequilibrium radiation is decreased with a reduction in density.

The spectral composition of the nonequilibrium radiation is determined by the shock-wave velocity and was studied in [280]. In Fig. 135 the nonequilibrium radiation intensity is depicted as a function of the wavelength for a shock-wave velocity of 6.8 km/sec and an initial pressure of 2×10^{-2} mm Hg [280]. The main contribution to the nonequilibrium radiation of air at temperatures above 5000 °K is made by the *1st* and *2nd* positive systems of N_2 and by the *1st* negative system of N_2^+. In addition, the β and γ bands of NO, radiation of O_2 in the Schumann-Runge region and radiation of electrons in free-free and free-bound transitions may make important contributions. The total intensity of the $N_2(1+)$ and $N_2^+(1-)$ bands substantially

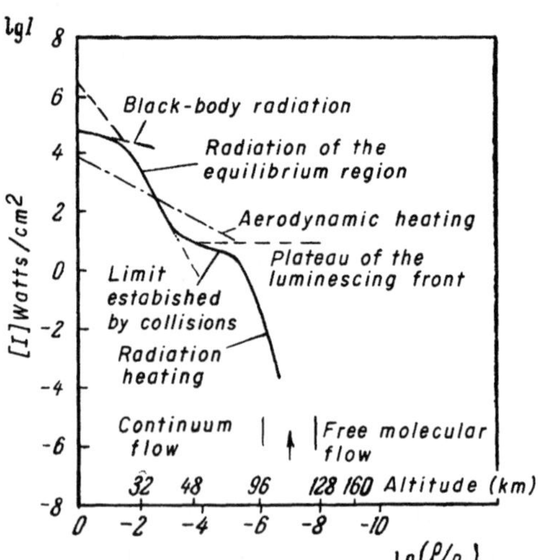

Fig. 136. Radiation and aerodynamic heating as a function of the reduced density (flight altitude) for a flight velocity of 7.5 km/sec and nose section radius of 3 meters.

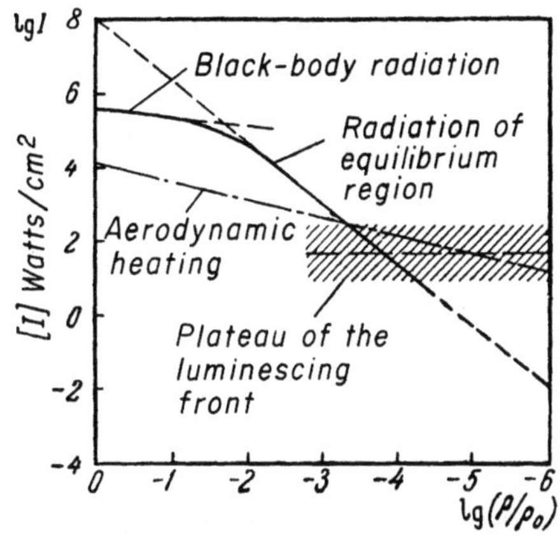

Fig. 137. Radiant and aerodynamic heating as a function of the reduced density for a flight velocity of 10.5 km/sec and nose section radius of 3 meters; ρ is the density ahead of the shock wave, ρ_0 is the density at $T = 288\,°K$ and $p = 1$ atm at sea level.

determines the nonequilibrium radiation of air in the range of 0.36 – 1 microns. When the shock-wave velocity is 10.5 km/sec the magnitude of the radiant flux in the direction of the stagnation point is included within the limits of 12–40 watts/cm^2 [569].

From the gasdynamics point of view, nonequilibrium radiation is of substance only in the case when its intensity exceeds that of the radiation from the equilibrium region and, in addition, the radiation-produced heating exceeds the aerodynamic heating. The intensity of nonequilibrium radiation behind the front of a plane shock wave in air exceeds the radiation intensity of the equilibrium region starting with altitudes of 45 km above sea level. However, the total radiant heating is small in comparison with aerodynamic heating up to shock-wave velocities of the order of 7 km/sec. At the same time, the radiant heating becomes comparable with aerodynamic heating starting with flight velocities of 10.5 km/sec.

The comparative role of aerodynamic and radiant heating for a body with a nose section radius of 3.05 m, moving with velocities of 7.5 and 10.5 km/sec is schematically represented in Figs. 136 and 137 [569].

At present, no calculations of nonequilibrium radiation for higher flight velocities are available. The difficulties in performing these calculations are due to the unavailability of reliable data on the behavior of air at the very high temperatures which are generated under these conditions. Thus, for example, at velocities of 12 km/sec the kinetic temperature directly behind the shock front is 75,000 °K. Under these conditions, not only the magnitudes of reaction rate constants but the reactions proper which result in establishing the equilibrium are unknown.

6

Flow of a Gas Undergoing Relaxation

[25] INTRODUCTION

It was pointed out in Sect. 1 that in the case when the characteristic time of relaxation processes taking place behind the front of a strong shock wave is much greater than the time of a mean-free molecular path, the gas flow in the relaxation zone can be studied by methods of macroscopic gasdynamics. However, the gasdynamics of a fluid undergoing relaxation differs in two important aspects from ordinary "equilibrium" gasdynamics. First, the equations of relaxation gasdynamics are based on a thermodynamic description of a fluid which is in the state of incomplete statistical equilibrium. This description requires the addition of new parameters which would characterize the fact that the fluid is not at equilibrium. Second, the equations of relaxation gasdynamics include equations which describe the kinetics of the relaxation processes.

Any flow in which the dimensions of the relaxation zone are of the order of a characteristic macroscopic length can be described by relaxation gasdynamics. Any complete consideration of the problems which thus arise is beyond the scope of this book. In this chapter, which actually is a continuation of Sect. 1, we shall restrict ourselves only to a short outline of the properties of a gas undergoing relaxation.

In this outline we shall formulate the equations of motions of fluids undergoing relaxation, note certain features peculiar to this motion and consider the limit transition from relaxation gasdynamics to "equilibrium" gasdynamics when the relaxation time goes to zero.

The last problem is of interest due to the fact that certain important properties of equations of relaxation gasdynamics do not become properties of equations of equilibrium gasdynamics when the relaxation time goes to zero. For example, the characteristics of the equations of nonsteady one-dimensional flow of a gas undergoing relaxation (without viscosity and thermal conductivity) are independent of the relaxation time and do not coincide with the characteristics of equilibrium gasdynamics. Hence the problem of limit transition to equilibrium gasdynamics requires special consideration.

Subsequently, as we shall see, the difficulties which arise here are eliminated by assuming that equilibrium gasdynamics is substantially an approximate description of only a special class of solutions of relaxation gasdynamics.

[26] EQUATIONS OF RELAXATION GASDYNAMICS

[1] Gaskinetic Methods for Obtaining Equations of Equilibrium and Relaxation Gasdynamics

As was pointed out, intermolecular collisions are an extremely effective mechanism for restoring the local Maxwellian distribution in a gas. This effectiveness is characterized by the relaxation time which is, by order of magnitude, equal to the time τ_0 of a mean-free path.

The nonequilibrium spatial distribution of flow variables (density, fluid velocity, temperature, etc.), is due to departure from the equilibrium distribution of thermal velocities, but this disturbing effect of spatial inhomogenity cannot compete with any degree of success with intermolecular collisions, which restore the disturbed statistical (local) distribution. Hence gasdynamic disturbances result in negligible disturbances of the Maxwellian distribution function. However, from the viewpoint of molecular kinetics, these small perturbations are the basis of all the dissipation processes in monatomic gases. The introduction into ordinary gasdynamics equations of terms with viscosity, thermal conductivity and diffusion coefficients are a method for taking into account the above-moderate disturbances of the statistical equilibrium; here the thermodynamic parameters of the fluid are assumed to be equal to their values in the state of total (local) thermodynamic equilibrium.

These results are obtained naturally within the framework of the Enskog-Chapman solution of Boltzmann's gaskinetic equations [220, 51]. The main idea behind this method (which to a certain extent is the culmination of works of Maxwell [439], Hilbert [371, 372] and Boguslavskiy [26]), consists in the following. It is assumed that the rapid (time τ_0!) process of establishing the local Maxwellian distribution (if the initial distribution was non-Maxwellian) is practically completed and the remaining small corrections to the equilibrium distribution, which owe their existence to the spatial inhomogeneity of the system, are at any given time completely determined by this inhomogeneity, *i.e.*, by the fields of the gas-flow variables, *e.g.* the density, flow velocity, temperature, etc. Thus, starting with this time, the distribution function is an implicit function of time (in terms of the time-dependent gasdynamic fields).

According to this, the solution $f(\boldsymbol{c}, \boldsymbol{r}, t)$ of Boltzmann's equations is sought in the form of a series expansion in terms of powers of the small parameter $\mu \sim l_0/L$, where l_0 is the mean-free path and L is a characteristic macroscopic length:

$$f(\boldsymbol{c}, \boldsymbol{r}, t) = f^{(0)}(\boldsymbol{c}, \boldsymbol{r}; n, \boldsymbol{c}_0, T) + \mu f^{(1)}(\boldsymbol{c}, \boldsymbol{r}; n, \boldsymbol{c}_0, T) + \\ + \mu^2 f^{(2)}(\boldsymbol{c}, \boldsymbol{r}; n, \boldsymbol{c}_0, T) + \cdots . \qquad (26\text{–}1)$$

Here the coefficients of powers of μ which are functions of $\boldsymbol{c}$ and $\boldsymbol{r}$, are functionals of the n, $\boldsymbol{c}_0$ and T fields, where n is the molecular number density, and $\boldsymbol{c}_0$ is the gasdynamic velocity. Substitution of Eq. (26–1) into Boltzmann's equation yields a series of equations for $f^{(0)}, f^{(1)}$, etc.

It follows from the physical considerations which serve as a basis of this method, that the zero approximation $f^{(0)}$ is a Maxwellian function which is defined by the values of n, $\boldsymbol{c}_0$ and T in the given point in space. Equations of gasdynamics of an ideal fluid are obtained in this approximation.

The first-order correction $f^{(1)}$ yields a tensor of viscous stresses and the thermal flux density vector in the Navier-Stokes form (and diffusion fluxes in multicomponent gases). Systematic consideration of higher-order terms in Eq. (26–1) yields Barnett and other systems of gasdynamic equations.

In the theory which is based on gaskinetic equations, the temperature T appears as a result of the definition:

$$\tfrac{3}{2}nkT = \int \frac{m(\boldsymbol{c} - \boldsymbol{c}_0)^2}{2} f d\boldsymbol{c} . \qquad (26\text{–}2)$$

For the case of statistical equilibrium, f is a Maxwellian function, Eq. (26–2) degenerates into a known theorem of statistical thermodynamics and the kinetic temperature of Eq. (26–2) becomes identical to the thermodynamic temperature.

The expansion (26–1) and definitions of the type of (26–2) make obvious the approximate character of the thermodynamic description of monatomic gases which are not at equilibrium and determine the limits of its applicability. Similar considerations can also be put forward for the case of polyatomic gases.

Let us simplify the picture by assuming that the establishing of an equilibrium energy distribution between the translational and internal degrees of freedom is characterized by a single relaxation time $\tau \gg \tau_0$. We shall also assume that the establishing of the Boltzmann energy distribution between the internal degrees of freedom (with a temperature T_1 which, in general, differs from temperature T of the Maxwellian distribution) is also a rapid process. The last assumption can follow either from the high relative effectiveness of internal energy transfer on collisions, or from the special properties of the probabilities of energy transfer from the translational to the internal degrees of freedom (this is the situation, for example, in the case of harmonic oscillators [451]).

First of all, let us consider the case when the characteristic macroscopic time t_m is great in comparison with τ, *i.e.*, $(t_m/\tau) \gg 1$. The ratio t_m/τ is a measure of effectiveness of collisions as a mechanism which restores the (local) statistical equilibrium in a system, including the equilibrium between the internal and translational degrees of freedom.

The method for generalizing the ideas of the Enskog-Chapman method in this case is quite obvious. We forego the description of the rapid ($\tau \ll t_m$!) process of establishing the local statistical distribution (with respect to all the degrees of freedom), we assume in the zeroth approximation that it is completed and we seek the solution of gaskinetic equations in the form of the following power series in $\mu \sim (\tau/t_m)$:

$$f\,(p_i, q_i, t) = f^{(0)}(p_i, q_i; n, c_0, T) + \\ + \mu f^{(1)}(p_i, q_i; n, c_0, T) + \mu^2 f^{(2)}(p_i, q_i; n, c_0, T) + \cdots . \qquad (26\text{–}3)$$

In the series (26–8) the first term of the series is the Maxwell-Boltzmann distribution which is determined by the local values of density, temperature and gas velocity. The terms of the first and second order of smallness in this series, which are functions of the generalized coordinates q_i and of the momenta p_i of the molecule are functionals of the gasdynamic fields and are not explicitly time-dependent.

As in the case of monatomic gases, it should be emphasized that the "flow variables" which are contained in Eq. (26–3) appear as a result of known gaskinetic definitions [such as (26–2)]. The possibility of identifying them with the corresponding quantities of phenomenological thermodynamics [classical thermodynamics; editor] arises within the framework of certain approximations which serve as a basis of expansion (26–3).

Retaining in Eq. (26–3) a given number of first terms, it is possible, in principle, to obtain from the gaskinetic equations, equations of gasdynamics in various approximations. In all these equations the thermodynamic properties of the fluid are identical to these properties in the state of complete thermodynamic equilibrium. The gasdynamics equations in the different approximations differ from one another only by the dependence of the viscous stresses tensor and the thermal flux density vector on the position derivatives of the flow variables.

Relaxation zones are characterized by the condition $\tau \sim t_m$. On this time scale in the assumed simplified scheme, the only rapid process is establishing the Maxwell and Boltzmann distributions (with the different temperatures T and T_1). Obviously, the considerations which yielded expansions of the type of Eqs. (26–1) and (26–3) can be also extended to the flow in the relaxation zone, if among the macroscopic fields which determine the corrections to $f^{(0)}$ we include the temperature field T_1 (of the internal degrees of freedom). Hence the solution of gaskinetic equations in the relaxation zone is represented in the form:

$$f(p_i, q_i, t) = f^{(0)}(p_i, q_i; n, c_0, T, T_1) + \\ + \mu f^{(1)}(p_i, q_i; n, c_0, T, T_1) + \cdots . \qquad (26\text{–}4)$$

Here μ is a small parameter of the order of τ_0/t_m, $f^{(0)}$ describes the Maxwellian distribution of translational velocities which corresponds to temperature T, and the Boltzmann energy distribution with respect to the internal degrees of freedom which corresponds to temperature T_1.

The expansion (26–4) makes it possible, in principle, to obtain equations of relaxation gasdynamics in various approximations. In these equations the state of the fluid is characterized by the temperatures T and T_1 and the system of equations of relaxation gasdynamics includes an equation which describes a slow relaxation process which is that of equalizing these temperatures.

There is no basis for regarding one of the two temperature characteristics as more fundamental and closer to the thermodynamic temperature. This becomes obvious in the case of a nonequilibrium mixture of two gases the particles of which have highly different masses. The state of such a fluid is described by two temperatures, one of the light and the other of the heavy component; each of them is related to the translational degrees of freedom. In both cases the difference in the temperatures T and T_1 characterizes the deviation of the system from the state of complete thermodynamic equilibrium.

[2] The Use of Methods of Thermodynamics of Irreversible Processes

In the general theory of the motion of gases undergoing relaxation (which can also be extended to liquids), which is based on the methods of thermodynamics of irreversible processes (see, for example [54, 54a]), it is convenient to introduce the nonequilibrium parameter ξ (or several such parameters), without specifying precisely its physical meaning [119].

The description of thermodynamic properties of systems in the state of incomplete statistical equilibrium can be given in terms of some characteristic functions which have been appropriately generalized. Within the framework of statistical physics, the concept of entropy is as fundamental and general as the energy concept. In particular, the statistically nonequilibrium states which are considered here can be assigned entropy values with the same degree of definiteness as for equilibrium systems. Hence as the characteristic function which would describe the thermodynamic properties of the fluid, it is natural to use the energy ε of a unit mass, considering it as a function of the density ϱ, entropy of a unit mass s and the parameter ξ

$$\varepsilon = \varepsilon(\varrho, s, \xi). \tag{26–5}$$

In the state of equilibrium ξ takes on the value $\xi_0(\varrho, s)$, which is defined by

$$\varepsilon_\xi(\varrho, s, \xi_0) = 0, \quad \varepsilon_\xi \equiv \left(\frac{\partial \varepsilon}{\partial \xi}\right)_{\varrho, s}. \tag{26–6}$$

As is known, in the case of systems in the state of complete thermodynamic equilibrium, the specification of ε as a function of ϱ and $s - \varepsilon(\varrho, s)$ defines all the thermodynamic properties of the fluid; in particular, for example, we have

$$\left(\frac{\partial \varepsilon}{\partial s}\right)_\varrho = T, \quad \left(\frac{\partial \varepsilon}{\partial \frac{1}{\varrho}}\right)_s = -p, \tag{26–7}$$

where T and p are the temperature and pressure.

In the case of a system in the state of incomplete statistical equilibrium with an energy ε, which is defined by Eq. (26–5), the physical meaning of the derivatives

$$(\partial \varepsilon / \partial s)_{\varrho, \xi} \quad \text{and} \quad \left(\frac{\partial \varepsilon}{\partial \frac{1}{\varrho}}\right)_{s, \xi}$$

depends on the choice of ξ. For example, in the cases of systems with the two temperatures T and T_1 which were considered below, the definition $\xi \equiv T_1$ results in

$$\left(\frac{\partial \varepsilon}{\partial s}\right)_{\varrho, \xi} = T, \tag{26–8}$$

where T is the temperature of the translational degrees of freedom in the one case and the temperature of the heavier component in the other. If we assume $\xi \equiv T$, then $(\partial \varepsilon / \partial s)_{\varrho, \xi}$ is identical to the temperature T_1 of the internal degrees of freedom or with the temperature T_1 of the lighter component.

It is also easy to prove that

$$\left(\frac{\partial \varepsilon}{\partial \frac{1}{\varrho}}\right)_{s, \xi \equiv T_1}$$

is equal to the pressure in the case when T_1 is the temperature of the internal degrees of freedom and is equal to the partial pressure of the heavier component when T_1 is the temperature of the lighter component.

In the simple case being considered, the state of the system as a whole is determined by the fields of ϱ, s, $\boldsymbol{c}_0$ and ξ. Obviously, a different selection of ξ yields different values of the thermal flux density vector and for the momentum flux density tensor, which includes the tensor of viscous stresses.

Usually, the best choice of ξ is indicated by the character of the relaxation process. Let us, for example, consider a chemically nonequilibrium gaseous mixture. If the establishing of the Maxwell-Boltzmann energy distribution with a temperature T common to all the particles is rapid in comparison with the chemical reaction, then selecting as the parameters ξ_i the relative concentrations of molecules of the ith kind, we get from $\varepsilon(\varrho, s, \xi_1, \xi_2, \ldots)$

$$\left(\frac{\partial \varepsilon}{\partial s}\right)_{\varrho, \xi} = T, \quad \left(\frac{\partial \varepsilon}{\partial \frac{1}{\varrho}}\right)_{s, \xi}. \tag{26–9}$$

The equilibrium values of ξ_i^0 are determined by the law of mass action.

In the general case, it is always possible to select the parameter ξ in a number such that the equality

$$\left(\frac{\partial \varepsilon}{\partial \frac{1}{\varrho}}\right)_{s, \xi} = -p, \tag{26–10}$$

where p is the hydrostatic pressure, will always be valid.

In fact, for the reversible isentropic change of volume during "frozen" relaxation processes, we have

$$d\varepsilon = \left(\frac{\partial \varepsilon}{\partial \frac{1}{\varrho}}\right)_{s, \xi} d\frac{1}{\varrho} + \varepsilon_\xi d\xi = -p d\frac{1}{\varrho}. \tag{26–11}$$

Assuming that p is a known function of ϱ, s and ξ, we can represent any solution of Eq. (26–11) in the form

$$\varphi(\varrho, s, \xi) = \text{const}. \tag{26–12}$$

Changing to a new parameter $\xi' \equiv \varphi(\varrho, s, \xi)$, we get from $\varepsilon(\varrho, s, \xi')$

$$\left(\frac{\partial \varepsilon}{\partial \frac{1}{\varrho}}\right)_{s, \xi'} = -p.$$

Henceforth, we assume that the choice of ξ satisfies Eq. (26–10).

The equations of continuity, momentum and energy have the usual form:

$$\left.\begin{aligned} &\frac{\partial \varrho}{\partial t} + \operatorname{div} \varrho \boldsymbol{v} = 0, \\ &\varrho\left(\frac{\partial \boldsymbol{v}}{\partial t} + (\boldsymbol{v} \cdot \nabla)\boldsymbol{v}\right) = -\operatorname{grad} p + \nabla \cdot \boldsymbol{\Pi}, \\ &\frac{\partial}{\partial t}\left(\frac{\varrho v^2}{2} + \varrho\varepsilon\right) = -\operatorname{div}\left\{\varrho \boldsymbol{v}\left(\frac{v^2}{2} + h\right) - (\boldsymbol{\Pi} \cdot \boldsymbol{v}) + \boldsymbol{q}\right\}. \end{aligned}\right\} \tag{26–13}$$

Here v is the fluid velocity, h is the enthalpy of a unit mass, $\boldsymbol{q}$ is the thermal flux density vector, $\boldsymbol{\Pi}$ is a tensor with components π_{ik}, which are determined from

$$P_{ik} = -\, p\delta_{ik} + \pi_{ik}\,, \tag{26–14}$$

where P_{ik} is the stress tensor. It follows from the subsequent discussion that $\boldsymbol{\Pi}$, in general, may also not be identical to the viscous stresses tensor.

From Eq. (26–13) we get the following form of the energy equation:*

$$\frac{\mathrm{d}\varepsilon}{\mathrm{d}t} = -\, p\frac{\mathrm{d}}{\mathrm{d}t}\frac{1}{\varrho} - \frac{1}{\varrho}\frac{\partial q_i}{\partial x_i} + \frac{\pi_{ik}}{\varrho}\cdot\frac{\partial v_i}{\partial x_k}$$
$$\left(\frac{\mathrm{d}}{\mathrm{d}t} \equiv \frac{\partial}{\partial t} + \boldsymbol{v}\boldsymbol{\nabla}\right). \tag{26–15}$$

From this and from the general expression for $\mathrm{d}\varepsilon$

$$\mathrm{d}\varepsilon = -\, pd\left(\frac{1}{\varrho}\right) + T ds + \varepsilon_\xi d\xi \quad \left(T = \left(\frac{\partial \varepsilon}{\partial s}\right)_{\varrho,\,\xi}\right) \tag{26–16}$$

we get

$$\frac{\mathrm{d}s}{\mathrm{d}t} = -\, \frac{1}{\varrho T}\frac{\partial q_i}{\partial x_i} + \frac{1}{\varrho T}\pi_{ik}\frac{\partial v_i}{\partial x_k} - \frac{\varepsilon_\xi}{T}\frac{\mathrm{d}\xi}{\mathrm{d}t}\,. \tag{26–17}$$

In Eqs. (26–16) and (26–17), use was made of condition (26–10) and the physical meaning of temperature T, as was pointed out, is determined by the choice of ξ.

In the general case, the parameter ξ in a given point can change as a result of the scalar flux r, as well as due to the vectorial flux $\boldsymbol{j}$. Here the physical meaning of ξ is not explicitly specified, which makes it possible to neglect the physical meaning of the fluxes r and $\boldsymbol{j}$.

In the particular case, when ξ defines the composition of a gas which undergoes dissociation according to the scheme $2A \rightleftarrows A_2$, the flux r will be the rate of dissociation and $\boldsymbol{j}$ will be the diffusion flux.

By properly defining r and $\boldsymbol{j}$ it is always possible to write an equation for the rate of change of ξ in the form:

$$\varrho\frac{\partial \xi}{\partial t} = r - \operatorname{div} j\,. \tag{26–18}$$

Substituting Eq. (26–18) into (26–17), after simple transformations we obtain the entropy balance equation in the form

$$\varrho\frac{\mathrm{d}s}{\mathrm{d}t} + \frac{\partial}{\partial x_i}\left(\frac{q_i}{T} - \frac{\varepsilon_\xi j_i}{T}\right) = \sigma\,, \tag{26–19}$$

* Here, as usual, we assume summation over the double subscripts.

where σ is defined by

$$T\sigma = -q_i \frac{1}{T}\frac{\partial T}{\partial x_i} + \pi_{ik}\frac{\partial v_i}{\partial x_k} - r\varepsilon_\xi - j_i T \frac{\partial}{\partial x_i}\frac{\varepsilon_\xi}{T}. \tag{26–20}$$

It is easy to see that

$$\left.\begin{aligned} \frac{\mathrm{d}(\varrho s)}{\mathrm{d}t} &= -\operatorname{div}(\boldsymbol{j}_s + \varrho s\boldsymbol{v}) + \sigma, \\ \boldsymbol{j}_s &= \frac{\boldsymbol{q} - \varepsilon_\xi \boldsymbol{j}}{T}. \end{aligned}\right\} \tag{26–21}$$

The vector $\boldsymbol{j}_s$ is the entropy flux density vector, or more precisely, that part of the entropy flux density which is not due to the macroscopic motion of the fluid; the total entropy flux density is equal to $\boldsymbol{j}_s + \varrho s\boldsymbol{v}$.

The right-hand side of Eq. (26–19) is the entropy generation, σ, *i.e.*, the rate of change of the entropy of a unit volume due to the irreversible processes: heat conduction, internal friction and processes described by the fluxes r and $\boldsymbol{j}$ (in a particular case these can be a chemical reaction and diffusion) which take place in it.

The fluxes I_α and forces X_α in terms of thermodynamics of irreversible processes here will be q_i, π_{ik}, r and j_i, and, correspondingly,

$$\frac{1}{T}\frac{\partial T}{\partial x_i}, \quad +\frac{\partial v_i}{\partial x_k}, \quad -\varepsilon_\xi, \quad -T\frac{\partial}{\partial x_i}\frac{\varepsilon_\xi}{T},$$

which satisfy the relationship

$$\sigma = \frac{1}{T}\sum_\alpha I_\alpha X_\alpha. \tag{26–22}$$

Assuming that the fluxes are linear functions of the forces, we can write for an isotropic medium

$$\left.\begin{aligned} q_i &= -A\frac{1}{T}\frac{\partial T}{\partial x_i} - BT\frac{\partial}{\partial x_i}\frac{\varepsilon_\xi}{T}, \\ \pi_{ik} &= \eta\left\{\left(\frac{\partial v_i}{\partial x_k} - \frac{\partial v_k}{\partial x_i}\right) - \frac{2}{3}\delta_{ik}\frac{\partial v_l}{\partial x_l}\right\} + \zeta\delta_{ik}\frac{\partial v_l}{\partial x_l} - M\delta_{ik}\varepsilon_\xi, \\ r &= -K'\varepsilon_\xi + L\frac{\partial v_l}{\partial x_l}, \\ j_i &= -C\frac{1}{T}\frac{\partial T}{\partial x_i} - DT\frac{\partial}{\partial x_i}\frac{\varepsilon_\xi}{T}. \end{aligned}\right\} \tag{26–23}$$

Here A, B, C, D, K', L and M are kinetic coefficients which, in general, are functions of the state, η and ξ are the shear and bulk viscosity coefficients.

The thermodynamic fluxes and forces in Eqs. (26–23) are tensor quantities of different rank—zero (scalars), first (vectors) and second. In the case of an isotropic fluid, the flow cannot be dependent on thermodynamic forces the rank of which differs by an odd number from the rank of the flow (the content of a partial formulation of the Curie principle or theorem [54, 54a]). For example, the vectorial fluxes $\boldsymbol{q}$ and $\boldsymbol{j}$ cannot be a linear function of the tensor force $(\partial v_i/\partial x_k)$ and scalar force ε_ξ, since in both these cases a vectorial kinetic coefficient would have been required, which is impossible for an isotropic medium. Here we refer to the isotropic state of a fluid in the state of thermodynamic equilibrium. On the same basis, the expressions for π_{ik} and r do not contain vector forces.

The selection of quantities which satisfy Eq. (26–22) as the fluxes and forces makes it possible to use the principle of symmetry of kinetic coefficients [54, 54a] [the "Onsager reciprocal relation of irreversible thermodynamics"; editor]

$$L_{\alpha\beta} = L_{\beta\alpha} \quad \text{or} \quad L_{\alpha\beta} = -L_{\beta\alpha} \quad \text{[see below]}, \tag{26–24}$$

where $L_{\alpha\beta}$ are kinetic coefficients in equations which define the linear dependence of fluxes on forces

$$I_\alpha = \sum L_{\alpha\beta} X_\beta \,. \tag{26–25}$$

In the case being considered, we get from the principle of symmetry of kinetic coefficients that

$$B = C\,, \quad L = -M\,. \tag{26–26}$$

We are assuming that ξ is an even function of the velocities of all the particles of the system (molecules, atoms, etc.). (Such a parameter is, for example, the concentration of any given component of the mixture.) Since the momentum is an odd function of the particle velocities, then in the second of equations (26–26) which relates the corresponding cross coefficients, there appears a minus sign, as this follows from the statistical derivation of the principle of symmetry of kinetic coefficients (see, for example [54, 54a]).

As we see, in the expression for π_{ik}, there appears a term which is proportional to ε_ξ and is not directly related to the deformation of the velocity field. The total stress tensor in a fluid undergoing relaxation is comprised of three parts: 1) "thermodynamic" pressure (with a minus sign)$p\delta_{ik} = -\varrho^2 \varepsilon_0 \delta_{ik}$; 2) viscous stresses; and 3) the term $M\delta_{ik}\varepsilon_\xi$. According to Eq. (26–26), in the expression for the scalar flux r there appears the term $L(\partial v_1/\partial x_1)$, which is determined by the deformation of the velocity field. The phenomenological theory leaves open the problem of the magnitude of these effects. They can be estimated in the molecular kinetics theory.*

* Here, we do not consider the problem of rotational viscosity, which appears when the average angular velocity of the natural molecular rotations $\boldsymbol{\omega}$ is the given point of the gas does not correspond to the value of the gasdynamic velocity rotor ($\boldsymbol{\omega} \neq \frac{1}{2}\,\text{curl}\,\boldsymbol{v}$). The rotational viscosity is described by the antisymmetric part of the stress tensor [214a, 54a].

The first two of equations (26–13), Eqs. (26–18) and (26–19), together with (26–23) and (26–26) form a closed system of equations of relaxation gas-dynamics. Here ϱ, s and ξ are taken as the independent variables which define the thermodynamic state of the fluid. The remaining thermodynamic variables are determined from the characteristic function $\varepsilon(\varrho, s, \xi)$.

In gasdynamic flows moving with hypersonic velocities, regions may be produced having large gradients of the flow variables, in particular, of ξ. Under these conditions, the consideration of the corresponding flows, *i.e.*, of the viscous stress tensor and of the vectorial fluxes $\boldsymbol{q}$ and $\boldsymbol{j}$, may become important. However, usually the main contribution to dissipation processes in the relaxation zone behind the front of a plane shock wave is made by scalar fluxes such as r. In this case the viscosity and $\boldsymbol{q}$ and $\boldsymbol{j}$ can be disregarded and the equation of relaxation gasdynamics takes on the form

$$\left.\begin{aligned} &\frac{\mathrm{d}\varrho}{\mathrm{d}t} + \varrho \operatorname{div} \boldsymbol{v} = 0, \quad &&\varrho\frac{\mathrm{d}\boldsymbol{v}}{\mathrm{d}t} + \operatorname{grad} p = 0, \\ &\varrho\frac{\mathrm{d}s}{\mathrm{d}t} = K'\frac{\varepsilon_\xi^2}{T}, &&\varrho\frac{\mathrm{d}\xi}{\mathrm{d}t} = -K'\varepsilon_\xi. \end{aligned}\right\} \tag{26–27}$$

Under these conditions $L = M = 0$ which follows, as can easily be shown, from the requirement that the quadratic form

$$L_{ik}X_iX_k \geqslant 0. \tag{26–28}$$

From Eqs. (26–27) we get a relationship between the coefficient K' and the relaxation time τ of small deviations from local thermodynamic equilibrium. Retaining in the power expansion in $(\xi - \xi)$ of the right-hand side of the last of Eqs. (26–27) only the linear terms and taking into account the fact that the entropy changes on the basis of Eq. (26–27) are a quantity of the second order of smallness, we get

$$\tau = \frac{1}{K\varepsilon_{\xi\xi}}, \quad K = \frac{K'}{\varrho}. \tag{26–29}$$

[27] CERTAIN PROPERTIES OF THE MOTION OF FLUIDS UNDERGOING RELAXATION. TRANSITION TO EQUILIBRIUM GASDYNAMICS

A number of properties common to equations of relaxation gasdynamics differ appreciably from properties of equations of equilibrium gasdynamics. Greater or less complete consideration of these problems is beyond the scope of this book. We shall restrict ourselves to only brief remarks concerning peculiar features of gasdynamics of fluids undergoing relaxation, and we shall consider in slightly more detail, the transition to equations of equilibrium gasdynamics in the case of small τ. As was already pointed out, the last problem is of importance due to the fact that certain important properties of equations of relaxation gasdynamics do not become identical to the properties of equations of equilibrium gasdynamics as τ approaches zero.

Calculating the curl of the right- and left-hand sides of the second of equations (26–13), we get

$$\left.\begin{aligned}&\frac{\partial \boldsymbol{R}}{\partial t} + \operatorname{curl}[\boldsymbol{R} \times \boldsymbol{v}] = \frac{1}{\varrho^2}\nabla p \times \nabla \varrho + \frac{1}{\varrho}\operatorname{curl}\boldsymbol{P} - \frac{1}{\varrho}\nabla \varrho \times \boldsymbol{P}, \\ &\quad \boldsymbol{R} = \operatorname{curl}\boldsymbol{v}, \quad \boldsymbol{P} = \nabla\cdot\boldsymbol{\Pi}.\end{aligned}\right\} \tag{27–1}$$

If we disregard the viscous forces and assume that the hydrostatic pressure is a function of the density only, then Eq. (27–1) is transformed into the known equation for an ideal fluid

$$\frac{\partial \boldsymbol{R}}{\partial t} + \operatorname{curl}[\boldsymbol{R} \times \boldsymbol{v}] = 0, \tag{27–2}$$

from which follows the circulation conservation theorem [98].

In a fluid undergoing relaxation $p = p(\varrho, s, \xi)$, where s and ξ are variable; hence circulation is not conserved even in the absence of viscous stresses and thermal conductivity.

Considering small deviations from the equilibrium state, we can show that in the first approximation the flow does not contain vorticity sources [197, 200]. Retaining in Eq. (27–1) quantities of the second order of smallness and denoting terms of first order of smallness by one prime and second order of smallness by a double prime, we obtain the equation for the curl

$$\frac{\partial \boldsymbol{R}''}{\partial t} - \frac{\eta}{\varrho_0}\Delta \boldsymbol{R}'' = \frac{p_\xi}{\varrho_0^2}\nabla \varrho' \times \nabla \xi' + \frac{\frac{4}{3}\eta + \zeta}{\varrho_0^3}\nabla \varrho' \times \nabla \frac{\partial \varrho'}{\partial t}. \tag{27–3}$$

Here the viscosity coefficients η and ζ are assumed to be constant. The physical meaning of the individual terms of Eq. (27–3) is obvious. The second term in the left-hand side defines the diffusion of the vortex through the fluid. The right-hand side of the equation describes the generation of the vortex. Its first term defines the rate of vortex generation in the given point as a result of relaxation processes and its second term defines the same quantity produced by the shear and bulk viscosity. The diffusion is determined solely by shear viscosity.

Propagation of small disturbances in fluids undergoing relaxation has a number of features [69, 196, 199, 200]. As was shown in [69], a sufficiently weak stationary shock wave in a fluid undergoing relaxation should not contain discontinuities (even in the absence of viscosity and heat conduction). Its thickness is determined in [60].

However, it is possible to have initial conditions which produce a low-intensity discontinuity in a fluid undergoing relaxation. Such, for example, is the disturbance which propagates in a quiescent gas in a cylindrical tube, if the piston instantaneously changes its velocity from zero to some small value. Analysis (within the framework of linearized equations) yields the following picture of propagation of this disturbance [196, 200]: the discontinuity which forms propagates with a velocity c_∞ of high-frequency sound [wave], being damped with the distance x according to the law $\exp(-x/l)$ and gradually yielding to a smoothly increasing disturbance which propagates with velocity c_0 of low-frequency sound [wave]. The characteristic length l is found to be equal to

$$l = \frac{2c_\infty^2}{c_\infty^2 - c_0^2} c_\infty \tau . \tag{27–4}$$

The relaxation processes are responsible for the quite complex structure of Mach lines in supersonic flow around obstacles [199, 200, 294]. Stationary discontinuities of intensity as small as desired are possible in these flows.

The equations of characteristics of one-dimensional flow which is described by Eqs. (26–27) have the form [272]

$$\frac{dx}{dt} = v \pm c_\infty , \quad \frac{dx}{dt} = v . \tag{27–5}$$

These equations do not contain the relaxation time τ and, consequently, the relaxation gasdynamic characteristics do not change when τ goes to zero and do not become characteristics of equilibrium gasdynamics which are determined by the velocity c_0. Hence from the viewpoint of relaxation gasdynamics, the experimentally observed (for small τ) Mach lines (along which the small disturbances propagate) do not coincide with the characteristics. Hence, transition to equilibrium gasdynamics which describes the motion of the fluid for small τ sufficiently well, requires special examination [197, 200].

Let us introduce the parameter $\mu = (c\tau/L)$, where c is the speed of sound (c_0 or c_∞), L is the characteristic macroscopic length (distance at which substantial changes occur in the flow variables); assuming that $\mu \ll 1$, we represent the solution of Eqs. (26–27) in the form of a series

$$y(\boldsymbol{r}, t) = y^{(0)}(\boldsymbol{r}, \ ) + \mu y^{(1)}(\boldsymbol{r}, t) + \cdots , \tag{27–6}$$

where $y(\boldsymbol{r}, t)$ is any of the flow variables.

Here it is found that the first approximation equation for $\xi^{(0)}$ describes the relaxation of parameter ξ with a characteristic time τ; quantities such as $\varrho^{(0)}$ remain constant in this approximation. Changes in these quantities are described only by the second approximation. However, then we obtain expressions which diverge with time. Hence this method of power expansion in μ results in equations of motion which are suitable only for small time intervals during which the flow variables do not appreciably change. The characteristics of equations which are thus obtained are determined by the velocity c_∞.

The difficulty of transition to equilibrium gasdynamics can be eliminated if we assume that the equations of equilibrium gasdynamics are, in substance, an approximate description of only a special class of solutions of relaxation gasdynamics. These solutions are characterized by the fact that the relative rapid relaxation of ξ is regarded as completed in the first approximation and the local nonequilibrium at any given time is fully determined by the fields of flow variables independently of the previous history of the fluid.

On the basis of these considerations, we assume the following expansion for ξ:

$$\xi = \xi^{(0)}(\boldsymbol{r}; \varrho, s, \boldsymbol{v}) + \mu \xi^{(1)}(\boldsymbol{r}; \varrho, s, \boldsymbol{v}) + \cdots; \tag{27–7}$$

where $\xi^{(0)} \equiv \xi_0(\varrho, s)$, and the coefficients of μ^n are functionals of the ϱ, s and $\boldsymbol{v}$ fields.

Under these conditions we obviously have the following expansions for the time derivatives of the flow variables:

$$\frac{\partial \varrho}{\partial t} = \mu N^{(1)}(\boldsymbol{r}; \varrho, s, \boldsymbol{v}) + \mu^2 N^{(2)}(\boldsymbol{r}; \varrho, s, \boldsymbol{v}) + \cdots, \tag{27–8}$$

$$\frac{\partial \boldsymbol{v}}{\partial t} = \mu \boldsymbol{U}^{(1)}(\boldsymbol{r}; \varrho, s, \boldsymbol{v}) + \mu^2 \boldsymbol{U}^{(2)}(\boldsymbol{r}; \varrho, s, \boldsymbol{v}) + \cdots, \tag{27–9}$$

$$\frac{\partial s}{\partial t} = \mu S^{(1)}(\boldsymbol{r}; \varrho, s, \boldsymbol{v}) + \mu^2 S^{(2)}(\boldsymbol{r}; \varrho, s, \boldsymbol{v}) + \cdots, \tag{27–10}$$

where the coefficients of μ^n also are functionals of $\varrho(\boldsymbol{r})$, $s(\boldsymbol{r})$, $v(\boldsymbol{r})$.

Expansions (27–7)—(27–10) are in substance entirely analogous to methods worked out in nonlenear mechanics and statistical physics in conjunction with the appearance of terms which increase with time in expansions such as (27–6). The coefficients of μ^n in (27–8)—(27–10) can be obtained in the explicit form by equating these equations to expressions for $\partial\varrho/\partial t$, etc., which are obtained from Eq. (26–27). For example, remembering that grad ϱ and div $\boldsymbol{v}$ are quantities of the order of μ, we find from Eqs. (27–8) and (26–26),

$$N^{(2)} = N^{(3)} = \cdots = 0. \tag{27–11}$$

Consulting the expansion for p

$$p = p_0(\varrho, s) + \mu\left(\frac{\partial p}{\partial \xi}\right)_{\xi_0} \xi^{(1)} + \cdots, \tag{27–12}$$

where p_0 is the equilibrium pressure, we find from Eqs. (27–9) and (26–27)

$$\mu \boldsymbol{U}^{(2)} = -\frac{1}{\varrho}\,\mathrm{grad}\,(p_\xi \xi^{(1)}). \tag{27–13}$$

From Eqs. (27–10) and (26–27) we find

$$\mu S^{(1)} = -(\boldsymbol{v}\cdot \mathrm{grad}\, s) \quad \mathrm{etc}. \tag{27–14}$$

Retaining in expansions (27–8)—(27–10) first-order terms for μ and using Eqs. (27–11)—(27–14), we obtain equations of motion of an ideal fluid with an equilibrium pressure. In order to obtain equations of motion in the next approximation, we must find $\xi^{(1)}$, which is contained in Eq. (27–13).

We find from Eqs. (27–8)—(27–10) that, with an accuracy to within μ^2,

$$\frac{\partial \xi^{(1)}}{\partial t} = \mu \frac{\partial \xi_0}{\partial \varrho} N^{(1)} + \mu \frac{\partial \xi_0}{\partial s} S^{(1)}. \tag{27–15}$$

Substituting into (27–15) Eqs. (27–11) and (27–14), we get

$$\frac{\partial \xi^{(1)}}{\partial t} = -\,\boldsymbol{v}\,\mathrm{grad}\,\xi_0 - \frac{\partial \xi_0}{\partial \varrho}\varrho\,\mathrm{div}\,\boldsymbol{v}. \tag{27–16}$$

On the other hand, in the same approximation

$$-K\varepsilon_\xi = -\mu \frac{1}{\tau}\xi^{(1)}. \tag{27–17}$$

Substituting Eqs. (27–16) and (27–17) into the last of Eqs. (26–27), we find $\xi^{(1)}$

$$\frac{\mu}{\tau}\xi^{(1)} = \frac{\partial \xi_0}{\partial \varrho}\varrho\,\mathrm{div}\,\boldsymbol{v}. \tag{27–18}$$

Substituting this value of $\xi^{(1)}$ into Eq. (27–13) we obtain an equation of motion in the form

$$\varrho \frac{\mathrm{d}\boldsymbol{v}}{\mathrm{d}t} = -\,\mathrm{grad}\,p_0 + \mathrm{grad}\,(\zeta\,\mathrm{div}\,\boldsymbol{v}), \tag{27–19}$$

where

$$\zeta = -\left(\frac{\partial p}{\partial \xi}\right)_{\xi=\xi_0} \tau \frac{\partial \xi_0}{\partial \varrho}\varrho \quad \text{or} \quad \zeta = \tau\varrho(c_\infty^2 - c_0^2) \tag{27–20}$$

plays the role of the second viscosity coefficient.

The derivation presented here assumes $\mu \ll 1$, however, it is not limited to small changes in flow variables and hence it is valid also outside the limits of the acoustic approximation.

[28] THE CASE OF SEVERAL NONEQUILIBRIUM PARAMETERS

The nonequilibrium states of a gas behind the front of a strong shock wave are usually defined by several nonequilibrium parameters ξ_i. The complex picture of nonequilibrium processes in this system is simplified appreciably if it is found that the characteristic times of the different processes are of different orders of magnitude (see Sect. 2).

Let us consider an element of a gas volume during a time period t after it has passed through a shock front. Let t be a quantity of the same order as τ_i, where τ_i is the characteristic time of one of the irreversible processes and let the inequalities

$$\tau_1 < \tau_2 < \tau_3 < \cdots \quad \text{etc.},$$

be valid. Then during a time t, excluding a relatively small initial time interval of the order of τ_{i-1}, only this process (with the characteristic time τ_i) takes place in the system. Processes with larger τ are too slow and those with a smaller τ are too rapid. Thus in the first approximation they can be regarded as completed.

Obviously, the motion is described by the equations of relaxation gasdynamics, where the rapidly relaxing parameters ξ_i are assumed to be equal to their equilibrium values, and the slow processes are regarded as being frozen. The formal method for obtaining these equations from the general system of equations in which the kinetics of all the relaxation processes is contained in the explicit form, is also obvious.

If the rapid processes are described by parameters $\xi_1, \xi_2, \ldots, \xi_{i-1}$, then the gasdynamic equations for the flow under study can be obtained by the natural generalization of expanding (27–7)

$$\xi_k = \xi_k^{(0)} + \xi_k^{(1)}, \tag{28–1}$$

where $\xi_k^{(0)}$ are determined by

$$\varepsilon_{\xi_k}(\varrho, s, \xi_1^{(0)}, \xi_2^{(0)}, \ldots, \xi_{i-1}^{(0)}, \xi_i, \xi_{i+1}, \ldots) = 0,$$

and $\xi_k^{(1)} = \xi_k^{(1)}(r\,; \varrho, s, v, \xi_i, \xi_{i+1}, \ldots)$ are functionals of the $\varrho, s, v, \xi_i, \xi_{i+1}, \ldots$ fields, etc.

Skipping the details of calculations, we only wish to note that in this approximation the rapid processes yield, as is the case of a single parameter, the bulk viscosity effect. The equations of motion describe the motion of a fluid undergoing relaxation. Its states of incomplete thermodynamic equilibrium are defined by an already smaller number of parameters: ξ_i, $\xi_{i+1}, \ldots$. Let us also note that the fact that the "slow" processes are frozen (which has not as yet been used) does not mean that the quantities $\xi_{i+1}, \xi_{i+2}, \ldots$, remain invariable with time, since, depending on their selection, they can vary with changes in ϱ and s.

Under the conditions being considered the overall picture of the structure of strong shock waves is represented in the form of a succession of relaxation zones, which are situated directly behind the shock front. One of the possible relaxation processes predominates in each of them. When a gas moves from one zone to another, its state is described by an increasingly smaller number of nonequilibrium parameters. These results are easily generalized to include the case when some of the characteristic times of irreversible processes are found to be quantities of the same order.

It is important to note the arbitrariness of the spatial separation of the zones. The passing from one zone to another, with other flow relationships, is in substance determined by changes on the time (or corresponding spatial) scale. The theory of shock waves (not only in this point) satisfactorily illustrates the situation, which is usual in physics, when physical concepts appear within the framework of certain approximations and, losing their definiteness, disappear when they are no longer contained within this framework [213, 69].

References

1. Alyamovskiy, V. N. and V. F. Kitayeva: *Optika i Spektrograf.*, **8**:152, 1960.
2. Angerer, E. von: *Technische Kunstgriffe bei physikalischen Untersuchungen*, Vieweg, Brunswick, 1952.
3. Anisimov, S. I.: *Zh. Teoret. Fiz.*, **31**:1941, 1961.
4. Artsimovich, L. A., S. Yu. Lukyanov, I. M. Podgornyy and S. A. Chuvatin: *Zh. Eksp. Teoret. Fiz.*, **33**, 3, 1957.
5. ———: *Upravlyayemyye termoyadernyye reaktsii (Controlled Thermonuclear Reactions)*, Moscow, Fizmatgiz, 1961.
6. Astrov, D. N. and A. S. Borovik-Romonav: *Prib. i Tekhn. Eksper.*, **1**:132, 1958.
7. Bazhenova, T. V.: In *Fizicheskaya gazodinamika i teploobmen (Physical Gasdynamics and Hear Transfer)*, Moscow, Acad. Sci., SSSR, 1961.
8. ———: In *Fizicheskaya gazodinamika i svoystva gazov pri vysokikh temperaturakh (Physical Gasdynamics and High-Temperature Gas Properties)*, Moscow, Acad. Sci., SSSR, 1963.
9. ——— and S. G. Zaytsev: In *Fizicheskaya gazodinamika, teploobmen i termodinamika gazov vysokikh temperatur (Physical Gasdynamics, Heat Transfer and Thermodynamics of High-Temperature Gases)*, Moscow, Acad. Sci., SSSR, 1962.
10. ——— and Yu. S. Lobastov: In *Fizicheskaya gazodinamika i teploobmen (Physical Gasdynamics and Heat Transfer)*, Moscow, Acad. Sci., SSSR, 1961.
11. ——— and ———: In *Fizicheskaya gazodinamika, teploobmen i termodinamika gazov vysokikh temperatur (Physical Gasdynamics, Heat Transfer and Thermodynamics of High-Temperature Gases)*, Moscow, Acad. Sci., SSSR, 1962.
12. ——— and ———: In *Fizicheskaya gazodinamika, i svoystva gazov pri vysokikh temperaturakh (Physical Gasdynamics and High-Temperature Gas Properties)*, Moscow, Acad. Sci., SSSR, 1963.
13. ——— and ———: Dokl. Acad. Sci., SSSR, **151**:519, 1963.
14. ——— and I. M. Naboko: *Ibid.*, **154**:401, 1964.
15. ———, O. A. Predvoditeleva and T. V. Nadezhdina: In *Fizicheskaya gazodinamika i svoystva gazov pri vysokikh temperaturakh (Physical Gasdynamics and High-Temperature Gas Properties)*, Moscow, Acad. Sci., SSSR, 1963.
16. Bao-Han-Lin and Yu. P. Lun'kin: *Zh. Tekhn. Fiz.*, **33**:234, 1963.
17. Bakhman, N. N.: *Inzh.-Fiz. Zh.*, **1**:94, 1960.
18. Biberman, L. M.: *Zh. Eksper. Teoret. Fiz.*, **19**:585, 1949.
19. ——— and B. A. Veklenko: *Ibid.*, **37**:164, 1959.
20. ———, V. S. Vorob'yev, E. G. Norman and I. T. Yokubov: *Kosmich. Issled.*, **2**, Issue 3, 1964.
21. ———, S. P. Yerkovich and V. N. Soshnikov: *Optika i Spektrograf.*, **7**:562, 1959.

22. ——— and E. G. Norman: *Ibid.*, **8**:433, 1960; **10**:565, 1961.
23. ——— and ———: *Quant. Spectr. Radiat. Transfer*, **3**, No. 3:221, 1963.
24. ———, V. G. Sevast'yanenko and I. T. Yakubov: *Teplofiz. Vysokikh Temp.*, **2**:333, 1964.
25. ——— and I. T. Yakubov: *Zh. Tekhn. Fiz.*, **33**:1244, 1963.
26. Boguslavskiy, S. A.: *Izbrannyye trudy po fizike (Selected Works on Physics)*, Moscow, Fizmatgiz, 1961.
27. Bond, J. W.: *Threshold of Space (Proc. of the Conference on Chem. Aeronomy, 1956)*, M. Zelikoff, Pergamon Press, 1957.
28. Borzunov, N. A., D. V. Orlinskiy and S. M. Osovets: *Zh. Eksper. Teoret. Fiz.*, **36**:717, 1959.
29. Borisov, A. A., S. M. Kogarko and A. V. Lyubimov: *Prikl. Matemat. i Teoret. Fiz.*, **3**:175, 1960.
30. Born, M.: *Optik*, Edwards, 1943.
31. Boronin, A. P. and T. G. Ignat'yeva: In *Fizicheskaya gazodinamika i svoystva gazov pri vysokikh temperaturakh (Physical Gasdynamics and High-Temperature Gas Properties)*, Moscow, Acad. Sci., SSSR, 1963.
31a. Brandt, N. B.: *Prib. i Tekhn. Eksper.*, **3**:189, 1961.
32. Brandt, A. A. and R. Kh. Kurtmulayev: *Ibid.*, **6**:941, 1958.
33. Busygin, E. P. and G. K. Tumakayev: *Zh. Tekhn. Fiz.*, **34**:122, 1964.
34. Bustlov, M. M.: *Uspekhi Nauchn. Fotograf.*, **6**:76, 1957.
35. Vaynshteyn, E. Ye.: *Rentgenovskiye spektry atomov v molekulakh khimicheskikh soyedineniy i v splavakh (X-Ray Spectra of Atoms in Molecules of Chemical Compounds and in Alloys)*, Moscow, Acad. Sci., SSSR, 1950.
36. Guthrie, A. and R. K. Wakerling: *Vacuum Equipment and Techniques*, 1st ed., New York, McGraw-Hill, 1949.
37. Vasil'yev, L. A., A. G. Galanin, I. V. Yershov and G. N. Suntsov: *Prib. i Tekhn. Eksper.*, **3**:195, 1964.
38. ———, S. S. Semenov and Ye. A. Tarantov: Izvestiya Acad. Sci., SSSR, Department of Technical Sciences, **11**:186, 1957.
39. Gaydon, A.: *Spectroscopy and Combustion Theory*, 2nd ed., London, Chapman & Hall, 1948.
40. Gaydon, A. G. and H. G. Wolfhard: *Flames, Their Structure, Radiation and Temperature*, London, Chapman & Hall, 1953.
41. Generalov, N. A.: Vestnik. Moscow State Univ., **2**:51, 1962.
42. ———: Dokl. Acad. Sci., SSSR. **148**:373, 1963.
43. ———: *Thesis*, Department of Physics, Moscow State Univ., 1963.
44. ——— and S. A. Kosev: *Prikl. Matemat. i Teoret. Fiz.*, **2**:64, 1960.
45. ——— and ———: Dokl. Acad. Sci., SSSR, **148**:552, 1963.
46. ——— and ———: *Prikl. Matemat. i Teoret. Fiz.*, **1**:145, 1963.
47. ——— and ———: Izvestiya Acad. Sci., SSSR, Fiz. Ser., **27**:1110, 1963.
48. ———, ———, V. D. Kosynkin and V. Ya. Ovechkin: Vestnik Moscow State University, No. 6, 1965.
49. ———, ——— and V. A. Maksimenko: *Optika i Spektrograf.*, **15**, Issue, 1:27, 1963.

49a. ——— and ———: *Journ. Quant. Spectr. Radiat. Transfer*, **5**, 1965.

50. Ginzburg, V. L.: *Rasprostraneniye elektromagnitnykh voln v plazme (Propagation of Electromagnetic Waves in Plasma)*, Moscow, Fizmatgiz, 1960.

51. Hirschfelder, J. O., C. F. Curtiss and R. B. Bird: *Molecular Theory of Gases and Liquids*, New York, John Wiley & Sons, 1954.

52. Golant, V. Ye.: *Zh. Tekhn. Fiz.*, **30**, Issue 11:1265, 1960.

53. Granovskiy, V. P.: *Elektricheskiy tok v gaze (Electrical Currents in Gases)*, Moscow, Gostekhizdat, 1952.

54. de Groot, S. R.: *Thermodynamics of Irreversible Processes*, Amsterdam, North-Holland Publishing Co., 1951.

54a. ——— and E. P. Mazur: *Non-Equilibrium Thermodynamics*, New York, John Wiley & Sons, 1962.

55. Gusev, V. N.: *Inzhen. Zh.*, **1**:161, 1961.

56. Dem'yanov, Yu. A.: *Priklad. Matemat. i Mekhan.*, **21**, No. 3:368, 1957.

57. ———: *Ibid.*, **21**, No. 4:473, 1957.

58. James, T. H. and G. C. Higgins: *Fundamentals of Photographic Theory*, 2nd ed., New York, Morgan and Morgan, 1960.

59. Dunayev, Yu. A., G. K. Tumakayev and A. M. Shukhtin: *Zh. Tekh. Fiz.*, **31**:1119, 1961.

60. D'yakov, S. P.: *Zh. Eksper. Teoret. Fiz.*, **27**, No. 6:728, 1954.

61. Yel'yashevich, M. A.: *Atomnaya i molekulyarnaya spektroskopiya (Atomic and Molecular Spectroscopy)*, Moscow, Fizmatgiz, 1962.

62. Yeremin, Ye. N.: *Zh. Fiz. Khim.*, **30**, Issue 7:1615, 1955.

63. Yeremina, B. G.: *Gazovyy analiz (Analysis of Gases)*, Goskhimizdat, 1955.

64. Zhigulev, V. N., Ye. A. Romishevskiy and V. K. Vertushkin: *Inzhen. Zh.*, **1**, Issue, 1:60, 1961.

65. Zaytsev, S. G.: *Prib. i Tekhn. Eksper.*, **6**:97, 1958.

66. ———, A. P. Shatilov, Ye, V. Lazareva and L. N. Trukhanova: In *Fizicheskaya gazodinamika i svoystva gazov pri vysokikh temperaturakh (Physical Gasdynamics and High-Temperature Gas Properties)*, Moscow, Acad. Sci., SSSR, 1963.

67. ———, ———, ———, ———, L. A. Averina and M. Yaichkov: *Ibid.*

68. Zel'dovich, Ya, B.: *Acta Physicochem.*, SSSR, **21**:577, 1946.

69. ———: *Teoriya udarnykh voln i vvedeniye v gazodinamiku (Theory of Shock Waves and Introduction to Gasdynamics)*, Moscow, Acad. Sci., SSSR, 1946.

70. ———, P. Ya. Sadovnikov and D. A. Frank-Kamenetskiy: *Okisleniye azota pri gorenii (Formation of Nitric Oxide Following Combustion)*, Moscow, Acad, Sci., SSSR, 1947.

71. ——— and Yu. P. Rayzer: *Fizika udarnykh voln i vysotemperaturnykh gidrodinamicheskikh yavleniy (Physics of Shock Waves and High-Temperature Gasdynamic Phenomena)*, Fizmatgiz, 1963.

72. Sommerfeld, A.: *Thermodynamics and Statistical Mechanics*, Academic Press 1955.

73. Ionov, V. P., G. N. Nikolayev, M. V. Gusev and O. I. Luneva: In *Fizicheskaya gazodinamika i svoystva gazov pri vysokikh temperaturakh (Physical*

Gasdynamics and High-Temperature Gas Properties), Moscow, Acad. Sci., SSSR, 1963.
74. Riddel, F. R.: *Hypersonic Flow Research*, Academic Press, 1962.
75. Ivanov, A. V.: *Prikl. Matemat. i Teoret. Fiz.*, **6**:99, 1964.
76. Kaptsov, N. A.: *Elektronika (Electronics)*, Moscow, Gostekhizdat, 1955.
77. Katushev, Ya. M. and V. I. Sheberstov: *Osnovy teorii fotograficheskikh protsessov (Fundamentals of the Theory of Photographic Processes)*, Moscow, "Isskustvo," 1964.
78. Kitayeva, V. F., V. V. Obukhob-Denisov and N. N. Sobolev: *Optika i Spektrograf.*, **12**, Issue 2:178, 1962.
79. Knorre, V. G. and G. I. Kozlov: *Inzhen.-Fiz. Zh.*, **6**, No. 2:109, 1963.
80. Kogan, V. I.: In *Fizika plazmy i problemy upravlyamykh termoyadernykh reaktsii (Plasma Physics and Problems of Controlled Thermonuclear Reactions)*, Acad. Sci. SSSR **4**:258, 1958.
81. Kozlov, G. I. and V. G. Knorre: *Inzhen.-Fiz. Zh.*, IV, No. 7:11, 1961.
82. ——— and ———: *Combustion and Flame*, **6**, No. 4:253, 1962.
83. Kolb, A.: In *Magnetohydrodynamics*, R. K. M. Landshoff, ed., Stanford Univ. Press, 1957.
84. Kolesnikov, V. N. and L. V. Leskov: *Uspekh: Fiz. Nauk*, **65**:3, 1958.
85. Compton, A. H. and S. K. Allison: *X-Rays in Theory and Experiment*, Van Nostrand, 1935.
86. Kondrat'yev, V. N.: *Spektroskopicheskoye izucheniye khimicheskikh gazovykh reaktsiy (Spectroscopic Study of Chemical Reactions in Gases)*, Moscow, Acad. Sci., SSSR, 1944.
87. ———: *Kinetika khimicheskikh gazovykh reaktsiy (Kinetics of Chemical Reactions in Gases)*, Acad. Sci., SSSR, 1958.
88. Kondrat'yev, K. Ya.: *Luchistyy teploobmen v atmosfere (Radiant Heat Transfer in the Atmosphere)*, Moscow, Gostekhizdat, 1956.
89. Korolev, Ye. I.: *Osnovy vakuumnoy tekhniki (Fundamentals of Vacuum Techniques)*, "Energiya," 1964.
90. Krivtsova, N. V. and Yu. P. Lun'kin: *Zh. Tekh. Fiz.*, **32**:69, 1962.
91. Kudryavtsev, Ye. M., N. N. Sobolev, L. N. Tunitskiy and F. S. Fayzullov: *Trudy FIAN*, **18**:159, 1962.
92. Kuznetsov, I. M.: *Inzhen.-Fiz. Zh.*, **9**:17, 1960.
93. ———: *Zh. Tekh. Fiz.*, **34**:624, 1964.
94. Lagar'kov, A. N. and I. T. Yakubov: *Optika i Spektroskopiya*, **14**:199, 1963.
95. Ladyzhenskiy, M. D. and L. G. Chernokova: *Prikl. Matemat. i Tekhn. Fiz.*, **2**:3, 1961.
96. Landau, L. D. and E. M. Lifshits: *Mechanics*, Addison-Wesley, 1960.
97. ——— and ———: *Quantum Mechanics – Non-Relativistic Theory*, Addison-Wesley, 1958.
98. ——— and ———: *Electrodynamics of Continuous Media*, Addison-Wesley, 1960.
99. Leontovich, M. A.: *Zh. Eksper. Teoret. Fiz.*, **6**:561, 1936.

100. Lobastov, Yu. S.: In *Fizicheskaya gazodinamika i svoystva gazov pri vysokikh temperaturakh (Physical Gasdynamics and High-Temperature Gas Properties)*, Moscow, Acad. Sci., SSSR, 1963.
101. Loytsyanskiy, L. G.: *Laminarnyy pogranichnyy sloy (Laminar Boundary Layer)*, Moscow, Fizmatgiz, 1962.
102. Losev, S. A.: Dokl. Acad. Sci., SSSR, **120**, No. 6:1291, 1958.
103. ———: *Nauchnyye doklady vysshey shkoly (Scientific Reports of High Schools)*, *Fizichesko-matematicheskiye nauki*, **5**:197, 1958.
103a. ———: *Thesis*, Physics Department, Moscow State Univ., 1958.
104. ———: Vestnik Moscow State Univ., **5**:47, 1960.
105. ———: Dokl. Acad. Sci., SSSR, **141**:894, 1961.
106. ——— and N. A. Generalov: *Prob. i Tekhn. Eksper.*, **3**:108; correction given in **5**:150, 1959.
107. ——— and ———: Dokl. Acad. Sci., SSSR, **133**:872, 1960.
108. ——— and ———: *Ibid.*, **141**:1072, 1961.
109. ——— and ———: In *Molekulyarnaya Spektroskopiya (Molecular Spectroscopy)*, Moscow-Leningrad, Acad. Sci., SSSR, II:15, 1963.
110. ——— and ———: In *Fizicheskiye problemy spektroskopii (Physical Problems in Spektroscopy)*, Acad. Sci., SSSR, **1**:114, 1962.
111. ———, ——— and V. A. Maksimenko: Dokl. Acad. Sci., SSSR, **150**, No. 4:839, 1963.
112. ———, ——— and L. B. Terebenina: *Optika i spektrograf.*, **8**:569, 1960.
113. ——— and A. I. Osipov: *Uspekhi Fiz. Nauk*, **74**:393, 1961.
114. Luk'yanov, S. Yu.: *Fotoelementy (Photoelements)*, Moscow, Acad. Sci., SSSR, 1947.
115. Lun'kin, Yu. P.: *Zh. Tekh. Fiz.*, **30**:622, 1960.
116. Mazing, M. A.: *Trudy FIAN*, **15**, 1961.
117. ——— and N. A. Vrubleskaya: *Optika i Spektroskopiya*, **13**, No. 3:308, 1962.
117a. Makarov. Yu, V. and A. M. Maksimov: *Zh. Tekh. Fiz.*, **35**:223, 1965.
118. Malyshev, G. M., G. T. Razdobarin and L. V. Sokolova: *Zh. Tekh. Fiz.*, **33**, No. 2:191, 1963.
119. Mandel'shtam, L. I. and M. A. Leontovich: *Zh. Eksper. Teoret. Fiz.*, **7**:438, 1937.
120. ——— and M. A. Mazing: Izvestiya Acad. Sci., SSSR, *Fiz. Ser.*, **23**, No. 8:1017, 1959.
121. ——— and N. K. Sukhodrev: Izvestiya Acad. Sci., *Fiz. Ser.*, **19**:11, 1955.
121a. Maslennikov, V. G., I. G. Pariyskiy, S. I. Rozov and A. M. Studenkov: *Prikl. Matemat. i Tekhn. Fiz.*, **1**:85, 1965.
122. Matveyev, V. V. and A. D. Sokolov: *Fotoumnozhiteli v stsintillyatsionnykh schetchikakh (Photomultipliers in Scintillation Counters)*, Moscow, Atomizdat, 1962.
123. Massey H. S. W. and E. H. S. Burhop: *Electronic and Ionic Impact Phenomena*, Fairlawn, N. J., Oxford University Press, 1952.
124. Millman, J. and H. Taub: *Pulse and Digital Circuits*, McGraw-Hill, 1956.

125. Mitchell, A. C. G. and M. W. Zemansky: *Resonance Radiation and Excited Atoms*, Cambridge University Press, 1934.
126. Model', I. Sh. *Zh. Eksper. Teoret. Fiz.*, **32**:714, 1957.
127. Nikitin, Ye. Ye.: Dokl. Acad. Sci., SSSR, **116**:585, 1957.
128. ———: *Ibid.*, **119** 526, 1958.
129. ———: *Optika i Spektrograf.*, **6**:141, 1959.
130. ———: Dokl. Acad. Sci., SSSR, **124**:1085, 1959.
131. ———: *Optika i Spektrograf.*, **9**:16, 1960.
132. ———: Dokl. Acad. Sci., SSSR, **132**:395, 1960.
133. ———: *Ibid.*, **135**:1442, 1960.
134. ———: *Kinetika i Kataliz*, **3**:322, 1962.
135. ———: *Ibid.*, **3**:380, 1962.
135a. ———: *Sovremennyye teorii termicheskogo raspada i izomerizatsiya molekul v gazovoy faze (Modern Theories of Thermal Disintegration and Isomerization of Molecules in the Gas Phase)*, "Nauka," 1964.
36.1 ——— and N. D. Sokolov: Dokl. Acad. Sci., SSSR, **124**:366, 1959.
37.1 Novikov, N. P.: *Ibid.*, **147**, No. 3:597, 1962.
38.1 ———: *Prikl. Matemat. i Tekhn. Fiz.*, 6:22, 1962.
39.1 Osipov, A. I.: *Nauchnyye doklady vysshey shkoly (Scientific Reports of High Schools), Ser. Fiz.*, 4:149, 1958.
140. ———: Vestnik Moscow State Univ., III, No. 4:45, 1958.
141. ———: Dokl. Acad. Sci., SSSR, **130**:523, 1960.
142. ———: Vestnik Moscow State Univ., III, No. 4:96, 1960.
143. ———: *Ibid.*, **1**:13, 1961.
144. ———: Dokl. Acad. Sci., SSSR, **137**:833, 1961.
145. ———: *Zh. Fiz. Khim.*, **35**:1524, 1961.
146. ———: Dokl. Acad. Sci., SSSR, **143**:1392, 1962.
147. ———: Vestnik Moscow State Univ., III, No. 2:41, 1962.
148. ———: *Zh. Fiz. Khim.*, **36**:1798, 1962.
149. ———: *Trudy XIII Vsesoyuznoy konferentsii po spektroskopii (Proc. of Thirteenth All-Union Conference of Spectroscopy)*, **1**:188, 1962.
150. ———: *Kinetika i Kataliz*, **4**:487, 1963.
151. ———: *Zh. Fiz. Khim.* (in print).
152. ———: Izvestiya Acad. Sci., *Fiz. Ser.* (in print).
153. ———: *Prikl. Matemat. i Tekhn. Fiz.*, 1:41, 1964.
154. ——— and Ye. V. Stupochenko: Izvestiya Acad. Sci., SSSR, *Fiz. Ser.*, 8:99, 1960.
155. ——— and ———: *Uspekhi Fiz. Nauk*, **79**:81, 1963.
156. Ferri, A.: *Fundamental Data Obtained from Shock-Tube Experiments*, Pergamon Press, 1961.
157. Patterson, G. N.: *Molecular Flow of Gases*, John Wiley & Sons, 1956.
158. Penner, S. S.: *Quantitative Molecular Spectroscopy and Gas Emissivities*, Addison-Wesley, 1959.
159. Petschek, H. I. and S. Byron: In *Threshold of space (Proc. of the Conference on Chem. Aeronomy)*, 1956, M. Zelikoff, ed., Pergamon Press, 1957.

160. Predvoditelev, A. S., Ye. V. Stupochenko, Ye. V. Samuylov, I. P. Stakhanov, A. S. Pleshanov and I. B. Rozhdestvenskiy: *Tablitsy termodinamicheskikh funktsiy vozdukha (dlya temperature ot 6000 do 12,000°K i davleniy ot 0.001 to 1000 atmosfer)* (Tables of Thermodynamic Functions of Air. For temperatures from 6000 to 12,000 °K and pressures from 0.001 to 1000 atm.) (Published in English translation by ATS, Glen Ridge, N. J.), Moscow, Acad. Sci. SSSR, 1957.

161. ———, ———, A. S. Pleshanov, Ye. V. Samuylov, and I. B. Rozhdestvenskiy: *Tablitsy termodinamicheskikh funktsiy vozdukha (dlya temperatur ot 12,000 to 20,000°K i davleniy ot 0.001 do 1000 atmosfer)* (Tables of Thermodynamic Functions of Air. For temperatures from 12,000 to 20,000 °K and pressures from 0.001 to 1000 atm.) (Published in English translation as [160]), Moscow, Acad. Sci., SSSR, 1959.

162. ———, ———, I. B. Rozhdestvenskiy, Ye. V. Samuylov and A. S. Pleshanov: *Tabilitsy gazodinamicheskikh i termodinamicheskikh velichin potoka vozdukha za pryamym skachkom uplotneniya (Tables of Gasdynamic and Thermodynamic Flow Variables for Air Behind a Normal Shock Front)*, Moscow, Acad. Sci., SSSR, 1960.

163. ———, ———, V. P. Ionov, A. S. Pleshanov, I. B. Rozhdestvenskiy and Ye. V. Samuylov: *Termodinamicheskiye funktsii vozdukha dlya temperatur ot 1000 do 12,000°K i davleniy ot 0.001 to 1000 (grafiki funktsii* (Charts of Thermodynamic Functions of Air. For temperatures of 1000 to 12,000 °K and Pressures of 0.001 to 1000 atm.) [English translation by ATS], Moscow, Acad. Sci., SSSR, 1960.

164. ———, ———, A. S. Pleshanov, Ye. V. Samuylov and I. B. Rozhdestvenskiy: *Tablitsy termodinamicheskikh funktsii vozdukha (dlya temperatur ot 200 to 6000°K i davleniy ot 0.00001 do 100 atmosfer* (Tables of Thermodynamic Functions of Air. For Temperatures from 200 to 6000 °K and Pressures from 0.00001 to 100 atm.), Moscow, VTs Acad. Sci., SSSR, 1962.

165. ———, ———, I. B. Rozhdestvenskiy, Ye. V. Samuylov and A. S. Pleshanov: *Tablitsy gazodinamicheskikh i termodinamicheskikh velichin potoka vozdukha za pryamym skachkom uplotneniya (dlya skorostey nabegayushchego potoka do 4500 m/sek)* (Tables of Gasdynamic and Thermodynamic Flow Variables for Air Behind a Normal Shock Front [for incoming flow velocities up to 4500 m/sec], Moscow, VTs Acad. Sci., SSSR, 1962.

166. Prinsgheim, P.: *Fluorestsentsiya i fosforestsentsiya (Fluorescence and Phosphorescence)* (Most likely a translation from German translation), IL, 1951.

166a. Prokof'yev, V. A.: *Uchenyye zapiski,* Moscow State Univ., *Mekhanika,* **172**:79, 1954.

167. Prokof'yev, V. K. *Fotograficheskiye metody spektral'nogo analiza* (Photographic Methods of Spectral Analysis), Gostekhizdat, 1951.

168. Rakhmatullin, Kh. A. and S. S. Semenov: *Vvodnaya stat'ya k sborniku "Udarnyye truby"* (Introductory article to the collection, *Shock Waves*), IL, 1962.

169. Riemann, B.: Sochineniya (Collected Works) translated from the German, p. 376, Moscow-Leningrad, OGIZ, 1948.
170. Resler, E. and B. Cary: In *Threshold of Space (Proc. of the Conference on Chem. Aeronomy)*, 1956, M. Zelikoff, ed., Pergamon Press, 1957.
171. Rozhdestvenskiy, D. S.: *Raboty po anomal'noy dispersii v parakh metallov (Works on the Anomalous Dispersion in Metal Vapors)*, Moscow-Leningrad, Acad. Sci., SSSR, 1951.
172. Rozhdestvenskiy, I. B.: In *Fizicheskaya gazodinamika (Physical Gasdynamics)*, Moscow, Acad. Sci., SSSR, 1959.
173. Rusanov, V. D.: *Sovremennyye metody izucheniya plazmy (Modern Methods of Plasma Study)*, Moscow, Atomizdat, 1962.
174. Ryabinin, Yu. N.: *Doctoral dissertation,* Inst. Khim. Fiz., Moscow, Acad. Sci, SSSR, 1950.
175. Salamandra, G. D., T. V. Bazhenova, S. G. Zaytsev, R. I. Soloukhin, I. M. Naboka ond I. K. Sevast'yanova: *Nekotoryye metody issledovaniya bystroprotekayushchikh protsessov i ikh primeneniye k izucheniyu formirovaniya detonatsii (Some Methods for the Study of Rapidly Occurring Processes and their Application to the Study of Denotation Formation)*, Moscow, Acad. Sci., SSSR, 1960.
175a. Safaryan, M. N. and Ye. V. Stupochenko: *Prikl. Matemat. Tekh. Fiz.*, **4**:29, 1964.
176. Safronov, B. G., G. G. Aseyev and Yu. S. Azovskiy: *Zh. Tekhn. Fiz.*, **31**:1066, 1961.
177. Sayasov, Yu. S.: Dokl. Acad. Sci., SSSR, **146**:409, 1962.
178. Sevast'yanenko, V. G. and I. T. Yakubov: *Optika i Spektroskopiya*, **16**:3, 1963.
179. Semenov, S. S.: Dokl. Acad. Sci., SSSR, **114**, No. 4:841, 1957.
180. Sobel'man, I. I.: *Vvedeniye v teoriyu atomnykh spektrov (Introduction to the Theory of Atomic Spectra)*, Moscow, Fizmatgiz, 1963.
181. ——— and L. A. Vaynshteyn: *Optika i Spektrograf.*, **6**:440, 1959.
182. Sobolev, N. N.: *Trudy FIAN*, **6**, 1956.
183. ———, A. V. Potapov, V. F. Kitayeva, F. S. Fayzullov, V. N. Alyamovskiy, Ye. T. Antropov and I. A. Isayev: Izvestiya Acad. Sci., SSSR, *Fiz. Ser.*, **32**:730, 1958.
184. ———, ———, ———, ———, ———, ——— and ———: *Optika i Spektrograf.*, **6**:284, 1959.
185. ———, Ye, T. Antropov, Ye. F. Gippous, A. P. Dronov, N. I. Krindach, Ye. M. Kudryatsev, A. I. Pechenov, A. G. Sviridov, L. N. Tunitskiy and V. P. Cheremisinov: *Report to the 15th Conference of Spectroscopy*, Minsk, 1963.
186. *Sovremennoye sostoyaniya aerodinamiki bol'shykh skorostey (The Present State of High-Velocity Aerodynamics)*, Vol 1 (Russian translation), II, 1955.
187. Solov'yev, V. A.: *Akusticheskiy zhurnal*, **7**:337, 1961.
188. ———: *Ibid.*, **9**, 1963.
189. Soloukhin, R. I.: *Udarnyye volny i detonatsiya v gazakh (Shock Waves and Detonation in Gases)*, Fizmatgiz, 1963.

190. ———: *Prikl. Matemat. Tekhi Fiz.*, **6**: 138, 1963.
191. Soshnikov, V. N.: *Optika i Spektrograf.*, **10**:448, 1962.
192. ———, *Uspekhi Fiz. Nauk*, **74**:61, 1961.
193. Stupochenko, Ye. V. and A. I. Osipov: *Zh. Fiz. Khim.*, **32**: 1673, 1958.
194. ——— and ———: *Ibid.*, **33**: 1526, 1959.
195. ———, Ye. V. Samuylov, A. S. Pleshanov and I. B. Rozhdestvenskiy: *Zh. Az. Khim*, **34**, No. 6, 1960.
196. ——— and I. P. Stakhanov: Dokl. Acad. Sci., SSSR, **117**, No. 1:65, 1957.
197. ——— and ———: *Ibid.*, **134**, No. 4:782, 1960.
198. ———, ———, Ye. V. Samuylov, A. S. Pleshanov and I. B. Rozhdestvenskiy: In *Fizicheskaya gazodinamika (Physical Gasdynamics)*, Moscow, Acad. Sci. SSSR. 1959.
199. Stakhanov, I. P. and Ye. V. Stupochenko: Dokl. Acad. Sci., SSSR, **134**, 5: 1044, 1960.
200. ——— and ———: *Prikl. Matemat. i Tekhn. Fiz.*, 2:2, 1963.
201. Stulov, V. P. and G. F. Telenin: Izvestiya Acad. Sci., SSSR, *Mekhanika*, 1:3, 1965.
202. Tuma*k*ayev, G. K. and V. R. Lazovskaya: *Zh. Tekh. Fiz.*, **34**, 1879, 1964.
203. *Udarnyye truby (Shock Tubes)*, collection of translations, IL, Moscow, 1962.
204. Unsöld, A.: *Physik die Sternatmosphären*, 1*st* ed., Russian translation published by IL, 1949.
205. Fayzullov, F. S.: *Trudy FIAN*, **18**, 1962.
206. Fayzullov, S. F. S., N. N. Sobolev and E. M. Kudryavtsev: Dokl. Acad. Sci., SSSR (Soviet Physics Doklady), **127**: 541, 1959.
207. ———, ——— and ———: *Optika i Spektrograf.*, **8**: 585, 1960.
208. ———, ——— and ———: *Ibid.*, **8**: 761, 1960.
209. Fayn, V. M.: *Uspekhi Fiz. Nauk*, **79**:641, 1963.
210. Fowler, R. H. and E. A. Guggenheim: *Statistical Thermodynamics*, Cambridge, Cambridge University Press, 1949.
211. Ladenburg, R. W., B. Lewis, R. N. Pease and H. S. Taylor: *Physical Measurements in Gas Dynamics and Combustion*, Vol. 9: *High-Speed Aerodynamics and Jet Propulsion*, Princeton University Press, 1954.
212. Filippova, T. I., N. V. Filippov, V. V. Zhurin and V. P. Vinogradov: *Yadernyy sintez*, **1**: 195, 1961.
213. Fok, V. A.: *Uspekhi Fiz. Nauk*, **16**: 1070, 1936.
214. Frenkel', Ya. I.: *Ibid.*, **20**:84, 1938.
214a. ———: *Kineticheskaya teoriya zhidkostey (The Kinetic Theory of Fluids)*, Acad. Sci., SSSR, 1945.
215. Frish, S. E.: *Uspekhi Fiz. Nauk*, **43**: 512, 1951.
216. ———: *Opticheskiye spektry atomov (Optical Spectra of Atoms)*, Moscow, Fizmatgiz, 1963.
217. Kholev, S. R.: In *Voprosy magnitnoy gidrodinamiki i dinamiki plazmy (Problems of Magnetodrohydynamics and of Plasma Dynamics)*, Riga, Latgosizdat, p. 93, 1959.
218. ——— and L. I. Krestnikova: *Izvestiya Vuzov, Fizika*, **1**:29, 1960.

219. ——— and D. S. Poltavchenko: *Dokl. Acad. Sci.*, SSSR, **131**:1060, 1960.
220. Chapman, S. and T. G. Cowling: *The Mathematical Theory of Non-Uniform Gases*, Cambridge, Cambridge University Press, 1953.
221. Chechik, I. O., S. M. Faynshteyn and T. M. Lifshits: *Elektronnyye umnozhiteli (Electronic Multipliers)*, 2nd ed., Moscow, Gostekhizdat, 1957.
222. Shpigel', I. S.: *Zh. Eksper. Teoret. Fiz.*, **36**:411, 1959.
223. Von Engel, A.: *Ionized Gases,* Fair Lawn, N. J., Oxford University Press, 1955.
224. Allen, R. A., J. C. Keck and J. C. Camm: Phys. Fluids, **5**, 3:284, 1962.
225. Alpher, R. A. and D. R. White: Ideal Theory of Shock Tubes with Area Change near Diaphragm, (quoted from [354]).
226. ——— and ———: J. Fluid Mech., **3**, 5:457, 1958.
227. ——— and ———: Phys. Fluids, **1**:452, 1958.
228. ——— and ———: Phys. Fluids, **2**:153, 1959.
229. ——— and ———: Phys. Fluids, **2**:162, 1959.
230. Amdur, I.: Planetary and Space Science, **3**:228, 1961.
231. Anderson, G. F.: J. Aero/Space Sci., **26**, 3:184, 1959.
232. Anderson, J. M.: Rev. Sci. Instr., **32**, 8:975, 1961.
233. Andersen, W. H. and D. F. Hornig: J. Chem. Phys. **24**:717, 1956.
234. ——— and ———: Mol. Phys., **2**:49, 1959.
235. ——— and ———: Phys. Fluids, **4**:650, 1961.
236. Atomic and Molecular Processes, D. Bates, ed,. Interscience, N. Y., 1962 (Russian translation published by "Mir" Publishing House, Moscow, 1964).
237. Atkinson, W. R.: One-Dimensional Fluid Flow Produced by Confined Sparks (quoted from [354]).
238. Back, T. A. and K. Andersen: Mat. Fys. Medd. Dan. Vid. Selsk, **33**:7, 1961.
239. Ballard, H. V. and D. Venable: Phys. Fluids, **1**:225, 1958.
240. Bartlett, M. S. and I. E. Moyal: Proc. Cambr. Phil. Soc., **45**:545, 1949.
241. Basco, N., A. Collear and R. Norrish: Proc. Roy. Soc., **A260**:459, 1961.
242. Basu, S.: Phys. Fluids, **3**, 3:456, 1960.
243. Bauer, H., H. Kneser and E. Sitting: J. Chem. Phys., **30**:1119, 1959.
244. Bauer, S. H., G. L. Schott and R. E. Duff: J. Chem. Phys. **28**, 6:1089, 1958.
245. Becker, R.: Z. Physik, **8**:321, 1922.
246. Beckerle, J. C.: J. Chem. Phys., **21**:2034, 1953.
247. Benson, S. W. and A. E. Axworthy: J. Chem. Phys., **26**:1718, 1957.
248. ——— and G. C. Berend: J. Chem. Phys., **38**:25, 1963.
249. ———, ——— and J. C. Wu: J. Chem. Phys., **37**:1386, 1962.
250. ———, ——— and ———: J. Chem. Phys., **39**:2777, 1963.
251. Bernstein, H.: J. Aeron. Sci., **20**, 11:790, 1953.
252. Bershader, D.: Rev. Sci. Instr., **20**:260, 1949.
253. Bird, G. A.: J. Fluid Mech., **5**:60, 1959.
254. Blackman, V. H.: J. Fluid Mech., **1**, Part 1, 61, 1956.
255. Bleakney, W. and A. H. Taub: Rev. Mod. Phys., **21**:584, 1949.
256. ———, D. K. Weiner and G. H. Fletcher: Rev. Sci. Instr., **20**:807, 1949.
257. Bloxsom, D. E.: J. Appl. Phys., **29**, 7:1128, 1958.

258. Blythe, P. A.: J. Fluid Mech., **10**, Part 1, 33, 1961.
259. Bodenstein, M.: Roszniki Chemie, **18**:374, 1938.
260. Bond, I.: Phys. Rev., **105**:1683, 1957.
261. ———: Jet Propulsion, **97**:228, 1958.
262. Bortner, M. H.: Planetary and Space Sci., **3**:99, 1961.
263. Boyd, R. L. F.: Proced. Roy. Soc., **64**:195, 1951.
264. Brabbs, T. A., F. E. Bells and S. A. Zlatarich: J. Chem. Phys., **38**:1939, 1963.
265. ———, S. A. Zlatarich and F. E. Bells: J. Chem. Phys., **33**:307, 1960.
266. Breadley, J. N.: Shock Waves in Chemistry and Physics, London, 1962.
267. ——— and G. B. Kistiakowsky: J. Chem. Phys., **35**:256, 1961.
268. ——— and ———: J. Chem. Phys., **35**:264, 1961.
269. Britton, D. and N. Davidson: Discussions Faraday Soc., **17**:58, 1954.
270. ———, ———, W. Gehman and G. Schott: J. Chem. Phys., **25**:804, 1956.
271. ——— and ———: J. Chem. Phys., **25**:810, 1956.
272. Broer, L. I. F.: J. Fluid Mech., **4**:276, 1958.
273. Bromberg, R.: Jet Propulsion, **26**:737, 1956.
274. Brout, R.: J. Chem. Phys., **22**:1189, 1954.
275. Byron, S. R.: J. Chem. Phys., **30**, 2:1380, 1959.
276. Camac, M. and A. Vaughan: J. Chem. Phys., **34**, 2:460, 1961.
277. ———: J. Chem Phys., **34**, 2:448, 1961.
278. ———, J. Camm, S. Feldman, Y. Kesk and C. Petty: Chemical Relaxation in Air, Oxygen and Nitrogen, IAS, N. Y., Preprint No. 802, (1958).
279. Camm, J. and I. Keck: Bull. Amer. Phys. Soc., ser. 2, **4**:19, 1959.
280. ———, B. Kivel, R. L. Taylor and J. D. Teare: J. Quant. Spectr. Radiat. Transfer, **1**:53, 1961.
281. Camm, J. C. and P. H. Rose: Phys. Fluids, **6**:663, 1963.
282. Cary, B.: Phys. Fluids, **8**:26, 1965.
283. Careri, G.: Nuovo cimento, **6**:94, 1949.
284. ———: Nuovo cimento, 7:155, 1950.
285. Carrington, T. and H. Davidson: J. Phys. Chem., **57**:418, 1953.
286. Chang, C. T.: Danish Atomic Energy Commission, Risö Report, No. 46, 1962.
287. Chartis, G., L. R. Doherty and T. D. Wilkerson: J. Chem. Phys., **27**:1415, 1957.
288. ——— and T. D. Wilkerson: Phys. Fluids, **2**:578, 1959.
289. Chen, C. J. and R. J. Emrich: Phys. Fluids, **6**:1, 1963.
290. Chester, W. Phil. Mag., **7**, 45:1293, 1954.
291. Chesik, I. P. and G. B. Kistiakowsky: J. Chem. Phys., **28**:956, 1958.
292. Chisnell, R. F.: J. Fluid Mech., **2**:286, 1957.
293. Christian, R. H. and F. L. Jarger: J. Chem. Phys., **23**:2042, 1955.
294. Clarke, J. F.: J. Fluid Mech., **7**, Part 4, 577, 1960.
295. Clouston, J. G., A. G. Gaydon and I. I. Glass: Proc. Roy. Soc., **A248**:429, 1958.
296. ——— and ———: Conf. Molecular Spectroscopy, London, Pergamon Press, 1959.

297. ——— and ———: Spectrochem. Acta, **14**:56, 1959.
298. ———, ——— and I. R. Hurle: Proc. Roy. Soc., **A252**, 1269:143, 1959.
299. Clyne, M. A. A. and B. A. Thrush: Nature, **189**:56, 1961.
300. Connor, J. V.: J. Acoust. Soc. Amer., **30**:297, 1958.
301. Cottrell, T. L. and J. C. McCoubrey: Molecular Energy Transfer in Gases, 1962.
302. ——— and N. Ream: Trans., Faraday Soc., **51**:159, 1955.
303. Cowan, G. R. and D. F. Hornig: J. Chem. Phys., **18**, 8:1008, 1950.
304. Cowling, T. C.: Phil. Mag., **33**:61, 1942.
305. Cunningam, S. P., F. R. Scott and R. F. Wenzel: Bull. Amer. Phys. Soc., **5**:124, 1960.
306. Curtiss, C. F. and F. T. Adler: J. Chem. Phys., **20**:249, 1952.
307. Daen, J. and P. C. T. de Boer: J. Chem. Phys., **36**, 5:1222–1228, 1962.
308. Davies, W. O.: J. Chem. Phys., **36**, 2:292, 1962.
309. De Boer, P. C. T.: Phys. Fluids, **6**:962, 1963.
310. Devins, J.: Electrochem. Soc., **103**:460, 1956.
311. Devonshire, A. F.: Proc. Roy. Soc., **A158**:269, 1937.
312. De Wette, F. W. and Z. I. Slawsky: Physica, **20**:1169, 1954.
313. Diesen, R. W.: J. Chem. Phys., **39**:2121, 1963.
314. ——— and W. J. Felmlee: J. Chem. Phys., **39**, 9:2115, 1963.
315. Donaldson, C. P. and R. D. Sullivan: The Effect of Wall Friction on the Strength of Shock Waves in Tubes and Hydraulic Jumps in Channels, NACA, TN 1942, 1949 (see [331, 435]).
316. Dorrance, W. H.: J. Aero/Space Sci., **28**:43, 1961.
317. Duff, R. E.: Phys. Fluids, **2**, 2:207, 1959.
318. ——— and N. Davidson: J. Chem. Phys., **31**, 4:1018, 1959.
319. ——— and J. L. Joung: Phys. Fluids, **4**, 7:812, 1961.
320. ——— and Webster: Bull. Amer. Phys. Soc., **4**:283, 1959.
321. Fairbeirn, A. R. and A. G. Gaydon: Nature, **175**:253, 1955.
322. ——— and ———: Proc. Roy. Soc., **A239**:464, 1954.
323. Fowler, R. G., W. R. Atkinson, W. D. Compton and R. J. Lee: Phys. Rev., **88**, 1:137, 1952.
324. Freedman, E. and J. W. Daiber: J. Chem. Phys., **34**, 4:1271, 1961.
325. Gardiner, W. C. Jr. and G. B. Kistiakowsky: J. Chem. Phys., **34**:1080, 1961.
326. Gaydon, A. G. and J. R. Hurle: Proc. Roy. Soc., **A262**:38, 1961.
327. ——— and I. R. Hurle: Eigth Symposium (Intern.) on Combustion, Baltimore, 309, 1962.
328. ——— and ———: The Shock Tube in High-Temperature Chemical Physics, London, Chapman & Hall Ltd., 1963.
329. ———, G. H. Kimbell and H. B. Palmer: Proc. Roy. Soc., **A276**:461, 1963.
330. Gilbarg, D. and Paolucci: J. Rat. Mech. and Analysis, **2**:617, 1953.
331. Glass, I. I.: Shock Tubes UTIA Review, No. 12, part 1, 1958.
332. ——— and G. N. Patterson: J. Aero Sci., **22**, 2:73, 1955.

333. Glansdorff, P.: Proc. of the Second International Symposium on Rarefied Gas Dynamics, N. Y., 1961.
334. ———: Phys. Fluids, **5**:371, 1962.
335. Glick, H. S.: Seventh Symp. (Intern.) Combust., Lnd. 98, 1958.
336. ———, J. J. Klein and W. Squire: J. Chem. Phys., **24**, 4:850, 1957.
337. ———, W. Squire and A. Hertzberg: Fifth Symp. (Intern.) Combust., Reinholds, 393, 1955.
338. ——— and W. H. Wurster: J. Chem. Phys., **27**:1224, 1957.
339. Gloersen, P.: Bull. Amer. Phys. Soc., **4**:283, 1959.
340. ———: Phys. Fluids, **3**:857, 1960.
341. ———: Phys. Fluids, **4**:790, 1961.
342. Grad, H.: Comm. Pure and Appl. Math., **5**:257, 1952.
343. Greene, E. F.: J. Amer. Chem. Soc., **76**:2127, 1954.
344. ———, G. R. Cowan and D. F. Hornig: J. Chem. Phys., **19**:427, 1951.
345. ——— and D. F. Hornig: J. Chem. Phys., **21**:617, 1953.
346. ——— and ———: J. Chem. Phys., **24**:767, 1956.
347. ——— and J. P. Toennies: Chemische Reaktionen in Stosswellen, Verlag, Darmstadt, 1959.
348. Greenspan, M.: J. Acoust. Soc. Amer., **31**:155, 1959.
349. Griem, H.: Astrophys. J., **132**:883, 1960.
350. Griem, H. R., A. C. Kolb and K. Y. Shen: Phys. Rev., **116**, 1:4, 1960.
351. Griem, H., A. C. Kolb and K. Y. Shen: Astrophys. J.
352. Griffith, W.: J. Appl. Phys., **21**:1319, 1950.
353. ———, D. Brickl and V. Blackman: Phys. Rev., **102**:1209, 1956.
353a. Gustafson, W. A.: Phys. Fluids, **3**:732, 1960.
354. Hall, J. G.: Shock Tubes UTIA Review, No. 12, p. II, 1958.
355. Hammerling, P., J. D. Teare and B. Kivel: Phys. Fluids, **2**:422, 1959.
356. Hand, C. W. and G. B. Kistiakowsky: J. Chem. Phys., **37**:1239, 1962.
357. Hansen, C. F., R. A. Early, F. E. Alzofon and F. C. Witteborn: Theoretical and Experimental Investigation of Heat Conduction in Air, NASA, TR R–27, 1960.
358. Hartunian, R. A.: Phys. Fluids, **4**, 9:1959, 1961.
359. ——— and P. V. Marrone: Phys. Fluids, **4**:535, 1961.
360. ———, A. H. Russo and P. V. Marrone: Jass, **27**, 8:587, 1960.
360a. Harwell, K. E. and R. G. Janh: Phys. Fluids, **7**:214, 1964.
361. Haught, A. F. Phys. Fluids, **5**:1337, 1962.
362. Hansen, K. and D. F. Hornig: J. Chem. Phys., **33**:913, 1960.
363. Henshall, B. D.: The Use of Multiple Diaphragms in Shock Tubes, ARS, Current Papers, No. 291, 1956. Translation: Collection "Shock Tubes," p. 190, Foreign Literature Press, 1962.
364. ———: On some Aspects of the Use of Shock Tubes in Aerodynamic Research, ARS Reports and Memoranda, No. 3044, Ind. 1957. Translation: Collection "Shock Tubes," p. 19, Foreign Literature Press, 1962.
365. Herman, R. and R. J. Rubin: Phys. Fluids, **2**:547, 1959.
366. ——— and K. E. Shuler: J. Chem. Phys. **29**:366, 1958.

367. Herzberg, A. and W. Smith: J. Appl. Phys., **25**:130, 1954.
368. Herzfeld, K. F. and T. A. Litovitz: Absorption and Dispersion of Ultrasonic Waves, N. Y., Acad. Press, 1959.
369. ———: Zeit. Phys., **156**:265, 1959.
370. Hey, J. S., J. T. Pinson and P. G. Smith: Nature, **182**:1220, 1958.
371. Hilbert, D.: Math. Ann., **72**:562, 1912.
372. ———: Grundzüge einer allgemeinen Theorie der linearen Integralgleichungen, Leipzig, 1912.
373. Hiraoka, H. and R. Hardwick: J. Chem. Phys., **36**:1715, 1963.
374. ——— and ———: J. Chem. Phys., **36**, 8:2164, 1962.
375. Hirsch, E. H.: J. Sci. Instr., **36**:477, 1959.
376. Hoenig, S. A.: Rev. Sci. Instr., **29**:704, 1958.
377. Holckes, R., G. R. Jounes, N. Pusat and W. Tempest: Trans. Farad. Soc., **58**:2342, 1962.
378. Hollyer: J. Appl. Phys., **27**:254, 1956.
379. Hooker, W. J.: Phys. Fluids, **4**, 12:1451, 1961.
380. ——— and R. C. Millikan: J. Chem. Phys., **38**, 1:214, 1963.
381. Hornig, D. F.: Phys. Rev., **72**:179, 1948
382. ———: J. Phys. Chem., **61**:856, 1957.
383. Howe, J. T. and J. R. Viegas: The Dissociative Relaxation of CO_2 behind Normal Shock Waves, Interplanetary Miss. Conf., Los Angeles, Calif., 1963.
384. Huber, P. W. and A. Kantrowitz: J. Chem. Phys., **15**:275, 1947.
385. Hugoniot, H.: J. École Polytechn., **57**, 1887; **58**, 1988.
386. Hurlbut, F.: J. Appl. Phys., **30**, 3:273, 1959.
387. Hurle, I. R. and A. G. Gaydon: Nature, **184**, Suppl., No. 24, 1854, 1954.
388. Jackson, J. M. and N. F. Mott: Proc. Roy. Soc., **A137**:703, 1932.
389. Jacobs, T. A. and K. R. Giedt: J. Chem. Phys., **39**:749, 1963.
390. ———, R. A. Hartunian, R. R. Geidt and R. Wilkins: Phys. Fluids, **6**:972, 1963.
391. Jahn, R. G.: Phys. Fluids, **5**, 6:678, 1962.
392. Jeans, J. H.: The Dynamical Theory of Gases, Cambridge, 1904 (4*th* ed,. 1925, also Dover, 1954).
393. Jellinek, K.: J. anorg. Chemie, **49**:261, 1906.
394. Johannesen, J.: Fluid. Mech., **10**:25, 1961.
395. Johnston, H. S. and W. Kornegay: Trans. Faraday Soc., **57**:1563, 1961.
396. ——— and E. Tschuikow-Roux: J. Chem. Phys., **36**, 2:463, 1962.
397. Jones, J. J.: Experimental Investigation of Attenuation of Strong Shock Waves in a Shock Tube with Hydrogen and Helium as Driver Gases, NACA, TN, 4072, 1957 (see [331, 354]).
398. Josephson, V.: J. Appl. Phys., **29**:30, 1958.
399. Karman, T. and N. S. Tsien: JAS, **5**:227, 1938.
400. Kaufman, F. and T. R. Kelso: J. Chem. Phys., **21**:751, 1953; **23**:1072, 1955.
401. Keck, J. C., J. Camm and B. Kivel: J. Chem. Phys., **28**:723, 1958.
401a. ———, ———, ——— and T. Wentink: Ann. of Phys., **7**, 1:1, 1959.
402. Kelly, B. T.: J. Acoust. Soc. Amer., **29**:1005, 1957.

403. Kirkwood, I. G. and W. W. Wood: J. Appl. Phys., **28**, 4:395, 1957.
404. Kistiakowsky, G. B. and G. G. Volpi: J. Chem. Phys., **27**:1141, 1957.
405. Kivel, B., P. Hammerling and J. D. Teare: Planetary and Space Sci., **3**:132, 1961.
406. Klein, A. F.: Phys. Fluids, **6**:310, 1963.
407. Knight, H. T. and D. Venable: Rev. Sci. Instr., **29**, 2:92, 1958.
408. Knorr, G.: Z. Naturforsch., **13a**:941, 1958.
409. Kolb, A.: Phys. Rev., **107**:345, 1957.
410. Kolb, A. C. and H. R. Griem: Atomic and Molecular Processes, Bates ed., p. 141, 1962 (see [236]).
411. Kornegay, W. and H. S. Johnston: Bull. Amer. Phys. Soc., **7**:183, 1962.
412. ——— and ———: J. Chem. Phys., **38**:2242, 1963.
413. Lamb, L. and S. C. Lin: J. Appl. Phys., **26**:754, 1957.
414. Landau, L. and E. Teller: Phys. Z. Sow., **10**:34, 1936.
415. Laporte and Wilkerson: JOSA, **50**:1293, 1960.
416. Lapworth, K. C. Hypersonic Flows, A. R. Collar and J. Tinkler eds., London, p. 316, 1960.
417. Levitt, B. P.: Trans. Farad. Soc., **58**:1789, 1962.
418. ———: Trans. Farad. Soc., **59**:59, 1963.
419. Lewis, B. and G. Elbe: Chem. Rev., **21**:247, 1937.
420. Light, J. C.: J. Chem. Phys., **36**:1016, 1962.
421. ———: J. Chem. Phys., **37**:2240, 1962.
422. Lin, S. C. and W. I. Fyfe: Phys. Fluids, **4**, 2:238, 1961.
423. ——— and B. Kivel: Phys. Rev., **114**, 4:1026, 1959.
424. ———, R. A. Neal and W. I. Fyfe: Phys. Fluids, **5**, 12:1633, 1962.
425. ———, E. L. Resler and A. R. Kantrovitz: J. Appl. Phys., **26**:95, 1955.
426. ——— and J. D. Teare: Phys. Fluids, **6**, 3:355, 1963.
427. Linzer, M. and D. F. Hornig: Physics of Fluids, **6**, 12:1661, 1963.
428. Lucasik, S. J. and J. E. Young: J. Chem. Phys., **27**:1149, 1957.
429. Mak, W. H.: Line-Reversal Temperature Measurements in Shock Tube Flows, Annual Progress Reports, UTIA, 1962.
430. Manheimer-Timnat, I. and W. J. Low: J. Fluid Mech., **6**:449, 1959.
431. Mark, H.: JAS, **24**:304, 1957.
432. Marlow, D., C. Nisenanger and W. Cady: J. Appl. Phys., **20**:771, 1949.
433. Marrone, P. V. and C. E. Treanor: Phys. Fluids, **6**, 3:1215, 1963.
334. Marshak, R. E.: Phys. Fluids, **1**:24, 1958.
435. Martin, W. A.: An Experimental Study of the Boundary Layer Behind a Moving Plane Shock Wave, UTIA, Report, No. 47, May 1957.
436. ———: JAS, **25**, 10:644, 1958.
437. Matthews, D. L.: Phys. Fluids, **2**:170, 1959.
438. Matthews, D.: J. Chem. Phys., **34**:639, 1961.
439. Maxwell, J. C.: Phil. Trans. Soc., **170**:231, 1879.
440. McChesney, M.: J. Sci. Instr., **38**, 12:496, 1961.
441. McCoubrey, J. C., R. C. Milward and A. R. Ubbelohde: Trans. Farad. Soc., **57**, Pt. 9, 1472, 1961.

442. McLean, E. A., C. E. Faneuff, A. C. Kolb, T. E. Milligan and H. Greim. Phys. Fluids, **3**:843, 1960.
443. Meador, W. E.: The Interaction between Nitrogen and Oxygen Molecules, NASA, Tech. Rep. R–68, 1960.
444. Mies, F. H.: J. Chem. Phys., **40**:523, 1964.
445. Millar, W. and N. H. Price: See [266].
446. Millikan, R. C. and D. R. White: J. Chem. Phys., **39**, 1:98, 1963.
447. ——— and ———: J. Chem. Phys., **39**:3209, 1963.
448. Mirels, H.: Attenuation in a Shock Tube due to Unsteady-Layer Action, NACA, Rep., No. 1333, 1957.
449. ———: The Wall Boundary Layer Behind a Moving Wave. Grenzschichtforschung, Symposium, Berlin, 1958. Translation: Collection "Shock Tubes," Foreign Literature Press, 1962.
450. ———: Phys. Fluids, **6**:1201, 1963.
450a. ———: AIAA-Journal, **2**:1, 1964.
451. Montroll, E. W. and K. E. Shuler: J. Chem. Phys., **26**:454, 1957.
452. ——— and ———: Advances in Chemical Physics, **1**:361, 1958.
453. Morduchow, M. and P. A. Libby: J. Aeronaut. Sci., **16**:674, 1949.
454. Mott-Smith, H. M.: Phys. Rev., **82**, 11:885, 1951.
455. Muckenfuss, C.: Phys. Fluids, **3**:320, 1960.
456. ———: Phys. Fluids, **5**, 2:165, 1962.
457. ———: Phys. Fluids, **5**, 11:1325, 1962.
458. Nagamatsu, H., K. Geiger and R. Sheer: ARS-Journal, **29**:332, 1959.
459. ———, ——— and ———: J. Aeron. Space Sci., **27**:241, 1960.
460. Nakai, S., C. Yamanaka and Y. Yamamura: Technol. Repts., Osaka Univ., **12**:217, 1962.
461. Nicholls, R.: Ann. geophys., **14**:208, 1958.
462. Niblett, B. and V. H. Blackman: J. Fluid Mech., **4**:191, 1958.
463. Palmer, H. B. and D. F. Hornig: J. Chem. Phys., **26**:98, 1957.
464. Parbrook, H. D. and W. Tempest: Acustica, **8**:345, 1958.
465. Parker, J. G.: Phys. Fluids, **2**, 4:449, 1959.
466. ———, C. E. Adams and R. M. Stavseth: J. Acoust. Soc. Amer., **25**:263, 1953.
467. Parkinson, W. H. and R. W. Nicholls: (See [205]).
468. Patch, R. W.: J. Chem. Phys., **36**, 7:1919, 1962.
469. Patterson, W. L. and E. F. Greene: J. Chem. Phys., **36**, 5:1146, 1962.
470. Penner, S. S.: Introduction to the Study of Chemical Reactions in Flow Systems, London, Butterworths Scientific Publications, 1955.
471. ———: J. Chem. Phys., **32**:617, 1960.
472. Petralla, S.: Nuovo cimento, **12**:241, 1955.
473. Petschek, H. E.: Phys. Rev., **84**:614, 1951.
474. ——— and S. Byron: Ann. of Phys., **1**, 3:270, 1957.
475. ———, P. Rose, H. Glick, A. Kane and A. R. Kantrovitz: J. Appl. Phys., **26**:83, 1955.
476. Pomerantz, J.: J. Quant. Spectrosc. Radiat. Transfer, **1**, No. 3–4, 185, 1962.

477. Present, R. P.: J. Chem. Phys., **31**:747, 1959.
478. Prigogine, I. and T. A. Bak: J. Chem. Phys., **31**:1368, 1959.
479. Pritchard, H.: J. Phys. Chem., **65**:504, 1961.
480. Rankine: Philos. Trans., **160**:277, 1870.
481. Rapp, D.: J. Chem. Phys., **32**:735, 1960.
482. ——— and P. Englander-Golden: J. Chem. Phys., **40**:573, 1964.
483. ——— and T. E. Sherp: J. Chem. Phys., **38**:2641, 1963.
483a. ———: J. Chem. Phys., **40**:2813, 1964.
484. Raw, C. J. G. and C. P. Ellis: J. Chem. Phys., **28**:1198, 1958.
485. Rayleigh: Proc. Roy. Soc., **A84**:247, 1910.
486. Resler, E. L., S. C. Lin and A. R. Kantrovitz: J. Appl. Phys., **23**, 12:1390, 1952.
487. ——— and M. J. Scheibe: J. Acoust. Soc. Amer., **27**:932, 1955.
488. ——— and ———: Ind. Eng. Chem., **47**:1182, 1955.
489. Rice, O. K.: Monatsch. Chemie, **90**:330, 1959.
490. ———: (quoted from [135]).
491. Rink, J. P.: J. Chem. Phys., **36**, 2:572, 1962.
492. ———: J. Ch. Phys., **36**:262, 1398, 1962.
493. ———, H. T. Knight and R. E. Duff: J. Chem. Phys., 34:1942, 1961.
494. Robben, F.: J. Chem. Phys., **31**:420, 1959.
495. Rosa, R. J.: Phys. Rev., **99**:633, 1955.
496. Rosciszewski, I. J.: J. Aero/Space Sci., **26**, 6:388, 1959; **28**, 2:164, 1961.
497. Rose, P. H. and W. Nelson: On the Effect of Attenuation on Gas Dynamic Measurements made in Shock Tubes, AVCO Res. Lab. RR–24, 1958. Also, Proc. *2nd* Shock Tube Symposium, Palo Alto, Calif., March 5–6, 1958.
498. Roshko, A.: Phys. Fluids, **3**, 6:835, 1960.
499. Rosen, P.: Appl. Phys., **25**:336, 1954.
500. ———: J. Chem. Phys., **22**:1045, 1954.
501. Roth, W.: J. Chem. Phys., **31**, 720:1683, 1959.
502. ———: J. Chem. Phys., **31**:844, 1959.
503. ———: J. Chem. Phys., **34**:999, 1961.
504. ———: J. Chem. Phys., **34**:2205, 1961.
505. ———: Phys. Fluids, **4**:788, 1961.
506. ———: J. Chem. Phys., **37**, 4:925, 1962.
507. ———: Fifth Intern. Conference on Ionization Phenomena in Gases, Munich, Germany, 1961; **1**:1101, Amsterdam, 1962.
508. ———: Bull. Amer. Phys. Soc., **7**:134, p. 2, 1962.
509. ———: Phys. Fluids, **5**, 7:870, 1962.
510. ——— and P. Gloersen: J. Chem. Phys., **29**:820, 1958.
511. Roy, A. S. and M. E. Rose: Proc. Roy. Soc., **A149**:511, 1935.
512. Rudinger, U. G.: Phys. Fluids, **4**:1463, 1961.
513. Russo, A. and A. Hertzberg: Jet Propulsion, **27**, 11:1191, 1957.
514. Sakurai, A.: Res. Rep. Tokyo Electr. Eng. College, **5**:39, 1957.
515. ———: J. Fluid Mech., **3**:255, 1957.
516. ———: Res. Rep. Tokyo Electr. Eng. College, **6**:49, 1958.

517. Salkoff, M. and E. Bauer: J. Chem. Phys., **29**:26, 1958.
518. Sandborn, V. A., H. Weisblatt and Flagg: AIAA-Journal, **1**, 5:1236, 1963.
519. Saporoschenko, M.: Phys. Rev., **111**:1550, 1958.
520. Sather, N. F. and J. S. Dahler: J. Chem. Phys., **35**:2029, 1961.
521. Schardin, H.: Die Grenzen der Hochfrequenzkinematographie, "Die Kinotechnik.," H. 3, 1932.
522. ———: Ergebnisse der Exakten Naturwissenschaften, **20**:304, 1942.
523. Svhexnayder, C. J.: J. Aero/Space Sci., **25**, 8:527, 1958.
524. Schopper, E. and B. Schumacher: Z. Naturforshung, **6a**:700, 1951.
525. Schott, G. and N. Davidson: J. Amer. Chem. Soc., **80**:1841, 1958.
526. Schultz, D. L.: Hypersonic Flow, A. R. Collar and J. Tinkler eds., London, p. 301, 1960.
527. Schultz, G. and S. Brown: Phys. Rev., **98**:1642, 1955.
528. Schumacher, B.: Ann. d. Phys., **13**:404, 1953.
529. Schwartz, R. N. and J. Eckerman: J. Appl. Phys., **27**:169, 1956.
530. ———, Z. I. Slawsky and K. F. Herzfeld: J. Chem. Phys., **20**:1591, 1952.
531. ——— and K. F. Herzfeld: J. Chem. Phys., **22**:767, 1954.
532. Sherman, F. S.: NACA TN 3298, 1955.
533. ———: J. Fluid. Mech., **5**:465, 1960.
534. ——— and L. Talbot: Rarefied Gas Dynamics, Pergamon Press, 1960.
535. Shreffler, G. and R. H. Christian: J. Appl. Phys., **25**:324, 1954.
536. Shuler, K. E.: J. Chem. Phys., **31**:1375, 1959.
537. ———: J. Chem. Phys., **39**:1692, 1960.
538. ———, G. Weiss and K. Anderson: J. Math. Phys., **3**, 3:550, 1962.
539. ———: Phys. Fluids, **2**, 4:442, 1959.
540. ——— and G. H. Weiss: J. Chem. Phys., **38**:505, 1963.
541. Sichel, M.: Phys. Fluids, **5**, 10:1168, 1962.
542. Skinner, G. B.: J. Chem. Phys., **31**:268, 1959.
543. ——— and E. M. Sokoloski: J. Phys. Chem., **64**:1028, 1960.
544. Smiley, E. and E. Winkler: J. Chem. Phys., **21**:1648, 1953.
545. ——— and ———: J. Chem. Phys., **22**:2018, 1954.
546. Spence, D. A. and B. A. Woods: Boundary Layer and Combustion Effects in Shock Tube Flows, Hypersonic Flows, London, 1960.
547. Srivastava, B. N. and A. K. Barna: Proc. Phys. Soc., **77**:677, 1961.
548. Stewart, E. S.: Phys. Rev., **69**:632, 1946.
549. ——— and I. L. Stewart: J. Acoust. Soc. Amer., **24**:194, 1952.
550. Strehlow, R. and A. Cohen: J. Chem. Phys., **30**, 1:257, 1959.
551. Stupochenko, E. V., E. V. Samuilov, A. S. Pleshanov, I. B. Rozdestvensky, and I. P. Stakhanov: Thermodynamic Properties of Air at High Temperatures, 8*th* Symposium (Intern.) Combustion, Pasadena, 1960.
552. Sulzer, P. and K. Wieland: Helv. Phys. Acta, **25**:653, 1952.
553. Sutton, E. A.: J. Ch. Phys., **36**, 11:2923, 1962.
554. Takayanagi, K.: Progr. Theor. Phys., Japan, **8**, 111:497, 1952.
555. ———: Prog. Theor. Phys., Japan, **11**:557, 1954.
556. ———: Proc. Phys. Soc., London, **A70**:348, 1957.

557. ———: Sci. Rep., Saitama Univ., **A3**:1, 1958.
558. ———: J. Phys. Soc., Japan, **14**:1458, 1959.
559. ———: Sci. Rep., Saitama Univ., **A3**:65, 1959.
560. ———: Sci. Rep., Saitama Univ., **A3**:87, 1959.
561. ——— and S. Kaneko: Science Reports, Saitama Univ., **1**:111, 1954.
562. ——— and ———: Sci. Reports, Saitama Univ., **A3**:167, 1960.
563. ——— and ———: Sci. Reports, Saitama Univ., **A4**:15, 1961.
563a. ———: Suppl. Progr. Theoret. Phys., **25**:1, 1963.
564. ——— and T. Kishimoto: Progr. Theor. Phys., **9**:578, 1953.
565. ——— and ———: Progr. Theor. Phys., **10**:369, 1953.
566. Talbot, L. and F. S. Sherman: NASA Memo 12–14–58W, 1959 (quoted from [266]).
567. Tanczos, F. J.: J. Chem. Phys., **25**:439, 1956.
568. Taylor, G. I.: Proc. Roy. Soc., **A84**:371, 1910.
569. Teare, J. D., S. Georgiev and R. A. Allen: Radiation from the Non-Equilibrium Shock Front, in Ionization in High-Temperature Gases, K. E. Shider and J. B. Fenn eds,. Academic Press, 1963.
570. Tempest, W. and H. D. Parbrook: Acoustica, **7**:354, 1957.
571. Thaler, W. J.: J. Acoust. Soc. Amer., **24**:15, 1952.
572. Toennies, J. and E. Greene: J. Chem. Phys., **26**:655, 1957.
573. Thomas, L. H.: J. Chem. Phys., **12**:449, 1944.
574. Treanor, C. E. and P. V. Marrone: Phys. Fluids, **5**, 9:1022, 1962.
575. Trimpi, R. L. and N. B. Cohen: A Nonlinear Theory for Predicting the Effects of Unsteady Laminar, Turbulent or Transitional Boundary Layers on the Attenuation of Shock Waves in a Shock Tube with Experimental Comparison, NASA Tech. Rep., R-85, 1961.
576. Turner, E. B.: Phys. Rev., **99**:633, 1955.
577. Turner, R. and D. Rapp: J. Chem. Phys., **35**:1076, 1961.
578. Unoul, T. and T. Takahashi: J. Phys. Soc. Japan, **9**:199, 1954.
579. Vanderslice, J. T., E. A. Mason and E. R. Lippincot: J. Chem. Phys., **30**:129, 1959.
580. ——— and ———: J. Chem. Phys., **33**:492, 1960.
581. ——— and ———: Rev. Mod. Phys., **32**:417, 1960.
582. ———, ——— and W. G. Maisch: J. Chem. Phys., **32**:515, 1960.
583. ——— and S. Weisman: J. Chem. Phys., **37**:2247, 1962.
584. Van Itterbeek, A.: Verhagen, Nature, **167**:478, 1951.
585. Venable, D. and D. Kaplan: J. Appl. Phys., **26**:639, 1955.
586. Vetter, K.: Zsch. für Electrochemie, **53**, 6:369, 1949.
587. Voorhees, H. G. and F. R. Scott: Phys. Fluids, **2**:576, 1959.
588. Wang Chang, C. S.: Rept. UMH–3–F, Univ. Michigan, Dept. Eng. Res., 1948.
589. ——— and G. E. Uhlenbeck: Univ. Michigan, Res. Inst. Rept. M999, 1953.
590. Watson, R.: Rev. Sci. Instr., **33** 10:1113, 1962.
591. Wecker, M. S.: J. Aero/Space Sci., **26**, 12:848, 1959.
592. Weitzsch, F.: Ann. Phys., **7**:403, 1961.

593. Weyman, H. D.: Bull. Amer. Phys. Soc., **4**:284, p. 2, 1959.
594. ———: Bull. Amer. Phys. Soc., **6**:212, p. 2, 1961.
595. ———: Phys. Fluids, **3**:545, 1960.
596. Wetzel, L.: Phys. Fluids, **5**, 7:824, 1962.
597. White, D. R.: J. Fluid Mech., **4**, 6:585, 1958.
598. ———: Phys. Fluids, **4**, 1:40, 1961.
599. ——— and R. C. Millikan: J. Chem. Phys., **39**:1803, 1963.
599a. ——— and ———: AIAA-Journal, 2:1844, 1964.
600. Widom, B.: J. Chem. Phys., **27**, 4:940, 1957.
601. ———: J. Chem. Phys., **31**:1027, 1959.
602. ———: J. Chem. Phys., **32**:913, 1960.
603. ———: J. Chem. Phys., **34**:2050, 1961.
604. ———: Disc. Faraday Soc., **33**:37, 1962.
605. ——— and S. H. Bauer: J. Chem. Phys., **21**:1670, 1953.
606. Wiese, W. H. F. Berg and H. R. Griem: Bull. Amer. Phys. Soc., II, **5**:251, 1960.
607. Windsor, M. W., N. Davidson and R. Taylor: J. Chem. Phys., **27**:315, 1957.
608. Tsang, W., S. H. Bauer and M. Cowparthwaite: J. Chem. Phys., **36**, 7:1768, 1962.
609. Wise, H. and M. F. Frech: J. Chem. Phys. **20**, 22:1724, 1952.
610. Witteman, W. J.: Rev. Sci. Instr., **32**:3, 1961.
611. ———: J. Chem. Phys., **37**, 3:655, 1962.
612. Wittliff, C. E. and M. R. Wilson: Shock Tube Driver Techniques and Attenuation Measurements (quoted from [354]).
613. Wray, K. L.: Chemical Kinetics of High Temperature Air, Intern. Hypers. Conf., ARS Rep. 1975–61, 1961 (also in Hypersonic Flow Research, F. R. Ridell, ed., Academic Press, 1962).
614. ———: J. Chem. Phys., **36**, 10:2597, 1962.
615. ———: J. Chem. Phys., **37**, 6:1254, 1962.
615a. ———: J. Chem. Phys., **38**, 7:1518, 1963.
616. ——— and J. D. Teare: J. Chem. Phys., **36**, 10:2582, 1962.
617. Wright, J. K.: Shock Tubes, Methuen Monographs, 1961.
618. Wurster, W. H.: J. Chem. Phys., **36**, 8:2111, 1962.
619. ———, C. E. Treanor and H. M. Thompson: J. Chem. Phys., **37**, 11:2560, 1962.
620. Zener, C.: Phys. Rev., **37**:556, 1931.
621. ———: Phys. Rev., **38**:277, 1931.
622. ———: Proc. Cambr. Phil., **29**:136, 1933.
623. Ziering, S. and F. Ek: Phys. Fluids. **4**:765, 1961.
624. ———, ——— and P. Koch: Phys. Fluids, **4**:975, 1961.
625. Zoller, K.: Z. Physik, **130**:1, 1951.
626. Zmuda, A. J.: J. Acoust. Soc. Amer., **23**:472, 1951.

Author Index

Subject Index

MIX
Papier aus verantwortungsvollen Quellen
Paper from responsible sources
FSC® C105338

If you have any concerns about our products,
you can contact us on
ProductSafety@springernature.com

In case Publisher is established outside the EU,
the EU authorized representative is:
Springer Nature Customer Service Center GmbH
Europaplatz 3, 69115 Heidelberg, Germany

Printed by Libri Plureos GmbH
in Hamburg, Germany